Australian Native Plants

Cultivation and Uses in the Health and Food Industries

Traditional Herbal Medicines for Modern Times

Each volume in this series provides academia, health sciences, and the herbal medicines industry with in-depth coverage of the herbal remedies for infectious diseases, certain medical conditions, or the plant medicines of a particular country.

Series Editor: Dr. Roland Hardman

Traditional Herbal Medicines for Modern Times

Australian Native Plants

Cultivation and Uses in the
Health and Food Industries

edited by
Yasmina Sultanbawa
University of Queensland, St Lucia, Queensland, Australia
Fazal Sultanbawa
Agrichem (Pty) Ltd., Yatala, Queensland, Australia

CRC Press
Taylor & Francis Group
Boca Raton London New York

CRC Press is an imprint of the
Taylor & Francis Group, an **informa** business

CRC Press
Taylor & Francis Group
6000 Broken Sound Parkway NW, Suite 300
Boca Raton, FL 33487-2742

First issued in paperback 2021

© 2016 by Taylor & Francis Group, LLC
CRC Press is an imprint of Taylor & Francis Group, an Informa business

No claim to original U.S. Government works

ISBN 13: 978-1-03-209788-6 (pbk)
ISBN 13: 978-1-4822-5714-4 (hbk)

Library of Congress Cataloging-in-Publication Data

Names: Sultanbawa, Yasmina, editor. | Sultanbawa, Fazal, editor.
Title: Australian native plants : cultivation and uses in the health and food industries / editors: Yasmina Sultanbawa and Fazal Sultanbawa.
Other titles: Traditional herbal medicines for modern times ; v. 17.
Description: Boca Raton, FL : CRC Press, Taylor & Francis Group, 2016. | Series: Traditional herbal medicines for modern times ; 17 | Includes bibliographical references and index.
Identifiers: LCCN 2015047155 | ISBN 9781482257144 (alk. paper)
Subjects: LCSH: Native plants for cultivation--Australia. | Native plant industry--Australia. | Endemic plants--Australia. | Medicinal plants--Australia.
Classification: LCC SB439.26A8 A97 2016 | DDC 635.9/5194--dc23
LC record available at http://lccn.loc.gov/2015047155

Visit the Taylor & Francis Web site at
http://www.taylorandfrancis.com

and the CRC Press Web site at
http://www.crcpress.com

Contents

Preface

INCREASING COMMERCIAL OUTCOMES AND AGRO-BIODIVERSITY THROUGH AUSTRALIAN INDIGENOUS FOOD PLANTS

The diminishing diversity of our food sources is an emerging issue that has raised concerns among all stakeholders in the food industry, from scientists to consumers. The Food and Agricultural Organisation (FAO) states that 75% of our food has come from just 12 plants and 5 animals, with a 75% loss of plant genetic diversity since the 1900s (www.fao.org/FOCUS/E/Women/Biodiv-e.htm). Rice, maize and wheat supply nearly 60% of the human intake of calories and protein. In contrast, the hunter-gatherer Palaeolithic diets consisted of over 200 plants and animals in a year.

The search for higher-yielding plant varieties has resulted in a significant reduction in nutritive value. Modern broccoli varieties have lost 63% of their calcium and 34% of iron over a 50-year period; potato has lost most of its calcium, iron and vitamins. The emerging paradigm of nutrition as the product of synergies among the different compounds in food rather than as effects of single ingredients ('nutritionism'; *Nutritionism* by Gyorgi Scrinis, Allen and Unwin) is gaining momentum.

Hippocrates' famous dictum 'let food be thy medicine' has never been truer than in indigenous Australian practice. The study of the health-giving properties of indigenous Australian food sources has long been neglected, and it is only in recent times that they are being systematically researched. The various contributing authors in this book have helped improve our understanding of the value and potential of some of these food sources.

The cultivation of 13 species of Australian native plants, namely Davidson plum (*Davidsonia* spp.), Kakadu plum (*Terminalia ferdinandiana*), riberry (*Syzygium luehmannii*), lemon myrtle (*Backhousia citriodora*), anise myrtle (*Syzygium anisatum*), bush tomato (*Solanum centrale*), native pepper (*Tasmannia lanceolata*), finger limes (*Citrus australasica*), desert limes (*Citrus glauca*), quandong (*Santalum acuminatum*), muntries (*Kunzea pomifera*), wattle seeds (*Acacia* spp.) and lemon aspen (*Acronychia* spp.), is discussed.

The processing of native plant food crops and changes occurring in quality, flavour and bioactivity during processing, packaging and storage are discussed for some species of native herbs and fruits. The importance of developing value chain analysis for commercial native foods is also discussed. Volatile and non-volatile phytochemicals as sources of flavouring agents and antibacterial and antifungal compounds for the food industry are reported. These compounds also have potential as natural preservatives that fulfil a current market need. Nutritional and bioactive compounds such as vitamins and antioxidants in the key species are reviewed, which are important for the healing and repair of damaged tissues and would be of value to an ageing population. The safety of using native foods as ingredients in the health and food sectors and new market opportunities are also addressed.

The departure from traditional foods and the predominance of processed foods in modern diets have been cause of the epidemic proportions of metabolic syndrome (obesity), diabetes in all its forms and several other diseases and conditions. It is as if the Hippocratic message is equally valid: 'Bad food can lead to ill health'. Many indigenous societies around the world have been plagued by obesity, diabetes and cardiovascular diseases, perhaps attributable, at least in large part, to the departure from their traditional food sources in favour of modern diets. Michael Pollan's suggestion (*In Defense of Food*, Penguin) that we 'avoid eating what our grandmothers would not recognise as food' is one trend that has gained momentum, as is the trend towards 'Paleo' and other diets. This sounds alarmist, even extreme and unrealistic in the short term, because we may have lost much of the germplasm to achieve this and may be akin to 'bringing back the dinosaurs' but does highlight how far we have moved away from our 'food roots'.

Like having many arrows in a quiver, diversity in our food sources is the best guarantee of food security. There are many instances where a single disease has wiped out a food source, best exemplified by the Irish potato famine, which claimed a million lives and led to the emigration of a similar number in the 1840s as a result of devastation of the potato crops by *Phytophthora* late blight. A narrow genetic base (most cereals) and clonal propagation of some of our key crops (e.g. potato, banana) put them in a highly vulnerable position to similar ravages. Therefore, Australian indigenous plants offer new possibilities that could add diversity to our germplasm, food experience and immense value as food and medicine.

The other opportunity that Australian indigenous plants offer is to design a path to utilisation that avoids the pitfalls that lead to nutritionism, where the industry and the media seek opportunities to exploit, even sensationalise, claims made by science. A holistic rather than reductionist approach is imperative. Commercialisation is the current economic paradigm and has to be managed equitably. Recognition and acknowledgement of indigenous knowledge and practice has to be addressed, as is consensus on intellectual property issues and distribution of the proceeds of commercialisation.

In this context, we are pleased to include a brief mission statement of an initiative by an aboriginal organisation to promote agribusiness opportunities, as an example of the support structure needed to foster indigenous participation and entrepreneurial activity in this sector.

GROWING HEALTHY COMMUNITIES

Dale Chapman is a renowned indigenous chef, born in Dirranbandi in southwest Queensland (Kooma tribal lands), and director of Coolamon Food Creations, and *Paul Keily* is also of indigenous origin and the chief executive officer of National Aboriginal Solutions. Both Dale and Paul represent Red Centre Enterprises Pty Ltd.

Red Centre Enterprises is an aboriginal-owned agribusiness organisation set up to expand regional agribusiness opportunities and development in communities across Australia. It is the first national company of its kind comprising an impressive board of women growing healthier communities. Associated network representatives' and

Red Centre Enterprises' commitment is to work collectively to provide the necessary support, skills, employment, market access, infrastructure improvement and assistance that make each participating community a viable and successful stakeholder.

There has been little access for Aboriginal and Torres Strait Islander community–driven agribusiness trade services, which employ local people and support local communities, to the domestic food supply market in Australia. Research has also shown that a growing population is building enormous momentum and will require combined leadership to secure food supplies and develop mechanisms that safeguard the interests of all stakeholders. From this opportunity, the agribusiness market potential in Australia is strengthening, as it is globally.

The Red Centre Enterprise collectives operate ethical systems of creating industry buy-in for community members, which can actively develop into sustainable agribusiness portfolios, raising opportunities locally, regionally, nationally and internationally for personal and community economic development while also generating positive social, environmental and cultural outcomes. Red Centre Enterprise initiatives allow improved co-ordination of assets, resources and market access for the supply of produce and value-added products in domestic and international markets. These strategies will expand the capabilities of local community stakeholders and enable independence and understanding towards Aboriginal and Torres Strait Islander communities in their region.

Red Centre Enterprises and associated network representatives are working to establish and integrate paddock-to-plate supply chains in multiple agribusiness sectors, as well as collectively delivering sustainable outcomes with culturally appropriate practices and integrity that will deliver genuine benefits to the community. Red Centre has developed a range of initiatives required to build capacity and capability, offers opportunities for local community member involvement throughout regional and national operations and provides access to multiple sectors of industry and community support through continued development of these integrated paddock-to-plate value chains—'a genuine win-win for all'.

Red Centre Enterprises understands the importance of developing initiatives through appropriate consultation and cultural consideration that genuinely meets the community's needs. We are committed to developing our relationship in the right way, with the highest integrity and transparency, exploring shared synergies and progressing these initiatives. The leaders of Red Centre Enterprises and Aboriginal and Torres Strait Islander have developed ongoing national engagements, working in unity with all stakeholders to develop mutually beneficial opportunities that avoid exploitation and create tangible, positive impacts at a local 'grass-root' level.

The idea is simple: Collective agribusiness leadership produces the highest-quality Australian products through shared innovation for mutual benefit. Red Centre Enterprises and associated network representatives are working collaboratively with aboriginal and non-aboriginal landowners, operators and industry leaders across Australia sharing a vision to establish long-term, sustainable agribusiness opportunities in national as well as international markets.

Acknowledgements

We thank Professor Roland Hardman, series editor, for guiding us from the invitation to write this book to the reading of the chapter contents and giving his valuable feedback and encouraging us through the course of this project. Our thanks also go out to the senior editor, John Sulzycki and to the project coordinator, Kathryn Everett from the CRC Press, Taylor & Francis Group. We would like to acknowledge the immense contribution that all the authors have made; they are all experts in their relevant fields and have been generous with their time and knowledge in contributing these chapters. Finally, we would like to thank our children Aamina and Fahim for their patience (but not for their cheeky comments about delayed meals) during the period of this project.

Editors

Yasmina Sultanbawa obtained a graduateship in chemistry from the Institute of Chemistry, Ceylon, in Sri Lanka, an MSc in food science from the University of Reading in the United Kingdom and a PhD in food chemistry from the University of British Columbia, Canada. She has worked as a senior research scientist in the area of food processing, preservation, food safety and nutrition for 15 years at the Industrial Technology Institute in Sri Lanka. For the past 8 years, she has worked as a senior food scientist at the Department of Agriculture and Fisheries, Queensland, the last 5 years of which have been as a senior research fellow at the Queensland Alliance for Agriculture and Food Innovation, University of Queensland, Australia. Her work on Australian native plant foods is focused on the incorporation of these plants into mainstream agriculture and diversification of diets. Working with the Australian native food industry and indigenous communities to develop nutritious and sustainable value-added products from native plants for use in the food, feed, cosmetic and healthcare industries is a key strategy. The creation of employment and economic and social benefits to these remote communities, as well as bringing indigenous ethnobotanical knowledge to the notice of the wider world, is an anticipated outcome. She considers it a privilege to engage with these communities and is very passionate that her work will have a positive socio-economic impact.

Fazal Sultanbawa obtained a BSc and an MPhil in agriculture and crop science from the University of Peradeniya, Sri Lanka, and a PhD from the University of Georgia, Athens, Georgia, focusing on biotechnology and the mass propagation of plants using tissue and protoplast culture. He later completed an assignment as a visiting scholar investigating heavy metal binding proteins at the University of British Columbia, Vancouver, Canada. His subsequent work over a 15-year period involved working with agribusiness companies in Sri Lanka, developing their commercial tissue culture labs and nurseries, analytical labs and plant nutrition programs, as well as production of a variety of field and greenhouse crops, for domestic and export markets. For the last 5 years, he has been technical manager at Agrichem Pty Ltd, Brisbane, Australia, a leading international plant nutrition company. He has been responsible for developing their crop nutrition programs, R&D programs and the conduct of field trials, as well as working closely with commercial growers across the country. He has held senior management positions in these companies and has been involved in projects and consultancies in several countries.

Contributors

Anoma Ariyawardana
School of Agriculture and Food
    Sciences
The University of Queensland
Gatton, Queensland, Australia

Chris Brady
Research Institute for the Environment
    and Livelihoods (Adjunct)
Charles Darwin University
and
EcOz Environmental Services
Darwin, Northern Territory, Australia

Mridusmita Chaliha
Innovative Food Technologies
Agri-Science Queensland
Department of Agriculture and
    Fisheries
Brisbane, Queensland, Australia

Vic Cherikoff
Australian Functional Ingredients
    Pty Ltd
Kingsgrove, New South Wales,
    Australia

Ray Collins
School of Agriculture and Food
    Sciences
The University of Queensland
Gatton, Queensland, Australia

Kim Courtenay
Kimberley Training Institute
Broome, Western Australia, Australia

Jock Douglas
Australian Desert Limes
Roma, Queensland, Australia

Lyle Dudley
Southern Flinders Ranges
Wilmington, South Australia, Australia

Frances Eliott
Southern Cross Plant Science
Southern Cross University
Lismore, New South Wales, Australia

Amanda Garner
Australian Native Food Industry
    Limited
Birregurra, Victoria, Australia

Rus Glover
Woolgoolga Rainforest Products
Sandy Beach, New South Wales,
    Australia

Julian Gorman
Research Institute for the Environment
    and Livelihoods
Charles Darwin University
Darwin, Northern Territory, Australia

Robert Henry
Queensland Alliance for Agriculture
    and Food Innovation
The University of Queensland
Brisbane, Queensland, Australia

L. Slade Lee
Division of Research
Southern Cross University
Lismore, New South Wales, Australia

La Vergne Lehmann
Australian Native Food Industry
    Limited
Woodbridge, Tasmania, Australia

Ben Lethbridge
South Australian Native Food
    Association
Clarendon, South Australia, Australia

Lilly Lim-Camacho
Land and Water
Commonwealth Scientific and
    Industrial Research Organisation
Pullenvale, Queensland, Australia

Hazel MacTavish-West
MacTavish West Pty Ltd
South Hobart, Tasmania, Australia

Jude Mayall
Outback Chef Melbourne, Victoria,
    Australia

Gary Mazzorana
Australian Rainforest Products Pty Ltd
Lismore, New South Wales, Australia

Melissa Mazzorana
Australian Rainforest Products Pty Ltd
Lismore, New South Wales, Australia

Tony Page
Verdien Pty Ltd
Innisfail, Queensland, Australia
and
Forests and People Research Centre
University of Sunshine Coast
Maroochydore DC, Queensland,
    Australia

Christopher D. Read
Diemen Pepper
Birchs Bay, Tasmania, Australia

Sheryl Rennie
Australian Fingerlime Caviar
Bangalow, New South Wales, Australia

Maurizio Rossetto
National Herbarium of New South
    Wales
The Royal Botanic Gardens and
    Domain Trust
Sydney, New South Wales, Australia

Donna Savigni
School of Anatomy, Physiology and
    Human Biology
The University of Western Australia
Crawley, Western Australia,
    Australia

Mervyn Shepherd
Southern Cross Plant Science
Southern Cross University
Lismore, New South Wales, Australia

Heather Smyth
Queensland Alliance for Agriculture
    and Food Innovation
The University of Queensland
Brisbane, Queensland, Australia

Fazal Sultanbawa
Agrichem Pty Ltd
Yatala, Queensland, Australia

Yasmina Sultanbawa
Queensland Alliance for Agriculture
    and Food Innovation
The University of Queensland
Brisbane, Queensland, Australia

Margo Watkins
Rainforest Heart Orchard
Millaa Millaa, Queensland, Australia

David J. Williams
Department of Agriculture and
    Fisheries
Agri-Science Queensland
Coopers Plains, Brisbane, Australia

CHAPTER 1

Overview of Australian Native Plants

Amanda Garner and La Vergne Lehmann

As the oldest living human culture on earth, Indigenous Australians were the first to harvest, process, prepare and consume the native species across the Australian continent. Fast forward to 2015 and what we can now see is an Australian Native Food Industry that is transitioning from being a novelty to a part of the mainstream Australian food culture. For Australian native foods, the world is starting to awaken to a new, but ancient, taste sensation.

Australian native plants are not just about food. We have just started to acknowledge and develop an understanding of indigenous plant species in Australia for their untapped medicinal, neutraceutical, cosmetic and pharmaceutical properties. In recent years, these plant species have also been recognised for their many ground-breaking bioactive, antimicrobial and antioxidant benefits. While many of us are barely aware of the research into these valuable properties, the long-term benefits for all of us are immeasurable.

If we are to tap into the potential of these plants, we also need to understand how we can develop a long-term viable and sustainable agricultural and horticultural industry in rural and remote Australia to ensure supply.

Increasing public awareness is our greatest tool to extend existing research into the practical application for Australian native plants in agriculture and horticulture, product development, supply chain management and marketing.

We know that existing native food markets are already experiencing expansion through new product development in many sectors. The result is a growing gap in the supply chain for many of our native food products.

One of the most important aspects of developing and expanding the Australian native food industry is finding a way to utilise the traditional knowledge of our Indigenous Australians. This means that forming partnerships and developing new types of business models are a priority if we are to successfully develop and promote the industry and achieve future success.

Creating networks for expanding existing knowledge to embrace new participants in the industry is currently being rolled out through a number of organisations successfully around Australia. Examples include programmes specifically targeting indigenous participation such as the Coles Indigenous Fund, agroforestry education models and indigenous traditional owner business models or adding to our current agricultural industry by working with Landcare Australia and Natural Resources Management bodies. Increasingly, there has also been interest shown in growing Australian native food plants in school kitchen gardens and community gardens. The groundswell of interest has become apparent in recent years, and it is now time to harness this growing national interest and focus on developing a cohesive long-term plan for the future.

The most critical aspect of developing the native food industry is engaging traditional owners in all regions to be part of this process. In achieving this, we will see strong industry partnerships forming while ensuring that we continue the process of documenting traditional knowledge of each particular species. At present, what we understand most about traditional knowledge is how much we still have to learn.

Traditional food farming practices, processing and uses with food, fodder, shelter, tools and medicines, wild harvesting, land management and propagation have been always successfully maintained by our Indigenous Australian families for thousands of years until European colonisation in 1788 started to displace traditional owners from their lands. The consequence was a critical disruption to important cultural ethnobotanical relationships for our Indigenous Australians and their food and plants.

Thus, it is quite clear that we need to recognise the knowledge of the traditional owners of this country while we seek to encourage and develop the modern scientific research that would lead to productive and beneficial outcomes for all involved.

The Australian Native Foods Industry Limited is in conjunction with the Rural Industries Research and Development Corporation, and many of our trailblazing growers around Australia have identified 13 priority species with commercial potential in recent years. While some of these have become well known through commercial product development and the creation of recipes featuring these species, others are still relatively unknown. These are species found in all kinds of Australian landscapes from the tropical north to the inland deserts to the cool climates of Tasmania and everything in between. Native Australian food species can be found all over the country. The priority species include Kakadu plum, Davidson's plum, wattleseed, lemon aspen, desert lime, finger lime, quandong, muntries, bush tomato, pepper berry, lemon myrtle, anise myrtle and riberry.

It should not be forgotten that there are already some globally successful Australian native food plants. Species such as Macadamia, sandalwood and tea tree have developed a significant industry presence.

While these are species that the industry is currently focussing on, there are innumerable other edible native food species about which very little is currently known. Some of these include Pindan walnut, native tamarind, saltbush, native rice, native pasture and grains, bloodroot, yam daisy, native lemon grass, sea parsley, samphire, warrigal greens, native grapes, native currants and seaweeds.

We cannot consider the future of agriculture or horticulture in Australia if we do not consider the issues relating to climate change. Understanding the impacts of climate change on the production of native foods is a highly agendum for all large and small landowners across our vast country, which surely focusses on what is truly an Australian food industry. At the same time, there is potential for native food production to move into mainstream agriculture. Whereas we now have grain farmers in the semi-arid regions of Australia from the Pilbara in western Australia to the Wimmera in western Victoria to the central west of NSW dealing with the vagaries of variable climatic conditions from year to year with varied results, there may be solutions in developing production of a native millet or sorghum as the solution to crop yield security.

Equally, in other productive regions of Australia, we have a significant horticultural fruit industry with produce being grown in all states across a diverse climate range. The challenge is now to include a range of native food horticultural species such as native citrus, quandongs and native nut species. We already have international health and medical corporations interested in many of our native plant species, making it important for us to promote these growing opportunities for Australian farmers.

Another focus for our agricultural industry is the growing farmgate sector where farmers can take the opportunity to take charge of their marketing and sales. As a growth sector in Australian agriculture, it is not unreasonable to suggest that governments across the country could assist in developing more viable farmgate options for farmers who want to develop native food production, whether they are involved in conventional agricultural practices or are developed as part of new agricultural models for traditional owners. In short, these are the species that we can develop for production in our country, and if we do not use them, then the potential is to lose them!

We know that the native food industry has a huge potential in Australia – both as a domestic product and as an export product. Right now, we know it is on the cusp of moving from a niche or novelty product to something we can recognise easily when we do a regular grocery shop. We know that consumers are becoming increasingly adventurous with food flavours and without doubt native foods will only add to the flavour sensations that people can experience.

Just a few years ago, the standard fare in Australia precluded many of the flavours that we now accept as a regular par to our diets. The advent of multiculturalism in Australia has allowed us to experience the full-bodied flavours we now associate with Thai and Indian curries, the delicate nuance of Japanese food and the fragrant spices associated with middle eastern dishes. Yet we are still to fully appreciate Australian native food flavours.

Our amazing native foods proteins like kangaroo, emu and crocodile have been a little slow on the uptake but are all receiving a lot of recognition in recent times. In contrast, native food herbs and spices are still relatively new to the market here. They are unique and educating consumers about how to utilise these products appropriately in our kitchens is going to take more time.

Consequently, we know that describing the flavours to the consumer will be a challenge. But does it have to be that hard? By using the descriptors that we are already familiar with, we can describe many of these flavours. Let us start with the self-describing lemon myrtle, a native plant that is already widely used in teas, biscuits, soaps, cosmetics and cleaning products with its unique aroma that continues to awake the senses with its citrus overtones. On the flip side, its cousin, the anise myrtle, is an amazing substitute for fennel with its strong anise flavours, making it a great match with pork.

Native citrus such as the finger lime produce spectacular colourful jewels that are widely accepted in high-end restaurants around the globe, with their amazing colours and zingy citrus orbs attractive to the eye and their use as a refreshing alternative to the lemon, with the appearance of a citrus caviar.

The riberry are little bombs of refreshing goodness with a hint of clove, making them an ideal fruit for preserves and other products. Riberry products are now found in many on-the-shelf products, but like many of our native food plants, they are limited by the current supply chain.

Wattleseed, widely acknowledged as an Australian culinary product since the 1980s, is revered for its multiple uses in the kitchen and potentially as a coffee flavour substitute.

The quandong and the muntrie are found growing naturally wild in SA, WA and northern Victoria and were both used widely by our early settlers in preserves and sweets. Today, there are commercial supplies available through many of our native food suppliers.

Then there is the humble and mostly overlooked culinary additive, saltbush. We know there are a number of varieties suitable for growing all around Australia, which also have culinary applications. Saltbush has been given much notoriety for its use as a stock feed in times of drought; however, it is a tasty additive as a leafy green as well as a viable salt alternative.

Finally, let us consider the Kakadu plum or gubinge, found primarily in the wild in coastal areas of the Kimberly and the Northern Territory and now showing great potential for cultivation. It has long been eaten as a whole fruit and used as a medicine also by traditional owners for many years.

To finish, it is appropriate to quote from an Aboriginal Elder from the Nyul Nyul people of Dampier Peninsula:

> Gulloord Irini, my old grandfather told me when I was a little boy- one day this tree will help the world, this tree will be medicine for a lot of people. Even though they don't notice it now, one day they will notice it.

Perhaps this is the strongest hint yet for our researchers to investigate the possibilities of this wonderful fruit that has so much potential, not only as a food but also as a medicine. We have, indeed, so much to learn.

As passionate advocates for our industry, please join us in encouraging research and traditional knowledge being brought together to the native plant industry to the forefront of agricultural and horticultural industries here in Australia.

PART I

Cultivation and Production

CHAPTER 2

Cultivation of Anise Myrtle (*Syzygium anisatum*)

Gary Mazzorana and Melissa Mazzorana

CONTENTS

INTRODUCTION

Anise myrtle (*Syzygium anisatum*) is a rare Australian rainforest tree native to the subtropical Mid-North Coast of New South Wales, with a natural distribution in the wild restricted to the Nambucca and Bellinger valleys, near Coffs Harbour. It is rather rare in the rainforest; however, its popularity within the native food and horticultural industries promises its security within our flora. *S. anisatum* owes its now commonly used name of 'anise myrtle' to the pleasant liquorice-like scent of aniseed, produced on crushing the leaves.

BOTANY

Anise myrtle, botanically named *S. anisatum*, was previously named *Backhousia anisata* after James Backhouse, the nineteenth century English nurseryman and Quaker missionary, as well as *Anetholea anisata*. It is a member of the family Myrtaceae and is closely related to cinnamon myrtle (*Backhousia myrtifolia*), curry myrtle (*Backhousia angustifolia*) and the much more commonly known lemon myrtle (*Backhousia citriodora*). The more recent botanical name *S. anisatum* is not yet widely accepted, and the former names such as *B. anisata*, *A. anisata*, native anise, ringwood or aniseed myrtle are still often used when referring to this species.

Anise myrtle is evergreen and may reach heights of 20–45 m with a spread of 4 m in a rainforest environment, but most often reaches 8–10 m as a small to medium tree in open garden situations and is usually harvested as a hedge to 2–3 m in native food plantations. It is a large tree in its natural habitat, but the timber has no commercial value. The plant has a dense cover of fine, lush, glossy, green foliage with wavy margins, with the young leaf flush being soft with pink to red colouring. It has creamy-white scented flowers in the spring. Fruit are small, white papery capsules (Figures 2.1 and 2.2).

TRADITIONAL USE AND ECONOMIC POTENTIAL

Little knowledge of traditional use is available, mainly because the tree is so rare and the traditional knowledge has largely been lost. Anecdotally, anise myrtle leaves were traditionally used as an Aboriginal tonic with vitalising effect. The trees were harvested during World War II, when aniseed flavouring was in short supply and then ceased again after the war ended. Native food pioneers rediscovered the plant and began selecting superior genetic varieties, then established commercial plantations in the northern New South Wales in the early 1990s.

Anise myrtle contains anethole and methyl-chavicol, which impart aniseed and licorice flavours, respectively. When crushed, the fresh leaves exude a lovely aniseed scent. It is considered a fine spice of Australia resembling the taste of the French Tarragon and has a soft, subtle, warm, sweet liquorice-like taste. It is an interesting alternative to aniseed and star anise and also makes a mild clove substitute. Anise flavours from other sources have been used for thousands of years in Europe, The Middle East and Asia in prized dishes and drinks.

Anise myrtle makes a delectable herb and seasoning, and readily transfers its subtle flavour to the other ingredients being cooked with it. This invigorating herb has a versatile range of application in sweet and savoury dishes. It can be used fresh, dried or as a ground herb in things like confectionary, liqueur, custards, sorbets or soups. It adds a touch of subtle, but exquisite taste sensation to your palate. Ideally suited to Mediterranean, Middle Eastern or Asian dishes, it goes well with seafood and pork and is great with Australian game meats. It can be used to flavour steamed rice and stocks or to add that extra something to biscuits, breads, pastries

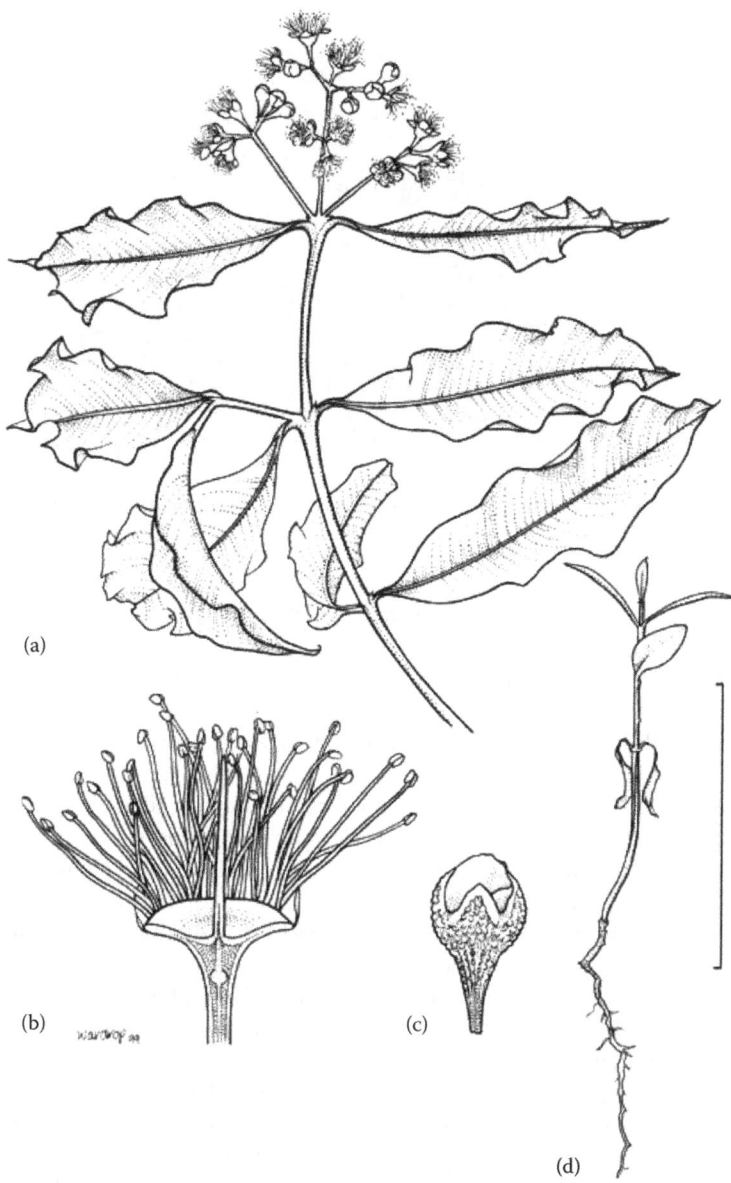

(a)

(b)

(c)

(d)

Figure 2.1   Botanical illustration of *Anetholea anisata* (*Syzygium anisatum*). (a) Habit, (b) longitudinal section (LS) of flower, (c) bud, (d) seedling. Scale bar: (a) 5.5 cm; (b and c) 1 cm; (d) 3 cm. (From Wilson, P.G. et al., *Aust. Syst. Bot.*, 13, 429, 2000; Illustrated by Catherine Wardrop. Copyright Royal Botanic Gardens and Domain Trust.)

Figure 2.2   (See colour insert.) Photo of an anise myrtle (*Syzygium anisatum*) tree. (Photo by Gary Mazzorana. Copyright Australian Rainforest Products.)

and desserts. Infused in warm milk, it can be added to yoghurt or ice creams, icings or cream fillings.

The flavour of anise myrtle is exquisite when added to Sauces or Marinades, and it makes a pleasant herbal infusion as a tea on its own, or as a tea blend. It can also be infused in oils and vinegars, setting a deep fragrant flavour to salad dressings and mayonnaises.

Anise myrtle has a superior antioxidant capacity compared to other native foods. It contains vitamin C and E, both powerful antioxidants and is rich in magnesium. Antioxidants in foods are important for our health and have been shown to help protect against heart disease, cancers and other diseases. Anise myrtle has been demonstrated as given in the RIRDC publication (09/133), to exhibit superior antioxidant capacity when compared to blueberry, which is recognised for its health enhancing properties. Anise myrtle has high levels of lutein. Lutein plays an important role in eye health. It improves visual function and symptoms in atrophic age-related macular degeneration that is the leading cause of vision loss in ageing Western societies. Lutein protects the retina from damage by inhibiting inflammation.

Given that the anise myrtle leaf is one of the highest known sources of the compound anethole, it is ideal for using as a flavour-masking agent, expectorant, sedative and stimulant in cough medicines. Plants containing anethole have traditionally been used to assist with weight loss, lactation and stomach complaints. Trans-anethole rich herbs have traditionally been used to treat conditions like anorexia, belching, hiccupping or reflux and persistent epigastric pain. It is regarded as an antiseptic, bactericide, cancer-preventative, carminative, dermatitogenic, expectorant, fungicide, gastro-stimulant and insecticide. Anecdotally, anethole cures intestinal cramps, colic and flatulence.

Table 2.1　Anise Myrtle Essential Oil Gas Chromatograph Analysis

Certificate of Analysis Method: Area Normalisation
Based on ISO 7609:1985

Sample: Oil of *Syzygium anisatum* (Vickery) Craven and Biffin (previously *Anetholea anisata* (L. Vickery) Peter G. Wilson; Backhousia anisata J. Vickery) anise myrtle oil

Client: Australian Rainforest Products

Components	Arena%; Batch No. 540
α-phellandrene	0.6
1,8-cineole	1.9
Estragole (syn. methyl chavicol)	9.4
Z-anethole (*cis-*)	0.4
E-anethole (*trans-*)	85.5

Source: Analysis report provided by Gary Mazzorana. Copyright Australian Rainforest Products.

Anise myrtle leaves can be steam distilled to produce oil. Although it is relatively new oil to the aromatherapy industry, it has the potential to supersede traditional Aniseed essential oils on the basis of a superior anethole content and a fresher Anise aroma. It also has applications within the fragrance industry for cosmetics. Table 2.1 show the composition of anise myrtle essential oil and a certificate of analysis.

Research indicates that anise myrtle has antimicrobial and antifungal properties. A study by Zhao and Agboola (2007) showed strong activity of anise myrtle in both methanol and water extract against the common food spoilage bacteria *Bacillus subtilis*. Anise myrtle methanol extract also demonstrated activity against the cholera-causing human pathogen *Vibrio cholerae*.

Antioxidant activity using ß-carotene bleaching method in this study showed 40.6% of inhibition and free radical–scavenging activity expressed as 1,1-diphenyl-2-picrylhydrazyl measured at 55.6%.

The anise myrtle leaf has one of the highest known concentrations of anethole, the compound that gives its aniseed flavour and aroma. Since the commercial establishment of the anise myrtle industry, growth has been steady. It started out as a small domestic industry, where growers were selling their handmade products locally at markets and stalls. From here, it became commercialised in the Australian domestic market and sold in supermarkets in a range of foods and in the form of confectionery.

Currently, the industry is producing up to 10 t per annum in anise myrtle leaf product and up to approximately 1 t in anise myrtle essential oil for the domestic and international markets.

Anise myrtle is now estimated to have an industry value of approximately $400,000–$500,000 per annum. Existing plantings have the capacity to produce much higher volumes; however, many growers have reported they are not harvesting due to lack of markets for this product.

Today, anise myrtle is still struggling to gain international recognition, mainly due to its lack of Novel Food status within European Union (EU) countries. Previous EU sales confirmed that this commodity is extremely well accepted. Should the Novel Food issue be overcome, the opportunity to sell anise myrtle into the EU will bring about a huge increase in sales, as Europeans are known for their preference for Aniseed flavoured products.

A large proportion of current sales are through the exportation of this commodity to the United States, who are the leading buyers of anise myrtle. Anise myrtle is being used as a nutraceutical product in an anti-arthritic formulation. The U.S. market has the potential to expand, but will require a strong marketing effort and further research into innovative new product applications. The Native Food industry considers anise myrtle to have the potential to become an important crop for the future. Continuing research indicates there may be many more health benefits associated with using anise myrtle, indicating that the economic future is promising.

AGROCLIMATIC REQUIREMENTS

Anise myrtle is a rare tree species, indigenous to the Mid-North Coast of New South Wales, Australia, being found along streams or on lower slopes. The tree is now grown commercially in plantations in many regions of Australia including the northern New South Wales, parts of South East Queensland, Victoria, South Australia and Western Australia.

Typically, anise myrtle is a hardy plant that handles all but the poorest drained soils. Plants prefer an acidic to neutral, well-drained soil, but will cope with some clay. They will tolerate full sun or a semi-shaded aspect. The trees are best suited to areas that are free of frost. However, it will also tolerate cooler climates, provided the plants are protected from frost when young. Once established, they will cope with −3°C to −4°C with minimal damage but should be protected from frost for the first couple of years. Although sensitive to dry conditions as a small plant, once established, it can tolerate and recover quickly from extremely dry conditions.

It takes several years for an anise myrtle plant to become mature enough to start producing commercial quantities and grades of foliage. The trunk is somewhat soft with a corky bark appearance, and new leaf growth has an attractive reddish tinge, making it a fine foliage plant with ornamental potential. The plants have fluffy, cream coloured flowers, forming in late spring.

AGRONOMY

The growth rate is highest in warmer subtropical climates, but leaf quality can vary with plant nutrition, irrigation regimes and climatic conditions. Anise myrtle is capable of being very productive and can thrive in a range of soils, provided that the soil preparation and nutrient management conditions are met. Although it can be grown in many soil types, it performs best in fertile, well-drained loamy-type soils for maximum biomass recovery.

Little research has been performed on the nutritional requirements of anise myrtle; however, the trees seem to respond when they are well mulched and are regularly fertilised with poultry manure–based fertiliser or compost. Liquid seaweed-, kelp- and fish-based fertilisers are also beneficial and have shown good results. Anise myrtle typically does not require huge nutrient inputs due to it being an Australian Native plant; however, in a plantation with dense plantings, with constant harvesting, they require a well-balanced fertiliser regime to maintain optimal production.

As in all good agricultural practices, a soil and leaf analyses should be undertaken prior to the application of any fertilisers.

Regular pruning of the trees from a young age is recommended to establish both lateral and vertical biomass. Pruning will also lessen the threat of wind damage.

Anise myrtle, although reasonably tolerant of dry conditions, grows best in areas with high rainfall with well-drained soils. In dryer areas, the implementation of an irrigation system is essential for optimal growth, particularly from first planting to an age of 2 years. Overirrigation should be avoided, as they are also sensitive to waterlogging.

As with most commercial plantations, layout needs to be carefully considered, taking into account maximisation of sunlight to both sides of the trees. Thus, a north-facing aspect with a north/south planting is ideal. Although many plantations have been established and reasonable biomass recovery has been achieved on south-facing aspects, this may be due to the plant having a natural distribution within moist, rainforest environments.

The planting layout for anise myrtle in the northern New South Wales is commonly a 3–3.5 m row spacing, with trees being planted 1–1.5 m along the row. Row spacing at 3 m × 1.2 m will give approximately 2500 trees per ha. Mechanical harvesting has been specifically designed to suit this spacing and tree configuration and cuts both sides of the tree, leaving the tree in an 'A' frame shape. Alternate planting designs can be used to suit different harvesting techniques, one of which is to plant the trees much closer together in multiple rows, of 2 or more, similar to that of tea production, where a harvester only cuts the top of the trees.

Once established, anise myrtle is very hardy; however, good weed management in newly established plantations is required as anise myrtle performs best without competition from weeds. There are several methods used for weed control. The most common is heavy mulching, which has a number of other benefits including moisture retention and improving soil fertility following the breaking down of this mulch by adding organic matter to the soil. The use of weed matting is another alternative. Once trees reach maturity, due to their dense canopy, they have little competition from weeds, and occasional hand weeding is required particularly prior to harvesting to eliminate any contamination. Slashing/mowing of the inter-row is recommended. The use of a side throw mower is preferred as the cut material is transferred back under the trees and breaks down to form organic matter (Figure 2.3).

Figure 2.3 **(See colour insert.)** Photo of a typical anise myrtle commercial plantation layout with lemon myrtle at the back. (Photo by Gary Mazzorana. Copyright Australian Rainforest Products.)

GENETIC IMPROVEMENT AND PROPAGATION

Anise myrtle can be grown by seed; however, commercially, they are propagated from tip cuttings, selected from specific, low methyl-chavicol and high trans-anethole varieties. Initially, there was little research on the different varieties, but during the mid-1990s, it was discovered that some plantations consisted of the high methyl-chavicol varieties, which was not ideal.

Research to identify different types of anise myrtle was conducted, and gas chromatograph analyses of these trees were undertaken, which in turn identified a superior, high trans-anethole variety. This selection is the most common and sought-after variety in commercial plantings.

Fortunately, this selection also happens to have a longer, broader and denser leaf, and in recent observations, this variety has proved to be more resistant against myrtle rust compared to other varieties.

BIOTIC CONSTRAINTS

Anise myrtle was extremely resistant to most pests and diseases, until 2010 when a new disease, myrtle rust, caused by the fungus (*Uredo rangelii* now identified as *Puccinia psidii*), reached Australia from South America. This disease affects the Myrtaceae family, which includes anise myrtle. Anise myrtle is highly susceptible to this fungal disease, which affects only the fresh, new, flush growth, and because of the way anise myrtle continually flushes; this causes severe leaf damage to the extent that young plants can have a high mortality rate. The rust has caused substantial damage in terms of production and raised costs due to the need for fungicides for its control. This in turn has made growing anise myrtle chemical-free or organically extremely difficult, as there can be a 50%–80% loss in production as a result of the rust.

On occasion, under specific conditions, anise myrtle will also be affected by sooty mould and psyllid gall, although these are easily controlled.

HARVESTING, PROCESSING AND MARKETING

In commercial plantations, anise myrtle trees are typically maintained as a shrub by regular harvesting from the top and sides. It is important to retain some lower branches when pruning for plant health and to eliminate soil contamination on the product.

Harvesting anise myrtle leaf can be undertaken all year round, but if it is harvested during warmer months, it leads to quicker plant regeneration and higher annual biomass recovery.

Harvesting techniques vary depending on plantation layout and size. Larger commercial plantations are mechanically harvested with specifically designed equipment to suit the layout, whereas smaller plantations are generally harvested manually.

Commercial situations for anise myrtle generally have mechanised harvesting systems in place, specifically designed to separate the leaf from the stem and then dry the leaf efficiently and hygienically without damaging the leaf's natural qualities. It is important to maintain low drying temperatures to avoid flavour loss. Best results are achieved through fully closed and automated, computerised drying systems, which provide a controlled environment, eliminating humidity of the ambient air. This allows for a high-quality, consistent product to be produced which has a pre-determined moisture level remaining in the leaf.

Small operations generally use rack-drying systems, which is subject to temperature fluctuations and humidity levels. This drying technique also can be time consuming in comparison to the fully automated systems and can lead to other issues such as higher microbial loadings.

Microwave drying techniques are also an option, although energy efficiency is much lower than using computerised fluid-bed air-drying systems. One of the benefits of microwave drying is its ability to reduce the microbial load within the product, lowering the microbial plate count.

Evaluation of combining both the fluid-bed air-drying system with microwave drying is being explored, as this gives the potential of making a drying system that is energy efficient along with giving lower microbial plate counts.

In large commercial processing plants, the equipment used ranges from custom-designed hammermills, herb/tea cutting equipment, vibratory separating equipment, conveyor systems and dust-extracting units, which allows the manufacturer to process in excess of 150 kg per hour. In smaller operations, the process is still performed manually and without the use of specialised equipment.

The packaging options for anise myrtle products vary. It is generally dependant on consumer requirements, but also must satisfy/preserve the expected/required quality of the product, which in turn will give the products an extended shelf life. The packaging must also be able to withstand shipping and handling conditions, taking into account export where the product will be subject to wide fluctuations in temperatures. Currently, the most common packaging for dried product is using plastic-lined cardboard cartons. Recent research conducted into alternate packaging for storage of dried product has indicated that the use of a multi-layered foil bag is a far superior material for packaging in regard to extending shelf life and quality retention. Shelf life of dried anise myrtle is very sensitive to correct temperature control and stores best in a dark, dry environment.

Good processors ensure that there are batch control procedures in place for traceability and that samples are taken for microbiological analysis to ensure it always meets the required standards for each customer.

The production of anise myrtle 100% essential oil is different from that of processing dried product, in that once the fresh leaf material is harvested, it can be directly placed into a distillation unit and no separating or drying is required. Anise myrtle essential oil is extracted from the leaves and young branches of anise myrtle trees by steam distillation. Leaves are freshly harvested and distilled immediately to ensure the highest quality oil. In large commercial operations, the harvested leaves are placed into large stainless steel vessels, and placed into the distillation units. Steam is introduced and the anise myrtle's volatile elements (essential oil) are extracted from the leaf with the steam. The condensation process turns this vapour mix into liquid (water) and essential oil. The essential oil floats on top of the water and is separated off. For more efficient distillation, the product can be cut or mulched into smaller particles, enabling more product to be placed into the distilling vessel and in turn yielding higher volumes. After distillation, the remaining leaf residue can be used as composted mulch, which can dramatically reduce input costs for fertiliser and soil conditioners. This by-product is also being evaluated for potential use in other areas, as only the volatile oils are removed and the water-soluble compounds remain.

The best storage and packaging option for anise myrtle oil is in stainless steel containers. The oil is generally sold and shipped in stainless steel or aluminium canisters. Table 2.1 gives the chemical composition of anise myrtle essential oil.

Anise myrtle is viewed as a traditional food of Australia by Food Standards Australia New Zealand (FSANZ) and has been exported to the EU prior to 1997. It is listed for inclusion in the Codex Alimentarius; however, to date, the work required to have Novel Food Status approval has not been completed. This has had severe implications for market growth in the EU, as large companies refuse to purchase anise myrtle until the Novel Food issue is resolved to avoid possible litigation. General consensus from prospective purchasers throughout Europe has been extremely positive and it indicates their willingness to use anise myrtle over other currently used aniseed flavoured products, due to its outstanding qualities.

At present, the United States is the largest market for anise myrtle internationally. This has potential to expand but will require an extensive marketing approach backed by innovative research findings, in order to open up new avenues of sale.

YIELD

Yields are not well documented, but current information suggests that in well-managed commercial plantations, green weight leaf production yields are in excess of 5 kg per tree. In plantations with planting densities of 2,500 trees per ha, this equates to a yield of approximately 12,500 kg per ha per annum: 12,500 kg of green leaf, once dried, will have a weight loss of over 50%; therefore, the dry weight recovery per hectare would be in the vicinity of 6,000 kg. Yields for anise myrtle oil production on the same basis as mentioned earlier, at a 1.5% recovery, should be just under 200 kg of essential oil per hectare per annum.

CONCLUSIONS AND RECOMMENDATION

Anise myrtle has not yet gained the recognition it deserves, mainly due to its lack of documented research, unlike other Australian Native plants. It is believed that the reason for this lack of information is due to the small size of the industry and limited finances. Anise myrtle has a long way to go in terms of becoming a well-recognised commodity, and although sales have indicated steady growth over the past 5 years, it is still considered a niche market.

Opportunities for the promotion of anise myrtle exist within the cosmetic, personal hygiene, perfume and aromatherapy industries, although currently the predominant market is as a food-flavouring agent or as a tea.

Another area to be capitalised upon would be to use the research findings into the health benefits associated with anise myrtle, such as its high lutein content and being rich in magnesium, which could lend itself to be marketed as a food supplement. Anise myrtle has great potential to be marketed for its microbial and antifungal properties, as these have not been utilised to their full extent. Future market growth will require major investments into research and marketing. However, the biggest hurdle to overcome is having anise myrtle listed as a Novel Food within the EU. This alone would see all available supply being expended.

REFERENCES

Wilson, P.G., O'Brien, M.M., Quinn, C.J., 2000. *Anetholea* (Myrtaceae), a new genus for *Backhousia anisata*: A cryptic member of the *Acmena* alliance. *Australian Systematic Botany* 13, 429–435.

Zhao, J., Agboola, S., 2007. *Functional Properties of Australian Bushfoods*. Rural Industries Research and Development Corporation, RIRDC Publication No. 07/030, Canberra, Australian Capital Territory, Australia.

BIBLIOGRAPHY

Australian Native Food Industry Limited (ANFIL). www.anfil.org.au.

Blewitt, M., Southwell, I.A., 2000. *Backhousia anisata* Vickery, an alternative source of (E)-anethole. *Journal of Essential Oil Research* 12, 445–454.

Brophy, J.J., Boland, D.J., 1991. The leaf essential oil of two chemotypes of *Backhousia anisata* vickery. *Flavour and Fragrance Journal* 6, 187–188.

Chaliha, M., Cusack, A., Currie, M., Sultanbawa, Y., Smyth, H., 2013. Effect of packaging materials and storage on major volatile compounds in three Australian native herbs. *Journal of Agricultural and Food Chemistry* 61, 5738–5745.

Clarke, M., 2012. *Australian Native Food Industry Stocktake*. Rural Industries Research and Development Corporation, RIRDC Publication No. 12/066, Canberra, Australian Capital Territory, Australia.

Craven, L.A., Biffin, E., 2005. *Anetholea anisata* transferred to, and two new Australian taxa of, *Syzygium* (Myrtaceae). *Blumea – Biodiversity, Evolution and Biogeography of Plants* 50, 157–162.

Konczak, I., Zabaras, D., Dunstan, M., Aguas, P., 2010. Antioxidant capacity and phenolic compounds in commercially grown native Australian herbs and spices. *Food Chemistry* 122, 260–266.

Konczak, I., Zabaras, D., Dunstan, M., Aguas, P., Roulfe, P., Pavan, A., 2009. *Health Benefits of Australian Native Foods: An Evaluation of Health-Enhancing Compounds*. Rural Industries Research and Development Corporation, RIRDC Publication No. 09/133, Canberra, Australian Capital Territory, Australia.

Smyth, H., 2010. *Defining the Unique Flavours of Australian Native Foods*. Rural Industries Research and Development Corporation, RIRDC Publication No. 10/062, Canberra, Australian Capital Territory, Australia.

Smyth, H.E., Sanderson, J.E., Sultanbawa, Y., 2012. Lexicon for the sensory description of Australian native plant foods and ingredients. *Journal of Sensory Studies* 27, 471–481.

Sultanbawa, Y., Williams, D., Chaliha, M., Konczak, I., Smyth, H., 2015. *Changes in Quality and Bioactivity of Native Foods during Storage*. Rural Industries Research and Development Corporation, RIRDC Publication No. 15/010, Canberra, Australian Capital Territory, Australia.

CHAPTER 3

Cultivation of Bush Tomato (*Solanum centrale*)
Desert Raisin

L. Slade Lee

CONTENTS

INTRODUCTION

Solanum centrale is a small Australian native plant adapted to arid environments which has a timeless tradition of use by Aboriginal peoples. In recent years, it has been incorporated in Australian bush food cuisine for its piquant flavour when used as a condiment in cooking and a flavouring in processed products. The fresh or dried fruit, although eaten in remote Aboriginal communities, is never sold for consumption as a whole fruit in commercial marketplaces. The flavour is considered by most to be rather acrid to be eaten whole in anything but small quantities, although as a flavouring ingredient, it imparts an attractive zest to food products. When used in processed products such as sauces, chutneys and dukkah, the *S. centrale* fruit content rarely exceeds 10%.

Commercially, the term 'bush tomato' has been adopted for marketing purposes, although the correct common name is 'desert raisin'; confusingly, the term 'bush tomato' officially pertains to several different edible species (*Solanum chippendalei*, *S. diversiflorum*, *S. cleistogamum*, *S. coactiliferum* and *S. esuriale*) and even one toxic species (*S. quadriloculatum*), but not to *S. centrale* (Latz, 1999).

Of the 13 main commercialised Australian native plant product species (including bush foods, flavourings and oils), Clarke (2012) ascertained that bush tomato was second only in production to the lemon myrtle industry; nonetheless, the total recorded farm-gate value in the year surveyed was little more than $0.5 M (AUD).

BOTANY AND AGROCLIMATIC REQUIREMENTS

Within the Solanaceae, the genus *Solanum* comprises approximately 1700 species globally (Jessop, 1981), with 200 *Solanum* species recorded in various parts of Australia (Black, 1965; Jessop, 1981; Symon, 1981), of which 132 are endemic (Purdie et al., 1982). Both diploid ($n = 12$) and tetraploid forms of the species *S. centrale* have been reported (Randell and Symon, 1976). *S. centrale* is a small sometimes prickly under-shrub, prostrate to erect in habit, rarely exceeding 0.5 m in height and breadth, with grey-green hirsute entire leaves. It bears flowers typical of the genus, medium in size with purple corolla and erect connivent (adjacent but not fused) yellow anthers, producing rounded pale yellow fruit up to 2 cm in diameter at maturity which shrivel to a dark-brown or red-brown raisin-like appearance typically retained on the plant (Bean, 2004; Symon, 1981; Whalen, 1984). From extensive research,

Figure 3.1 (See colour insert.) Photo of foliage, flowers and young fruit of bush tomato (*Solanum centrale*). (Photo by Slade Lee. Copyright Southern Cross University.)

Figure 3.2 Photo of developing fruit of bush tomato (*Solanum centrale*). (Photo by Slade Lee. Copyright Southern Cross University.)

Collins (2002) ascertained that *S. centrale* flowers are partially self-incompatible, exhibiting pin and thrum heterostyly, cyclical floral opening, and stigma receptivity, resulting in a high level of outcrossing mediated by insects, particularly native bees (Figures 3.1 and 3.2).

    S. centrale occurs throughout large swathes of arid Australia predominantly in Western Australia, South Australia and Northern Territory (Jessop, 1981; Symon, 1981; Whalen, 1984), with limited distribution also in Queensland (Bean, 2004), but the region of greatest prevalence is the extensive Central Australian desert areas on either side of the South Australia or Northern Territory border (Albrecht et al., 2007; Barker et al., 2005).

    A notable characteristic of *S. centrale* is its great propensity for clonal regeneration from roots (Dennett, 2006; Latz, 1995; Peterson, 1979), exhibited as aggressive suckering under cultivation, but which, in fact, is an adaptation to regular fire events and the arid conditions of Central Australia (O'Connell et al., 1983). This attribute has ramifications on horticultural production, both in its utility for clonal propagation and for inter-row management in cropping situations.

TRADITIONAL USE

    S. centrale has a well-reported tradition of use as a food by Aboriginal people of the arid regions, such as Alyawarr speakers from Ampilatwatja (Ampilatwatja et al., 2009; Latz, 1995; O'Connell et al., 1983; Peterson, 1979). The plant's use was widespread as reflected by its variety of names in Aboriginal languages from different regions in Central Australia: *akatyerr, katyerr, yakajirri,* and *kampurarpa* (Alyawarr speakers from Ampilatwatja et al., 2009); also *akatyerr* and *katyerr* (Anmatyerr language, according to Latz, 1999); *akatyerre* (East Arrernte language); *katyerre* (West Arrernte language); *arlkerre* (Kaytetye language); *kampurarrpa, kanytjilyi, katarapalpa* and *kintinyka* (Pintupi language); *kampurarpa* and *kati-kati* (Pitjantjatjara language); and *yakajirri, jungkunypa, kampurarrpa, kararrpa, karturu* and *mulyu* (Warlpiri language) (Latz, 1999).

O'Connell et al. (1983) reported that the energy return from collection and consumption of *S. centrale* fruit in traditional Aboriginal culture, at almost 6000 kcal/hour, was second only and just marginally so to that of the root vegetable *Ipomoea costata*. Of such significance was *S. centrale* as a food to traditional societies that it is reported to have been traded by Aboriginal clans from areas where it grew in abundance, with those less fortunate (Bonney, 2006).

There are significant issues concerning cultural acknowledgment and benefit sharing in regard to the commercial exploitation of Aboriginal traditional knowledge enshrined within long-used bush food plants, and many questions are raised about the aspects of intellectual property and customary rights (Lee, 2012; Lingard, 2015), which are beyond the scope of this chapter.

ECONOMIC POTENTIAL

Definitive validated data on mean bush tomato production and price are difficult to obtain, but variability in production, prices and supply reliability is apparent. Fruit is most commonly traded in the dried form, though fresh fruit is also traded. Bryceson (2008) indicated that a 15–20 t/year crop commanded a price range of $4–$14 per kg (AUD) in 2007. Somewhat consistent with that, 4 years earlier, Foster et al. (2005) reported a 2003 production of 13 t (8 t of dried fruit plus 5 t of fresh fruit) commanded $24 per kg dried and $12 per kg fresh and a total 2003 production value of $252,000. Robins and Ryder (2008) reported a 2004 farm-gate price of $20–$24 per kg dried fruit with production at 8–10 t/year, and Vincent (2010) explained that variability in production from year to year, due to weather factors, has caused prices to fluctuate from as low as $10 per kg (2005) to $50 per kg (2008/2009 – drought years when demand outstripped supply). However, Morse (2005) reported harvest in approximately the same period of 2–5 t/year at $12 per kg but pointed out that production was steadily increasing.

The most recent data come from the Rural Industries Research and Development Corporation Australian Native Food Industry Stocktake in August 2012 (Clarke, 2012). In this survey, fruit was reported by some buyers to be unavailable for the 4 years up to 2010; short supply was credited with pushing prices up to $32–$40 per kg (AUD) reported for 2011, with some reports of prices as high as $45 per kg, while small specialty packaged product retailed for up to $20/100 g ($200 per kg). For 2010/2011, total bush tomato production (includes wild harvesting) was 15 t with a mean farm-gate price of $36 per kg giving a total farm-gate value of $540,000 (Clarke, 2012). Anecdotal information (pers. comm.) suggests that this quoted farm-gate value may be conflated with the higher 'free-in-store' (FIS) value and thus slightly inflated. Reports from processors and growers (pers. comm.) indicate that market demand cannot be satisfied by current supply, and this has kept the 2012/2013 farm-gate price above $25 per kg with the average delivered price of $32 per kg FIS Melbourne, and the highest reported price paid of $64 per kg for dried fruit.

In experimental bush tomato field plots, Ryder et al. (2009) achieved yields of 96–360 kg/ha depending upon source of seedlings, but yields as high as 8000 kg/ha have been cited (Collins, 2002; Miers, 2003). Ryder et al. (2008) achieved a best yield equivalent to 5000 kg/ha. Industry participants (pers. comm.) indicate that 2000 kg/ha sustainable commercial production is easily achievable.

Based on the aforementioned data and stakeholder discussions, conservative assumptions for bush tomato horticultural production as at 2012/2013, for the purposes of generating estimates, indicate

- Mean farm-gate price – $30 per kg (AUD)
- Mean yield – 2000 kg/ha
- Mean annual production – 12 t
- Mean annual market growth – 8%

Through deduction, the 2012/2013 total area approximated 6 ha grossing $60,000 per ha; thus, the 2012/2013 estimated industry farm-gate value was $360,000 (AUD). The higher value reported by Clarke (2012) coincided with a period of scarcity and concomitant elevated prices (plus the aforementioned conflated FIS price component).

Wild harvest complements horticultural production and is highly variable due to seasonal factors. Further, the merit of point-of-harvest value addition is convoluted because of the costs associated with the range of other ingredients and processes involved. For these reasons, estimation of additional value to the wild harvest sector of the bush tomato industry and to the value-added processing sector is too speculative to reliably estimate, but it would be considerable – possibly more than half as much again (and up to fourfold the value for speciality products) as the dollar value of the farm-gate price attributable to the bush tomato horticulture component alone (Clarke, 2012). The economics of value addition is difficult to estimate, because while wholesalers are reported to apply markups of 250%–300%, product is often sold directly to processors who attract a value-added margin of just 15%–20% (Bryceson, 2008). The value of other ingredients and processing costs in creating the processed product makes it difficult to ascertain the margin that applies to the bush tomato component per se.

Furthermore, there is a complex supply/demand relationship with an unknown upper price limit in the value chain where purchasing resistance becomes a factor. One of the key drivers in this relationship is the seasonal climate variability in the arid zone where fruit is produced. To some extent, this is mollified by irrigation, but not entirely – fully half of the market supply typically comes from wild harvesting which declines almost completely in adverse years, while irrigation water for horticultural producers becomes less available or faces competitive uses (e.g. watering livestock) during drought.

AGRONOMY

As seen earlier, the bush tomato industry is not large. Unsurprisingly, very little formal research has been conducted into optimising the agronomic management of the crop. Where better guidance is absent, the abundant information on agronomic practices of regular commercial tomatoes is applied (Vincent, 2010). Vincent's handbook is the only comprehensive guide to bush tomato production.

Australian native plants have in general evolved under nutrient-deficient conditions and particularly so with regard to phosphorus nutrition (Handreck, 1997). Accordingly, it is often opined that restricted fertilizer application is preferable for such plants during the course of cultivation. This practice has been used with no reported ill effects (Dennett et al., 2011; Miers, 2003); however, information from several growers (pers. comm.) indicated the fertilizer applications typical of normal horticultural situations are clearly beneficial. Furthermore, it has been definitively demonstrated that S. centrale is amenable to arbuscular mycorrhizal symbiosis and that this results in enhanced phosphorus uptake and enhanced plant vigour (Dennett et al., 2011).

Bush tomato is a perennial crop and, once planted, can be expected to last many years by taking advantage of the plants' propensity for copious suckering. Plants are transferred to the field as potted tubestock propagated usually as seedlings, although ideally, vegetatively propagated clonal material is preferred (see section 'Genetic Improvement and Propagation'). Planting is generally recommended in rows spaced at 1.5 m with plants 30–50 cm apart (13,300–22,200 plants/ha) (Vincent, 2010). Miers (2003) used a density of 12,500 plants/ha but suggested that plantings as high as 37,000 plants/ha could be considered. As a compromise between optimising the cost of plants, maximising space utilisation and early yield, and minimising competition, a target density of approximately 20,000 plants/ha is suggested.

Bush tomato fruit yield was enhanced almost 700% by removing weed competition in experimental plots and irrigation produced a significant positive response in yield (Ellis et al., 2010), although a maximum threshold of irrigation application was observed, above which there was no yield benefit. Whereas seasonal conditions and local circumstances need to be accommodated, Ellis et al. (2010) recommend an irrigation regime for Central Australia using drippers spaced at 300 mm with a single application of 2 L/hour for 2 hours each week in autumn and spring, twice weekly in summer and no irrigation through most of winter.

An unusual feature of the bush tomato plant insofar as horticultural management is concerned is the propensity for adventitious sprouting from roots, resulting in copious suckering. This inevitably occurs in bare inter-rows and presents a dilemma for growers – on the one hand, it represents regeneration/replacement of older plants, additional fruit bearing and optimal utilisation of space, but on the other hand, suckers can interfere with access of machinery and personnel, can be problematic for weed control and creates an untidy crop. Consequently, there are two schools of thought on crop management – some growers practise inter-row cultivation, which risks damaging plants and encouraging root pathogens; conversely, others do not intervene, resulting in the aforementioned management difficulties. Individual circumstances and preferences will dictate the grower's preferred approach.

An alternative approach to bush tomato production is the 'enrichment planting' strategy. The practice of enrichment planting is widely used in forestry to enhance natural stands whereby existing vegetation is interplanted with plants of a desired species to augment the production of the crop of interest. The method has had only limited application in Australian bush food cultivation (Cunningham et al., 2009; Merne Altyerre-ipenhe Reference Group et al., 2011), and not at all with bush tomato to date.

Growing plants horticulturally bears little relevance to Australian Aboriginal culture; however, collecting plants for food is traditionally a very significant activity, steeped in social significance and rich in intergenerational transfer of knowledge and sharing of cultural practices. The establishment of enrichment plantings may serve the dual purpose of supporting and encouraging these valuable cultural pursuits, on the one hand, and providing ready access to bush food products for local consumption and for generating income (Lee and Courtenay, 2016).

As it is estimated that half of the bush tomato production in a normal season comes from wild harvesting, establishment of enrichment plantings could serve to improve reliability of product supply and provide an opportunity for Aboriginal people to benefit from increased participation in the industry. The author is experimenting with the technique as a means of increasing bush tomato production and to provide an income opportunity in a manner that is culturally relevant to remote Aboriginal communities.

BIOTIC CONSTRAINTS (PESTS AND DISEASES)

Insect pests identified in bush tomatoes include aphids (*Macrosiphum* sp.), crusader bugs (*Mictis profana*), mealy bugs (*Pseudococcus* sp.), native moth (*Alloeocysta* sp.), Sphingidae caterpillars and weevils (Curculionidae) (Collins, 2002). Cowpea aphid (*Aphis craccivora*), Rutherglen bug (*Nysius vinitor*) and silver-leaf whitefly (*Bemisia tabaci*) have also been found to cause plant damage (Vincent, 2010). Chemical applications with products registered for these pests are the usual form of recommended control.

Postharvest pests of bush tomatoes include eggplant caterpillar (*Sceliodes cordalis*) (Vincent, 2010), Indian meal moth (*Plodia interpunctella*), rust red flour beetle (*Tribolium castaneum*), saw-toothed grain beetle (*Oryzaephilus surinamensis*) and warehouse beetle (*Trogoderma variabile*) (De Sousa Majer et al., 2009). In order to minimise postharvest fruit infestations, recommendations are provided for cleaning and grading, washing, drying, milling and storing (De Sousa Majer et al., 2009; McDonald et al., 2006; Vincent, 2010). Satisfactory control of postharvest pests, without any adverse effect to bush tomato fruit quality, has been demonstrated using high-temperature and low-pressure disinfestation treatments (De Sousa Majer et al., 2009).

Few pathogens appear to afflict bush tomatoes; the only ones reported are root fungi (*Pythium* and *Phytophthora* spp.) and sooty mould on aerial parts (Collins, 2002). There is also potential for nematode infestations under cultivation (Vincent, 2010).

HARVESTING, PROCESSING AND MARKETING

The bush tomato supply chain is not well developed; supply is typically limited in all but the most abundant years, and sources are often not divulged by supply-chain participants, but it is known that approximately half of the raw product supply in a normal year comes from wild harvest. This supply diminishes significantly in drought seasons (Bryceson, 2008; Clarke, 2012; Cleary, 2009). The value-chain tends to be short with food processors often sourcing bush tomato supplies directly from the small number of growers, from wild harvesters or alternatively from a single middleman acting as a collection and resale point for wild harvested fruit. Wild harvesting is conducted by Aboriginal women in arid Central Australia, and it serves a significant cultural and social function at least as much as the pursuit of profit (Alexandra and Stanley, 2007; Holcombe et al., 2011; Morse, 2005; Walsh and Douglas, 2011) (Figures 3.3 and 3.4).

Limited value adding at the point of harvest is conducted apart from the removal of contaminants and the drying of fruit. There is thus an opportunity for harvesters to increase the price and/or quality of their product by means of simple steps such as grading, grinding and postharvest pest control (McDonald et al., 2006). Australian bush tomato (*S. centrale*) fruit is almost exclusively used as a condiment to flavour

Figure 3.3 (See colour insert.) Photo of hand harvesting of bush tomato (*Solanum centrale*) fruit. (Photo by Slade Lee. Copyright Southern Cross University.)

Figure 3.4 Photo of harvested bush tomato (*Solanum centrale*) fruit. (Photo by Slade Lee. Copyright Southern Cross University.)

processed products such as chutney, sauces, dukkah and meat products and as a dried seasoning; therefore, the most significant value adding occurs in processing.

In addition to manual wild harvesting, two approaches are applied in farmed operations – manual picking and mechanical harvesting. The ideal plant phenotype is influenced by the harvesting method. Many desirable features are common to both manual and mechanical harvesting, while other ideal features depend on the harvesting approach. Consequently, genetic selection ideotype characters are affected by harvesting method. For mechanical harvesting, concentrated maturity is required to minimise the number of harvester passes required and reduces the number of undesirable immature fruit, but for manual harvesting, a prolonged maturity period facilitates the continuous harvesting required due to ubiquitous labour shortages in remote parts of Australia where the crop grows. The presence of prickles is of no consequence to mechanical harvesters, but is a major problem for hand harvesting as they are a serious irritant and can result in infections to fruit pickers' hands. Mechanical harvesting is typically performed using small modified grain harvesters (Dessart, 2008), and although these are not ideal, the size of the bush tomato industry does not warrant the costly development of a purpose-designed machine.

GENETIC IMPROVEMENT AND PROPAGATION

The amount of genetic improvement undertaken in bush tomato is minimal to date (Bryceson, 2008), but there is research evidence of the potential benefits that could accrue (Ellis et al., 2010; Ryder et al., 2009). Several horticultural growers are known to select superior plants as parents when re-propagating in order to develop the quality of their stocks; however, due to prevalent outcrossing, the results appear to be inconsistent and variable, although improvements are certainly being made. At the time of publication, the only known institutional genetic improvement program is the Plant Business project of the Cooperative Research Centre for Remote Economic Participation.

Molecular techniques have been applied to understand the genetics of Australian *Solanum* species (Collins, 2002; Waycott, 2010; Waycott et al., 2011). High levels of genetic diversity were detected in *S. centrale* (HT 0.59–0.89), and phylogenetic analysis of 14 *Solanum* taxa from Central Australia using five chloroplast DNA loci indicated hybridisation and phylogenetic complexity. Collins (2002) identified a trait-specific molecular marker (RAPD) for prickliness, which could prove useful in genetic selection of improved varieties.

Propagation is almost universally conducted via seedling production; however, the benefits of clonal propagation have been recognised, particularly when improved genetic lines have been identified (Lee, 2012). The propagation of root cuttings appears to offer great potential, due to the propensity for *S. centrale* to produce suckers from roots (Dennett, 2006; Vincent, 2010).

S. centrale is notorious for poor seed germination under nursery conditions. A range of pregermination treatments for *S. centrale* seeds has been investigated (reviewed by Lee, 2012), and various methods are presented by Vincent (2010). The highest germination percentage is consistently achieved by the use of gibberellic acid treatments at a concentration of 1.4–2.9 mM (Commander et al., 2008; Vincent, 2010).

Tissue culturing protocols have also been established for *S. centrale* (Johnson and Ahmed, 2005; Johnson et al., 2003), and the author is aware of one commercial tissue culture laboratory that maintains bush tomato cultures. However, given the small size of the industry, there seems little justification for the routine use of tissue culture for bush tomato propagation; it may, nonetheless, prove highly valuable for rapidly increasing stocks of selections of superior genetic lines.

CONCLUSIONS AND RECOMMENDATION

Australian bush tomato (*S. centrale*) has origins as a food plant deep in the Aboriginal culture thousands of years ago, yet it is only in the last few decades that its potential as a commercial crop has been recognised. Accordingly, bush tomato production is still a nascent industry of very small size and managed only by small-scale producers. However, there is consistently high demand for fruit, and there is a clear market opportunity. A significant challenge to tapping this opportunity, however, is the fact that the plant is adapted to grow in a quite specific environment – namely the arid conditions and sandy soils typical of Central Australia. The plant does not appear to flourish in locations where it is exposed to regular rain periods. The ideal locations coincide with remote sparsely populated regions and are disadvantaged by vast distances to markets and to support services. Moreover, notwithstanding the market potential, it is very likely that a situation of oversupply, and resultant price impact, could easily occur for a product which is used only as a flavouring agent in what is at present a niche market.

Another major challenge to development of a bush tomato industry is the current heavy reliance on the variable wild harvested supply. Due to regular seasonal adversity and the unreliable supply of wild harvester labour, considerable variability of bush tomato supply results and unpredictability in the value chain occurs. Horticultural production, concurrently with wild harvesting, should result in stabilisation of supply for buyers and processors, yet opens an opportunity for high-value indigenous enterprises trading on authenticity and/or capitalising on a fair trade theme. The enrichment planting approach would enhance this value.

These constraints suggest that a managed coordinated approach would best serve the development of the industry to the benefit of all involved. Partnership arrangements amongst business people are being developed and cooperative ventures are being promoted. There will always be sole operators in this restricted marketplace, but it appears that the best prospect for a sustainable Australian bush tomato industry will come through cooperation amongst industry stakeholders.

ACKNOWLEDGEMENT

The author's research and development work with Australian bush tomato is undertaken with the financial support of the Australian Government Cooperative Research Centre's Program through the Cooperative Research Centre for Remote Economic Participation (http://crc-rep.com/) and its parent organisation Ninti One Limited (http://www.nintione.com.au/). The views expressed here do not, however, necessarily reflect the views of all Ninti One Limited partners or participant organisations of the CRC for Remote Economic Participation.

REFERENCES

Albrecht, D., Duguid, A., Coulson, H., Harris, M., Latz, P., 2007. *Vascular Plant Checklist for the Southern Bioregions of the Northern Territory: Nomenclature, Distribution and Conservation Status*, 2nd ed. Department of Natural Resources, Northern Territory Government, Alice Springs, Northern Territory, Australia.

Alexandra, J., Stanley, J., 2007. *Aboriginal Communities and Mixed Agricultural Businesses – Opportunities and Future Needs*. Rural Industries Research and Development Corporation, RIRDC Publication No. 07/074, Canberra, Australian Capital Territory, Australia.

Alyawarr speakers from Ampilatwatja, Walsh, F., Douglas, J., 2009. Angka Akatyerr-akert: A Desert raisin report. Desert Knowledge Cooperative Research Centre. Alice Springs, Northern Territory, Australia.

Barker, B., Barker, R., Jessop, J., Vonow, H., 2005. Census of South Australian vascular plants – Ed. 5. *Journal of the Adelaide Botanic Gardens Supplement* 1, 1–397.

Bean, A.R., 2004. The taxonomy and ecology of *Solanum* subg. *Leptostemonum* (Dunal) Bitter (Solanaceae) in Queensland and far northeastern New South Wales, Australia. *Austrobaileya* 6, 639–816.

Black, J., 1965. *Flora of South Australia*. South Australian Branch of the British Science Guild Handbooks Committee. Government Printer, Adelaide, South Australia, Australia.

Bonney, N., 2006. *Adnyamathanha and Beyond – Useful Plants of an Ancient Land*. Australian Plants Society, Unley, South Australia, Australia.

Bryceson, K., 2008. Value chain analysis of bush tomato and wattle seed products. DKCRC Research Report No. 40. Desert Knowledge Cooperative Research Centre. Alice Springs, Northern Territory, Australia.

Clarke, M., 2012. *Australian Native Food Industry Stocktake*. Rural Industries Research and Development Corporation, RIRDC Publication No. 12/066, Canberra, Australian Capital Territory, Australia.

Cleary, J., 2009. Perspectives on developing new cooperative arrangements for bush-harvested bush tomatoes from desert Australia. DKCRC Working Paper No. 48. Desert Knowledge Cooperative Research Centre. Alice Springs, Northern Territory, Australia.

Collins, C., 2002. A study into the domestication of *Solanum centrale*, Australian bush tomato. PhD thesis, University of Adelaide, Waite Campus, South Australia, Australia.

Commander, L.E., Merritt, D.J., Rokich, D.P., Flematti, G.R., Dixon, K.W., 2008. Seed germination of *Solanum* spp. (Solanaceae) for use in rehabilitation and commercial industries. *Australian Journal of Botany* 56, 333–341.

Cunningham, A., Garnett, S., Gorman, J., Courtenay, K., Boehme, D., 2009. Eco-Enterprises and *Terminalia ferdinandiana*: 'Best Laid Plans' and Australian policy lessons. *Economic Botany* 63, 16–28.

Dennett, A., 2006. Underground structures and mycorrhizal associations of *Solanum centrale* (the Australian bush tomato). Honours thesis, University of Sydney, Sydney, New South Wales, Australia.

Dennett, A., Burgess, L., McGee, P., Ryder, M., 2011. Arbuscular mycorrhizal associations in *Solanum centrale* (Bush Tomato), a perennial sub-shrub from the arid zone of Australia. *Journal of Arid Environments* 75, 688–694.

Dessart, B., 2008. Mechanical harvesting of bush tomato/desert raisin, *Solanum centrale*. Desert Knowledge Cooperative Research Centre. Alice Springs, Northern Territory, Australia.

De Sousa Majer, M., Singh, Z., De Lima, F., Ryder, M., 2009. Sustainable bush produce systems – Post-harvest storage of *Solanum centrale* and impact on produce quality. DKCRC Report No. 46. Desert Knowledge Cooperative Research Centre. Alice Springs, Northern Territory, Australia.

Ellis, G., Oliver, G., Vincent, A., Raghu, S., 2010. The effect of irrigation and weed competition on yield of a commercial scale *Solanum centrale* (bush tomato) production system: A preliminary investigation. DKCRC Working Paper No. 74. Desert Knowledge Cooperative Research Centre. Alice Springs, Northern Territory, Australia.

Foster, M., Jahan, N., Smith, P., 2005. *Emerging Animal and Plant Industries – Their Value to Australia*. Rural Industries Research and Development Corporation, RIRDC Publication No. 05/154, Canberra, Australian Capital Territory, Australia.

Handreck, K., 1997. Phosphorus requirements of Australian native plants. *Australian Journal of Soil Research* 35, 241–289.

Holcombe, S., Yates, P., Walsh, F., 2011. Reinforcing alternative economies: Self-motivated work by central Anmatyerr people to sell Katyerr (Desert raisin, Bush tomato) in central Australia. *The Rangeland Journal* 33, 255–265.

Jessop, J., 1981. *Flora of Central Australia*. A.H. and A.W. Reed, Sydney, New South Wales, Australia.

Johnson, K., Ahmed, A., 2005. *Developing Propagation Techniques for Bush Tomato.* Australian Flora Foundation, Sydney, New South Wales, Australia.

Johnson, K., Ahmed, A., Armstrong, G., 2003. Investigations into *in vitro* manipulation of *Solanum centrale* (Bush tomato). *Acta Horticulturae* 616, 169–175.

Latz, P., 1995. *Bushfires & Bushtucker – Aboriginal Plant Use in Central Australia.* IAD Press, Alice Springs, Northern Territory, Australia.

Latz, P., 1999. *Pocket Bushtucker.* IAD Press, Alice Springs, Northern Territory, Australia.

Lee, L.S., 2012. Horticultural development of bush food plants and rights of Indigenous people as traditional custodians – The Australian Bush Tomato (*Solanum centrale*) example: A review. *The Rangeland Journal* 34, 359–373.

Lee, L.S. and Courtenay, K. 2016. Enrichment plantings as a means of enhanced bush food and bush medicine plant production in remote arid regions – A review and status report. *Learning Communities – International Journal of Learning in Social Contexts* Special Issue 19 (in press).

Lingard, K. 2015. Legal support for the interests of Aboriginal and Torres Strait Islander peoples in the commercial development of new native plant varieties: Current status and future options. *Australian Intellectual Property Journal* 26, 39–57.

McDonald, J., Caffin, N., Sommano, S., Cocksedge, R., 2006. *The Effect of Post Harvest Handling on Selected Native Food Plants.* Rural Industries Research and Development Corporation, RIRDC Publication No. 06/021, Canberra, Australian Capital Territory, Australia.

Merne Altyerre-ipenhe Reference Group, Douglas, J., Walsh, F., 2011. Aboriginal people, bush foods knowledge and products from central Australia: Ethical guidelines for commercial bush food research, industry and enterprises. DKCRC Report No. 71. Desert Knowledge Cooperative Research Centre. Alice Springs, Northern Territory, Australia.

Miers, G., 2003. *Cultivation and Sustainable Wild Harvest of Bushfoods by Aboriginal Communities in Central Australia.* Rural Industries Research and Development Corporation, RIRDC Web-only Publication No. W03/124, Canberra, Australian Capital Territory, Australia.

Morse, J., 2005. Bush resources: Opportunities for Aboriginal enterprise in central Australia. DKCRC Report No. 2. Desert Knowledge Cooperative Research Centre. Alice Springs, Northern Territory, Australia.

O'Connell, J., Latz, P., Barnett, P., 1983. Traditional and modern plant use among the Alyawara of central Australia. *Economic Botany* 37, 80–109.

Peterson, N., 1979. Aboriginal uses of Australian Solanaceae. In: Hawkes, J.G., Lester, R.N., Skelding, A.D. (eds.), *The Biology and Taxonomy of Solanaceae.* Linnean Society Symposium Series 7. Academic Press, London, U.K., pp. 171–189.

Purdie, R., Symon, D., Haegi, L., 1982. Solanaceae. In: George, A.S. (ed.), *Flora of Australia*, Vol. 29. Australian Government Publishing Service, Canberra, Australian Capital Territory, Australia, pp. 1–2.

Randell, B., Symon, D., 1976. Chromosome numbers in Australian *Solanum* species. *Australian Journal of Botany* 24, 369–379.

Robins, J., Ryder, M., 2008. *New Crop Industries Handbook – Native Foods.* Rural Industries Research and Development Corporation, RIRDC Publication No. 08/021, Canberra, Australian Capital Territory, Australia.

Ryder, M., Latham, Y., Hawke, B., 2008. *Cultivation and Harvest Quality of Native Food Crops.* Rural Industries Research and Development Corporation, RIRDC Publication No. 08/019, Canberra, Australian Capital Territory, Australia.

Ryder, M., Walsh, F., Douglas, J., Waycott, M., Robson, H., Singh, Z., de Sousa Majer, M., Collins, T., White, J., Cheers, B., 2009. Sustainable bush produce systems – Progress Report 2004–2006. DKCRC Working Paper No. 31. Desert Knowledge Cooperative Research Centre. Alice Springs, Northern Territory, Australia.

Symon, D., 1981. A revision of the genus *Solanum* in Australia. *Journal of the Adelaide Botanic Gardens* 4, 1–367.

Vincent, A., 2010. *The Bush Tomato Handbook.* Desert Knowledge Cooperative Research Centre, Alice Springs, Northern Territory, Australia.

Walsh, F., Douglas, J., 2011.No bush foods without people: The essential human dimension to the sustainability of trade in native plant products from desert Australia. *The Rangeland Journal* 33, 395–416.

Waycott, M., 2010. Genetic diversity, trait variation and plant improvement. DKCRC CP2.1 Final Report. Desert Knowledge Cooperative Research Centre. Alice Springs, Northern Territory, Australia.

Waycott, M., Jones, B., Van Dijk, J., Robson, H., Calladine, A., 2011. Microsatellite markers in the Australian desert plant, *Solanum centrale* (Solanaceae). *American Journal of Botany* 98, e81–e83.

Whalen, M., 1984. Conspectus of species groups in *Solanum* subgenus *Leptostemonum.* *Gentes Herbarum* 12, 179–282.

CHAPTER 4

Cultivation of Davidson's Plum (*Davidsonia* spp.)

Tony Page and Margo Watkins

CONTENTS

BOTANY

Family: Cunoniaceae

Until recently the genus *Davidsonia* was listed in Davidsoniaceae, but since its revision in 2000, the genus is considered to be part of the Cunoniaceae family (Harden and Williams, 2000). The genus was also formerly considered a monotypic genus; however, *Davidsonia* comprises of three species:

1. *Davidsonia pruriens* described by F. Muell. and a type of species from north-eastern Queensland
2. *Davidsonia jerseyana* described by F. Muell. (ex F.M. Bailey) and G. Harden and J.B. Williams and is a known variety from north-eastern New South Wales (NSW)

3. *Davidsonia johnsonii* described by G. Harden and J.B. Williams which is a well-known and newly described species from north-eastern NSW and South East Queensland

A detailed description of the two principle commercial species (*D. pruriens* and *D. jerseyana*) is provided below.

Scientific name: *Davidsonia pruriens* F. Muell.

Common names: Davidson's plum, Davidsonian plum, North Queensland plum, ooray plum and sour plum.

Description: Small slender rainforest tree, which grows up to 18 m high; bark is brown to dark grey that becomes flaky. Tree form is generally a single trunk with a few branches. Fine hairs are prominent on all leaves and stems, which are long and cause irritation (Isaacs, 2002). New vegetative growth flushes are pink in colour. Adult leaves are compound and are positioned on the stem in an alternate arrangement. The rachis is slightly winged and is 60–1500 mm long, toothed with small leaf-like growths. Each leaf comprises of 7–19 leaflets, and these leaflets are 50–460 mm wide and 25–160 mm long. The upper side of the leaves are densely hairy with the underside less so. Lateral petioles are 1–5 mm long and the stipules are 5–8 mm long.

Inflorescences: An axillary or mostly ramiflorous panicle or raceme, often borne on the axis of new branch stems.

Flowers: Diameter 3–5 mm, petals absent, sepals 4 or 5 and green to pink in colour.

Fruit: A drupe, purple, blue or blackish in colour when ripe, with a bloom (Figure 4.1). Its size is 30–55 mm long × 32–60 wide. The surface of the fruit is sparsely

Figure 4.1   Photo of *Davidsonia pruriens* fruit. (Photo by Margo Watkins. Copyright Rainforest Heart Orchard.)

covered in fine rusty-coloured hairs with firm pink-red flesh. The fruits often contain 2, sometimes 1 and rarely 3, seeds. The shape of the seed is strongly laterally compressed, 3–6 mm long with fringing edge of radiating soft fibres attached to the seed body. In a two-seeded fruit, one seed is typically viable (larger of the two) and the other smaller seed is infertile. Detailed illustration of the leaves, flower, seed and fruit is presented in Figure 4.2.

Habitat and distribution: Widespread in tropical rainforest in north-eastern Queensland, chiefly from Cardwell to Cooktown and inland to near Atherton. It grows from near sea level to altitudes of over 1000 m and is cultivated in both northern NSW and tropical north Queensland near Atherton.

Phenology: Generally flowers between February and July, but flowers have been recorded in most months, with fruit ripe mostly between March and June. Small quantities of fruit are also recorded throughout the year in the north Queensland plantings.

Conservation status: Due to its widespread distribution, it is not considered to be under threat. It is known to occur in a number of conservation areas, for example Lake Eacham National Park and Wooroonooran National Park, and is widely planted for domestic and commercial purposes.

It should be noted that there is a recognised variant of *D. pruriens* among the north Queensland plantations. This variant, locally described as the 'smooth' or 'coastal' plum, is described further in the 'Genetic Improvement and Propagation' section of this chapter. This variant is also recognised in NSW plantations and is known as the 'lowland' variety.

Scientific name: *Davidsonia jerseyana* (F. Muell. ex F.M. Bailey) G. Harden and J.B. Williams.

Common names: Davidson's plum, Southern Queensland plum and sour plum

Description: *D. jerseyana* is a slender tree growing up to 10 m high, with few branches or with several stems from the base. The leaves are compound with an alternate arrangement and mostly 35–80 cm long. Each leaf has 11–17 oblong-shaped leaflets, mostly 6–30 cm long and 3–10 cm wide and more or less covered with fine hairs.

Flowers: The species has small dark pink- to red-coloured flowers that are arranged on panicles.

Inflorescences: The panicles are usually cauliflorous, borne on the main stem or older branches, and 4–10 cm long.

Fruit: Fruit are blue-black in colour, pear to oval shaped and lightly covered with golden-brown hairs (Figure 4.3). The fruit has a dark-red flesh and usually has two large seed cases that are fibrous on the surface, each containing a single seed. A detailed illustration of the leaves, flower, seed and fruit is found in Figure 4.4. See 'A revision of Davidsonia' by Harden and Williams (2000) for a full taxonomic description of the species.

Habitat and distribution: Usually grows in subtropical and riverine rainforest, in moderately high rainfall from 1100 to over 2000 mm, at altitudes less than 300 m

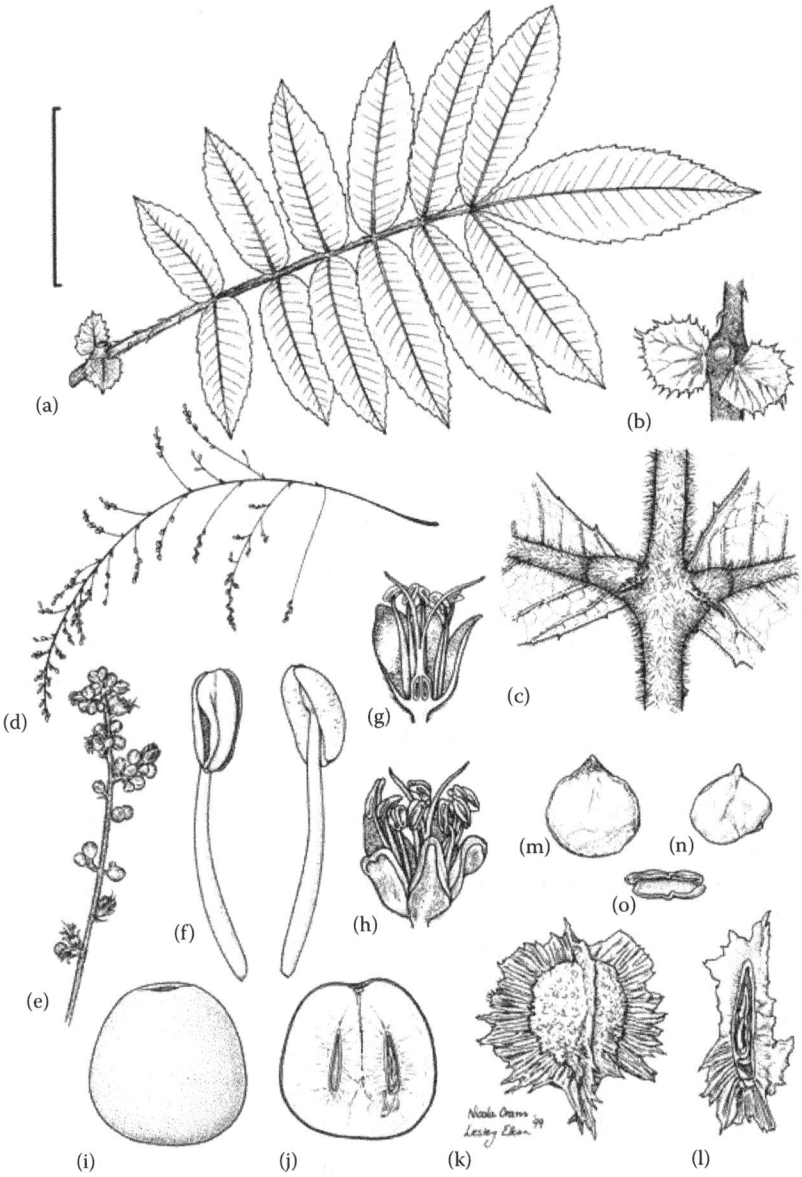

Figure 4.2 Botanical illustration of *Davidsonia pruriens*. (a) Adult leaf; (b) stipules; (c) rachis, pinnae bases and stipellae; (d) inflorescence and diagrammatic; (e) part inflorescence; (f) stamens; (g) flower LS; (h) flower; (i) drupe; (j) drupe L.S.; (k) pyrene; (l) pyrene in LS; (m) seed; (n) embryo; and (o) embryo showing 2 cotyledons. Scale bar: (a and d) 30 cm, (b, e, i and j) 6 cm, (c) 2 cm, (k, l, m–o) 3 cm, (f) 4 mm and (g and h) 1 cm. (From Harden, G.J. and Williams, J.B., *Telopea*, 8, 413, 2000; Illustrated by Lesley Elkan and Nicola Oram. Copyright Royal Botanic Gardens and Domain Trust.)

Figure 4.3   (See colour insert.) Photo of *Davidsonia jerseyana* fruit. (Photo by Margo Watkins. Copyright Rainforest Heart Orchard.)

above sea level. Occurs mostly on red and yellow podsolic soils of clay texture, over fine-grained meta sediments, greywacke, slate, phyllite or quartzite, also on alluvial deposits. It is confined in north-eastern NSW, chiefly from the Brunswick River and Tweed River catchments, from Mullumbimby north to Urliup, Upper Crystal Creek and Settlement Road. The species has been established as plantations in conjunction with *D. pruriens* plantings in both subtropical and tropical climates.

TRADITIONAL USE

The name 'ooray' came from the Tully River Aborigines (Bailey, 1900 cited in Harden and Williams, 2000), and so it is documented very early during Australia's colonisation that the Aborigines used this fruit for food. In the Malanda region of tropical north Queensland, the plum is called *Wiraa* in the local Ngadjon Jii language (Andrew Morta, Ngadjon man, pers. comm., 2014) and was consumed as a fresh fruit. Mayall (2014) noted that Aboriginal women advised that the flesh was also used topically to heal wounds and kill bacteria. Several descendants of Atherton Tableland settlers have stated that these settlers considered that the hairs of the tree could cause

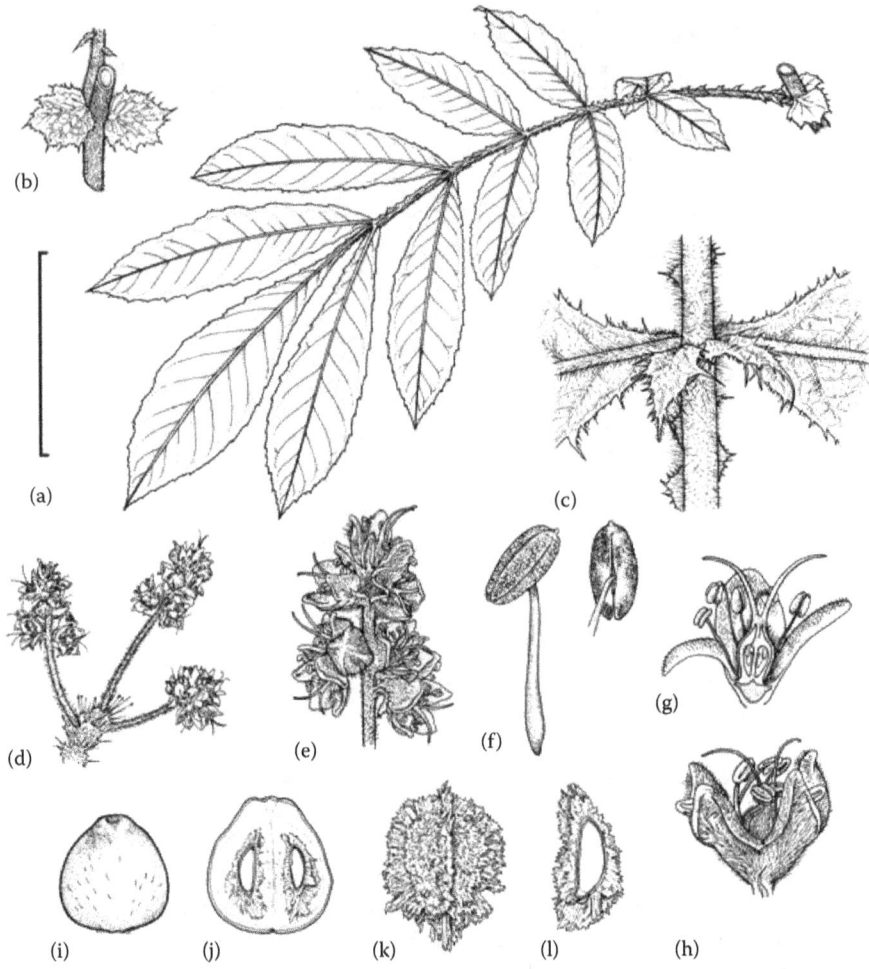

Figure 4.4 Botanical illustration of *Davidsonia jerseyana*. (a) Adult leaf; (b) stipules; (c) rachis, pinnae bases and stipellae; (d) inflorescence; (e) part inflorescence; (f) stamens; (g) flower LS; (h) flower; (i) drupe; (j) drupe in LS; (k) pyrene; and (l) pyrene in LS. Scale bar: (a) 15 cm, (b, d, i and j) 6 cm, (c) 2 cm, (e, k and l) 3 cm, (f) 4 mm and (g and h) 1 cm. (From Harden, G.J. and Williams, J.B., *Telopea*, 8, 413, 2000; Illustrated by Lesley Elkan and Nicola Oram. Copyright Royal Botanic Gardens and Domain Trust.)

blindness (Ian Thompson, pers. comm., 2011). While the settler's perception is inaccurate, they cut down *D. pruriens* on specific properties across the southern Tablelands. Interestingly, numerous large living specimens (*D. pruriens*) occur as isolated individuals or small clusters within open grazing paddocks across a band of country between the localities of Upper Barron and Lake Eacham on the Atherton tableland. This appears to indicate that in specific localities, early settlers were aware of the tree's food value (probably informed by indigenous workers in agriculture and forestry) and thus, settlers retained these individuals.

Records from the nineteenth century (Bailey, 1895 cited in Harden and Williams, 2000; Bailey, 1898) recommended this species for cultivation because of its edible fruits. *D. pruriens* is described as suitable for preparation as a full flavoured, dry red wine (SGAP, 1988). Both *D. pruriens* and *D. jerseyana* are widely cultivated for their edible fruits and planted as ornamental trees in parks and gardens.

ECONOMIC POTENTIAL

Wild harvested Davidson's plum has provided the basis for the development of a small niche industry in Australia. Over the past 15 years, the supply of Davidson's plum has made an almost complete transition from wild harvest to plantations. In north Queensland, the size of plantings ranges from the very small (50–200 trees) to modest plantings (1000–5000 trees). Recent research (Konczak et al., 2009, 2012) has highlighted the health properties of the Davidson's plum including its high potential as a functional food, with particular reference to the high levels of total phenolics and antioxidants. Functional foods, also known as nutraceuticals, are those foods of man which contain biologically active components, which offer the potential of either enhanced health or reduced risk of disease. Examples of functional foods include foods that contain specific minerals, vitamins, fatty acids or dietary fibre and foods with high biologically active phytochemicals or other antioxidants and probiotics that have live beneficial cultures. This research information has the potential to raise the fruit's market profile as a health product and increase demand. This positive market outlook has led several growers to expand their orchards and mechanise their processing to meet the demands of large-scale food manufacturers. Production estimates therefore increased from 10 tons in 2012 (Clarke, 2012) to a projected 12–15 tons in 2016 (ANFIL and RIRDC, 2014).

A large percentage of producers value-add their own produce and service farmer's markets, online sales and a range of domestic and international retailers. Some producers have sufficient quantity of minimally processed fruit, and can offer competitive prices, to service medium-scale processors and food manufacturers. However, given that just one boutique yogurt maker currently uses 2.6 tons per annum, it is clear that an increase in production will be required if large-scale food and health sector manufacturers take up native functional foods. This projected increase in demand should help to make Davidson's plum a commercially attractive orchard crop. In addition, research currently (RIRDC project PRJ-9026) underway aims to identify the best producing individual trees (for reproduction) and horticultural practices (in tropical north Queensland conditions), which will further guide the viable production of this species. Mechanising post-harvest processing and improved post-harvest storage is also fundamental in ensuring the commercial viability of this species, and both of these factors are considered important areas of development as part of the broader native food industries' ongoing R&D priorities.

AGROCLIMATIC REQUIREMENTS

D. pruriens grows well in a tropical to subtropical environment with a minimum annual rainfall of 1200–2500 mm, particularly where distributed evenly through the year. The species can grow across a range of soil types and textures but performs best on well-drained friable, slightly acidic (pH 5.5–6.0) soil types with a high level of organic matter and good water-holding capacity. Trees prefer sheltered sites, particularly with surrounding vegetation to provide protection from persistent winds. This shelter is particularly important during the period of flowering and fruiting to ensure good fruit set and development (Hotson, 2008).

D. jerseyana is endemic to the Northern Rivers region of NSW and wild populations are currently only known from the Tweed and Brunswick River catchments (DEC-NSW, 2004). However, there are plantation plantings of the species near Port Macquarie on the mid-north coast of NSW.

Wild specimens have been recorded at altitudes up to 300 m; however, most records reported specimen at altitudes below 100 m and the elevation of the Port Macquarie area is under 100 m AHD. Slope and aspect vary, but the majority of known wild subpopulations occur in south- and east-facing slopes, with only a few individuals recorded on north-facing slopes (A. Benwell pers. comm.). Plants are generally located in clusters around sheltered lower hill slopes, and this would suggest that *D. jerseyana* prefers protected sites. In north Queensland plantations, the species is successfully grown at elevations between 450 and 900 m Australian Height Datum (AHD). Higher elevation plantings can perform well, and this is most likely due to congruence in temperature between high elevation tropical climate of north Queensland and the subtropical climate of coastal mid-north NSW. This species can also grow across a range of soil types but will perform best on fertile, well-drained soils with evenly distributed annual rainfall of 1200–2500 mm, similar to that of *D. pruriens*.

AGRONOMY

The agronomy of *D. pruriens* is not well documented, although the importance of maintaining adequate soil moisture throughout the growing and fruiting season is gaining wide acceptance among growers. Irrigating the crop is essential for many sites to mitigate variability in rainfall, particularly during the period of fruit set and development. Much benefit is also gained in the general health and vigour of the tree by maintaining high soil organic matter, and protecting the soil from high temperatures. Regular applications of small amounts of fertiliser (either organic or inorganic) over the entire growing season are important for increasing overall tree vigour, which supports higher flower and fruit production.

Canopy structure is an important modifying factor affecting fruit yields, where multi-stemmed trees have higher yields than single stemmed trees. Trees with multiple stems have a greater number of sites where inflorescences can be formed (particularly in leaf axils for *D. pruriens*), which has an overall positive effect on fruit numbers and thus yield. Much of the multi-stemmed trees in north Queensland have come about after natural wind damage, particularly the weather events of Cyclone Larry (2006) and Yasi (2011). A multi-stemmed habit can also be promoted by pruning the central leader during the first few years after establishment. The crowns of most cultivated trees are upright and erect, which contrasts to the spreading canopies of much older remnant trees observed in pasture areas of the Atherton Tablelands. It is not yet clear whether the difference in canopy structure between wild and cultivated trees is due to higher tree density of plantations, or the difference in age between the plantations and remnant trees.

Protecting the crop from prevailing winds through the use of planted or natural windbreaks will reduce leaf-to-air vapour pressure deficit, allowing leaf stomata to remain open and tree growth to continue. This protection combined with ensuring adequate soil moisture will contribute to improved growth and reproduction. The reduction in wind velocity that windbreaks afford may also promote efficient transfer of pollen between trees to affect successful fertilisation and subsequent fruit set.

D. pruriens trees typically attain sexual maturity at around 3 years of age, although consistent production of commercial quantities of fruits occurs at around 5–8 years of age. Following cyclone damage, mature trees can take a further 2–3 years to recover condition and begin producing commercial volumes. In north Queensland, the timing of fruiting varies between sites, but typically occurs between March and October. At some sites, a low level of fruit set occurs continuously over most of the season, whereas other sites can experience a more intense season lasting around 2 months. The species produces a mass of flower buds that open gradually over a period of months. During the season, flowering is promoted by rain events followed by warm sunny conditions. Flowers open at a rate of 1–2 per day with generally no more than 10 flowers open at a given time on an inflorescence with up to 200 flower buds. This type of flowering phenology may be described as opportunistic where a small number of flowers are available for pollination at any given time and only set fruit when the environmental conditions are conducive for pollination and/or fruit set.

The pollen of *D. pruriens* is minute in size and clearly observed to drift on air currents. This observation combined with a consistent lack of flower-tending insects across all orchards in north Queensland suggests wind pollination may be a feature of this species. Most growers report poor fruit set when flowering coincides with heavy and/or persistent rain. Systematic observations of flowers reveal that pollen shedding is inhibited during periods of rain when the anthers are wet, which may reduce pollen transfer and fertilisation. Despite these reports and observations, Eliott et al. (2014) proposed that apomixis (clonal seed set without the need for fertilisation) was widespread in *D. pruriens*. Therefore, heavy and/or persistent rain may be a physical mechanism for flower drop, rather than causing a reduction in pollen transfer between flowers.

The rate of flower development is positively associated with temperature and flower opening and pollen shed occurs primarily in the mid-afternoon particularly during periods of warmer temperature and lower humidity. Flowers can open over a period of approximately 12 hours, followed by pollen shed for up to 24 hours, but under warm and dry conditions, this may occur within 6–8 hours. Flowers can be abscised (shed) immediately after pollen shed, or later after the anthers too have been abscised.

Agronomic requirements of *D. jerseyana* are similar to those of *D. pruriens* in that plantings require soil moisture and fertiliser to ensure vigour and productivity. In north Queensland, the onset of sexual maturity appears to occur when trees are at least 4–5 years old. It is noted, however, that NSW growers report that the *D. jerseyana* fruits earlier (at 3 years) than the *D. pruriens* (at 5–6 years), demonstrating a possible climatic effect on the length of juvenile period for both species. Further work on the reproductive biology of this species is required to determine flower and fruiting phenology, mode of pollination and weather effects on fruit set and development.

GENETIC IMPROVEMENT AND PROPAGATION

Two broad variants of *D. pruriens* are recognised in cultivation: (1) 'hairy' or 'typical' form and the (2) 'smooth' or 'coastal' form which has less hairy fruit and leaves, softer fruit skin, smaller fruit size but a higher yield and a lighter crown than the 'typical' variant (Cornelius, 2011). In north Queensland, the 'smooth' variant sets fruit earlier (between March and June) in the year compared with the 'typical' (between May and October). The broad morphology of the 'typical' variant is found throughout the natural populations of the species in north Queensland, whereas the origins of the 'smooth' variant are yet to be determined. Growers report that seedlings derived from both the 'smooth' and 'typical' variants breed 'true-to-type', resembling their parents in broad morphology. These observations are consistent with *D. pruriens* having a predominantly clonal breeding system where 89% of the progeny were identical to the heterozygous maternal parent (Eliott et al., 2014). Therefore, the smooth variant may potentially represent a selection from a single genotype and clonally propagated through seed. Sultanbawa et al. (2015) reported that the highland or 'typical' variety of *D. pruriens* had significantly higher total phenolics level and antioxidant capacity than the lowland or 'smooth' variety. The total phenolic levels and FRAP values of the lowland (smooth) fruits from both NSW and QLD orchards were similar, which indicates that the variation between the lowland and the highland varieties might be genetic, rather than agronomic. Depending on the target market, this is another factor that perhaps should be considered in any evaluation of the best cultivars to propagate.

Development of varieties that produce consistently sized large fruit is important for the development of the industry as this will increase labour efficiencies in harvesting and processing operations. The average weight of *D. pruriens* fruit is 50–60 g; however, significant variation between fruits exists within a tree where an average of a two- to fivefold difference in the weight has been recorded between the smallest and largest fruits. The current fruit size variation within a tree will also present challenges for mechanised harvesting and processing protocols, most likely requiring fruit to be graded according to size before processing.

Commercial plantings of Davidson's plum in north Queensland are typically established using seedlings collected initially from unselected wild sources. The fruit of *D. pruriens* generally contains two seeds, one of which is filled and viable and the other is reduced in size and often non-viable (Nand et al., 2004). A high level of germination can be attained by sowing freshly collected seeds, with close to 100% for the filled seed and 20% for the smaller seeds. Seeds can be readily stored wet in a sealed plastic bag for up to 1 month without a substantial loss in viability. As already mentioned, most (89%) of the seedlings produced are genetically identical to parental plant (Eliott et al., 2014), which will facilitate the development of clonal cultivars derived from seed.

D. pruriens can be clonally propagated by stem cuttings, marcots (Stanger and Page, 2012) and micropropagation (Nand et al., 2004). While the percentage root induction in juvenile (18%–40%) and mature cuttings (6%–16%) and marcots (28%) was relatively low, the results of these experiments offer the potential for further refinement for more routine operation. Generally, the application of IBA was positive on adventitious root induction in cuttings and marcots of mature tissues, although greater success was found without IBA in juvenile cuttings (Stanger and Page, 2012). Nand et al. (2004) demonstrated very good results with micropropagation of *D. pruriens* and *D. jerseyana* juvenile stem tissues with up to 80% root initiation and 60% survival after acclimation. Similar results were obtained using mature tissues for *D. jerseyana*, but *D. pruriens* was more problematic. While these results for the vegetative propagation of *D. pruriens* are important, given the clonal nature of most seedlings (Eliott et al., 2014), propagation of clonal seed would probably be the most efficient way to develop stable genotypes suited to cultivation.

In north Queensland plantings of *D. jerseyana*, the growth habit of the species has two distinct structures, referred to as the 'bushy' type and the 'bare trunk' type. The 'bare trunk' type comprises a bare single or multi stem trunk, with little or no leaf formation along the main stem/s. The crown of the 'bare trunk' type starts where branches form on the trunk/s at approximately 1–1.5 m from the ground (Figure 4.3). The other growth habit, the 'bushy' type, is a smaller multi-stemmed habit, with a very bushy continuous canopy, extending from the top of the crown down the stems to near-ground level. This variation in habit does not appear to be a result of wind damage but possibly a distinct variety or subspecies; however, the origins of these two variations are unknown.

BIOTIC CONSTRAINTS (PESTS AND DISEASES)

D. pruriens is affected by several animal pests including stem-boring jewel beetle (*Cyriodes cincta*); an as-yet-unidentified moth borer (possibly one of the Xylorictine moth borers); red-shouldered leaf beetle (*Monolepta australis*) (Ahmed and Johnson, 2000); sulphur-crested cockatoos (*Cacatua galerita*); endemic rats, such as the giant white-tailed rat (*Uromys* spp.); rats (*Rattus* spp.); and feral pigs (*Sus scrofa*). Northern growers consider three of these pests to have a high potential to cause significant harm to their orchards and hence cost substantial time and effort in managing the risks of impact. The sulphur-crested cockatoos, which often visit orchards in small flocks, strip immature fruit and young branches when fruits are nearing maturity. Sound cannon use is the principal method for deterring the birds. The giant white tailed rat is a nocturnal pest and consumes green immature fruit in copious amounts and can strip a tree of its crop if not trapped and relocated. The third most serious pest is the stem boring jewel beetle (*Cyriodes cincta*), which favours the Davidson's plum as a host for its larvae. Severe infestations can ring-bark numerous branches and even the main trunk. Infestations can be fatal to the tree particularly when the main stem is affected close to the ground. Treatment with systemic insecticides is not registered and is, in any case, not considered viable given the high cost of treatment and low rate of success, with the insecticide often not reaching the targeted larvae. Therefore, physical removal of the infested wood is the current recommended management of this pest. This may involve a major prune, resulting in yield depression in subsequent years and take up to 3–4 years for the trees to recover their yield potential. No stem borer pest problems have been reported for *D. jerseyana* in NSW, but in north Queensland plantations, the stem borer also attacks *D. jerseyana*.

The 'typical' variant of *D. pruriens* does not appear to be susceptible to fruit fly infestation, probably due to the tough skin and dense cover hairs on the fruit. The 'smooth' forms of *D. pruriens* and *D. jerseyana* are susceptible to fruit fly infestation, and this is a significant constraint for orchard management. This is probably due to softer skin and less irritant hairs on the fruit to deter the fly. Growers in northern NSW report that other pests include the king parrot (*Alisterus scapularis*) together with the rodent and bird pest species already mentioned for *D. pruriens*.

HARVESTING, PROCESSING AND MARKETING

D. pruriens fruits are often borne on the axils of branches, high in the tree or where new branches are produced from lower on the main trunk. Harvesting is currently performed manually after ripe fruits fall to the ground. This method is suitable, since the fruits are considerably firm and thus, the fall does not harm the fruit. Frequent collection from the orchards ensures minimal losses to wild fauna such as native rodents. Due to the relatively hardy nature of the *D. pruriens*, it would seem that mechanical collection similar to macadamia nut harvest systems could be considered for this species.

After harvesting, the fruits are tumble-washed to remove the fine hairs and any debris, and sorted according to their ripeness (those ready to process or freeze, are separated from those that require between 1 and 4 days to further ripen). Fruits to ripen are arranged in a single layer on large trays in screened open-air conditions and monitored daily until mature.

D. jerseyana harvest occurs in a more intensive period, over a maximum of 2 months, generally between November and January. The fruits are picked directly from the tree as soon as a blue/purple blush is visible and the fruit will readily ripen thereafter. This early picking helps to minimise fruit fly populations. Care in picking the fruit is required as they are quite soft-skinned and more delicate than *D. pruriens*. It has been found that it is best to twist the fruits to remove them, rather than pull the fruits, to reduce the incidence of damage to the fruit and at its point of attachment on the trunk. It is likely that mechanised harvesting of *D. jerseyana* would require customised machinery and processes to accommodate the delicate nature of the fruit.

Post-harvest handling of *D. jerseyana* fruits is similar to that of *D. pruriens* where they are washed and sorted according to their ripeness. Near-ripe fruits are set aside in trays to ripen over a slightly shorter period than those of *D. pruriens* of 1–2 days.

The ripe fruits of both species are washed in a mild-sanitising solution before the seed is removed, or if being kept whole, before being packed into vacuum sealed bags of desired weight (varies from 1 to 10 kg sizes).

Removal of the seeds is done primarily by hand; however, some larger growers have invested in paddle finishers and process the whole fruits through these pulping machines to extract the seeds and produce a homogenous puree from the pulp and skin. The puree is then frozen for future manufacturing use. The health properties (total phenolics and antioxidant activity) of frozen fruit puree are not maintained for as long as for frozen fruit that has been cut to remove the seed (Sultanbawa et al., 2015). More research is needed to further understand the reasons for this loss and if the reduction in health benefits is significant enough to reduce potential demand. If whole fruit is preferred by the market then identifying an alternative mechanical cutting process will be important.

Most Davidson's plum production goes into value-added products such as condiments and is also increasingly used as an additive for dairy products (yogurts and dips) and beverages.

Processing the fruits for a larger variety of products made by large-scale food manufacturers would expose a broader cross section of the consumer market to Davidson's plum, and thus raise the general public's awareness of the fruit.

In a RIRDC report on marketing the native food industry (Cherikoff, 2000), it was suggested at that time that a focus on growing the market could be achieved by targeting the food service and manufacturing markets so as to increase consumer awareness. Since 2000, uptake by manufacturers has been marginal and while large-scale manufacturers are interested in using native fruit flavours, they have concerns about supply limitations and higher prices than alternative additives. This is where producers with the capacity to efficiently grow, harvest and process Davidson's plum (whether themselves or in partnership with processors) and can capitalise with regard to bulk supply opportunities.

Davidson's plum producer's service the following markets: farmer's markets, online sales, processors, wholesalers and a range of domestic retailers. Export markets are accessed by some grower/value-adding entities, but it is not yet a major avenue for sales. However, given the raised awareness of the functional qualities of the fruit, export is likely to become a stronger opportunity. Market constraints to exporting value-added product include the economics of food manufacturing in Australia (high cost comparative to developing countries) and currency fluctuations.

YIELD

Initial fruit set in *D. pruriens* is dependent upon settled weather (warm and low wind) that possibly promotes wind pollination, where pollen floats on slow-moving air currents. Sufficient rain is however required to support further fruit development to maturity. These seemingly particular conditions are likely to contribute to the variation in the timing of yield between locations. A survey of 30 individual trees of the 'typical' variety across two plantings over 12 months revealed similar mean yield of 4.5 kg at one site and 4.9 kg per tree at the other, but fruits were collected over a 3- and 10-month season, respectively. Substantial variation in fruit yield also occurs between individual trees ranging from 1.5 to 8.7 kg and 0.1 to 13.5 kg and at each of the two respective sites. The trees at the former site were found to produce a greater number of inflorescences per tree (10–12) and flowers per inflorescence (80–90) during the season compared with the latter site (4–5 inflorescences and ~50 flowers respectively). Difference in yields between the sites may therefore be due to relative differences in tree productivity rather than issues with pollination. The differences in productivity may be attributed to genetic variations between trees, and the environmental conditions and crop management between the sites. Comparison between the 'typical' and 'smooth' varieties at one site revealed a significantly higher mean yield for the 'smooth' (8.8 kg/tree) compared with the 'typical' (4.5 kg/tree). Given the clonal nature of seed-derived plants in *D. pruriens* (Eliott et al., 2014), it is important that the relative influence of genotype and environment be determined; this can be used as the basis for further cultivar development. This work is currently underway in north Queensland.

In north Queensland, *D. jerseyana* commences flowering in mid-July, with some ripe fruit ready by October and continues to produce until late summer (end of February). To date, there has not been any systematic study of the fruit set factors and yield of *D. jerseyana*, but north Queensland growers report yields ranging from 4 to 13 kg per tree.

CONCLUSIONS AND RECOMMENDATIONS

The development of the native foods industry has come a long way since its emergence as a unique food experience in the late 1970s and early 1980s. Recent research on potential health benefits has shifted the emphasis of marketing away from novelty towards a focus on functional food. Ongoing studies which aim to improve planting stock and cultivation practices, together with more grower uptake of processing and production efficiencies, coupled with sophisticated marketing strategies focused on functional food values, should see an improvement in the potential for success of the Davidson's plum fruit industry.

ACKNOWLEDGEMENT

The authors acknowledge Ernie Raymont, Drew Morta, Sue and Ken Pyke, Peter Lawlor, Andrew Lilley, Richard Horsburgh, Geraldine McGuire, Bob Peever, Gillian and John Russell, Jenny and John Rogers, Kris Kupsch, Kim Kido, Volker Stanger, Jonathan Cornelius, Kevin Blackman, and Ellie Bock for their assistance in gathering knowledge on the Davidsons plum.

REFERENCES

Ahmed, A.K., Johnson, K.A., 2000. Turner review No. 3: Horticultural development of Australian native edible plants. *Australian Journal of Botany* 48, 417–426.

ANFIL, RIRDC, 2014. *Focus on Davidson Plum: Davidsonia jerseyana, Davidsonia pruriens*. Rural Industries Research and Development Corporation, RIRDC Publication No. 14–112, Canberra, Australian Capital Territory, Australia.

Bailey, F.M., 1895. Peculiarities of the Queensland flora. *Botanical Bulletin of Department of Agriculture, Queensland* 12, 11–26.

Bailey, F.M., 1898. Edible fruits indigenous to Queensland No. 1. Davidsonian plum. *Queensland Agricultural Journal* 2, 471.

Bailey, F.M., 1900. *Davidsonia pruriens* var. *jerseyana*. *The Queensland Flora* 2, 538.

Cherikoff, V., 2000. *Marketing the Australian Native Food Industry*. Rural Industries Research and Development Corporation, RIRDC Publication No. 00/61, Canberra, Australian Capital Territory, Australia.

Clarke, M., 2012. *Australian Native Food Industry Stocktake*. Rural Industries Research and Development Corporation, RIRDC Publication No. 12/066, Canberra, Australian Capital Territory, Australia.

Cornelius, J., 2011. Davidsonia cultivation: Summary and synthesis of FNQ grower responses to question on productivity, germplasm, and research priorities. RIRDC-funded project – 'Davidsonia domestication: Productivity constraints in Far North Queensland'. James Cook University, Cairns, Queensland, Australia.

DEC-NSW, 2004. *Recovery Plan for the Davidson's Plum (Davidsonia jerseyana)*. Department of Environment and Conservation (NSW), Hurstville, New South Wales, Australia.

Eliott, F.G., Shepherd, M., Rossetto, M., Bundock, P., Rice, N., Henry, R.J., 2014. Contrasting breeding systems revealed in the rainforest genus *Davidsonia* (Cunoniaceae): Can polyembryony turn the tables on rarity? *Australian Journal of Botany* 62, 451–464.

Harden, G.J., Williams, J.B., 2000. A revision of *Davidsonia* (Cunoniaceae). *Telopea* 8, 413–428.

Hotson, A., 2008. The Davidson plum. In: Salvin, S. (ed.), *The New Crop Industries Handbook: Native Foods*. RIRDC, Canberra, Australian Capital Territory, Australia, pp. 40–47.

Isaacs, J., 2002. *Bush Food: Aboriginal Food and Herbal Medicine*. New Holland Publishers, Sydney, New South Wales, Australia.

Konczak, I., Sakulnarmrat, K., Bull, M., 2012. *Potential Physiological Activities of Selected Australian Herbs and Fruits*. Rural Industries Research and Development Corporation, RIRDC Publication No. 11/097, Canberra, Australian Capital Territory, Australia.

Konczak, I., Zabaras, D., Dunstan, M., Aguas, P., Roulfe, P., Pavan, A., 2009. *Health Benefits of Australian Native Foods – An Evaluation of Health-Enhancing Compounds*. Rural Industries Research and Development Corporation, RIRDC Publication No. 09/133, Canberra, Australian Capital Territory, Australia.

Mayall, J., 2014. *The Outback Chef: Cooking with Native Australian Ingredients*. New Holland Publishers, Sydney, New South Wales, Australia.

Nand, N., Drew, R.A., Ashmore, S., 2004. Micropropagation of two Australian native fruit species *Davidsonia pruriens* and *Davidsonia jerseyana* G. Harden & J.B. Williams. *Plant Cell, Tissue and Organ Culture* 77, 193–201.

SGAP, 1988. *North Queensland Native Plants, The Society for Growing Australian Plants NSW, The Tablelands Branch*. Kangaroo Press, Kenthurst, New South Wales, Australia.

Stanger, V., Page, T., 2012. *Vegetative Propagation of Davidsonia pruriens*. School of Marine and Tropical Biology, James Cook University, Cairns, Queensland, Australia.

Sultanbawa, Y., Williams, D., Chaliha, M., Konczak, I., Smyth, H., 2015. *Changes in Quality and Bioactivity of Native Foods during Storage*. Rural Industries Research and Development Corporation, RIRDC Publication No. 15/010, Canberra, Australian Capital Territory, Australia.

CHAPTER 5

The Reproductive Systems of Davidson's Plum (*Davidsonia jerseyana, Davidsonia pruriens* and *Davidsonia johnsonii*) and the Potential for Domestication

Frances Eliott, Mervyn Shepherd, Maurizio Rossetto and Robert Henry

CONTENTS

BACKGROUND

The species in the *Davidsonia* genus, commonly called Davidson's plums, are important edible fruit crops for the Australian native food industry and are also of conservation concern, because two of the three species in the genus (*Davidsonia jerseyana* and *D. johnsonii*) are endangered. Understanding the reproductive systems in the genus is a crucial step in developing strategies for both plant improvement and conservation management.

The reproductive systems in both *D. jerseyana* and *D. pruriens* were investigated by analysing progeny arrays with microsatellite markers (Eliott et al., 2014) and the following description (unless otherwise referenced) is summarised from the results of that study along with personal observations and an analysis of the genetic diversity and structure occurring in natural populations (Eliott et al., unpublished data). The genus is interesting in that the reproductive systems are quite different among the three species, and vary from highly self-fertilising to putative apomixis (asexually derived seeds) with polyembryony (more than one embryo per seed) and clonality through root suckering. These reproductive systems have contributed to low genetic variation within each species and have important implications for cultivation, crop improvement and germplasm conservation.

Botany

The *Davidsonia* species are small rainforest trees with edible, albeit sour-tasting, fruit which are mainly used for making jams and sauces or as a flavouring agent. The taxonomy and a detailed botanical description of *D. jerseyana* and *D. pruriens* have been provided (Page and Watkins, 2016).

D. jerseyana and *D. pruriens* are quite similar in morphology, whereas *D. johnsonii* differs considerably despite having a distribution which is sympatric with *D. jerseyana*. Both *D. pruriens* and *D. jerseyana* are small, slender trees, usually single-stemmed (sometimes multi-stemmed in *D. jerseyana*), each with a terminal tuft of large pinnate leaves covered with irritant hairs. Their inflorescences are cauliflorous (occasionally ramiflorous), with panicles that are large and open in *D. pruriens* but often condensed and racemose in *D. jerseyana*. In contrast, the habit of *D. johnsonii* is bushy and spreading with many branches and smaller pinnate, glabrous leaves, and the inflorescences, which are terminal or subterminal panicles, are found among the leaves on branchlets. The flowers and fruit of *D. johnsonii* are otherwise similar to *D. jerseyana* and *D. pruriens*. All three species have small (~5 mm diameter) pink to red hermaphrodite flowers with 4–5 sepals. The sepals are hirsute (hairy) on the outer and pubescent on the inner surface in *D. pruriens* and *D. jerseyana* but finely pubescent on both surfaces in *D. johnsonii*. Likewise, the ovary is hirsute in *D. jerseyana* and *D. pruriens* but finely pubescent in *D. johnsonii*. Many *D. johnsonii* flowers have been found to be infertile, and seeds are not known to develop in the pyrenes within the fruit; consequently, reproduction is only known to occur through root suckering (Harden and Williams, 2000).

Distribution

D. pruriens is fairly widespread in the wet tropics of North Queensland and is not considered to be threatened. In contrast, *D. jerseyana* and *D. johnsonii* are both endangered and occur in small fragmented populations restricted to far north-eastern New South Wales (NSW), with *D. johnsonii* extending just over the border into southern Queensland. The southern *D. jerseyana* and *D. johnsonii* are listed as endangered at both national and NSW state level under the *Environment Protection and Biodiversity Conservation Act 1999* and the NSW *Threatened Species Conservation Act 1995*, with *D. johnsonii* also listed under the *Queensland Nature Conservation (Wildlife) Regulation 2006 – Schedule 2*. Recovery plans have been prepared for both species (DEC-NSW 2004a,b), and these should be consulted for guidelines concerning the collection of material for produce or propagation as these activities may require approval via a licence or certificate.

THE REPRODUCTIVE SYSTEMS

Davidsonia jerseyana

The reproductive system in *D. jerseyana* was found to be predominantly (if not entirely) self-fertilising. A progeny array analysis was conducted on the progeny from two mature mother trees, growing approximately two metres apart. The trees were grown in cultivation (*ex-situ* plantings) and were from genetically distinct populations. *Ex situ* plantings were chosen, because previous studies had shown that there was little genetic variation within populations (Eliott et al., unpublished data) and therefore, it would be impossible to determine whether progeny in a natural population resulted from self-fertilisation or mating between closely related individuals (bi-parental inbreeding). The segregation of microsatellite alleles in the progeny was consistent with self-fertilising in a diploid plant, and there was no evidence of any cross-pollination despite the proximity of the mother trees and the inclusion of progeny from fruiting seasons over several years. Alleles at a heterozygous locus segregated in the expected pattern for a selfing individual, that is half the progeny were homozygous and half were heterozygous for the mother tree alleles. These results are congruent with observations of single, isolated trees which consistently produce fruit crops.

The results of the genetic analysis of natural populations also support self-fertilisation in this species (Eliott et al., unpublished data). Within natural populations, there were very few polymorphic loci, almost all individuals were homozygous and most of the genetic variation was found among rather than within populations. Heterozygosity, in an exclusively selfing population, is lost at the rate of 50% per generation; thus, heterozygosity would be almost absent in such a population after seven generations. Given that trees are thought to live more than 100 years (Watson, 1987), the extreme loss of heterozygosity in *D. jerseyana* populations suggests that the species has been selfing for a prolonged period of time.

Another reason to suspect long-term selfing in *D. jerseyana* is that inbreeding depression was not apparent in offspring included in fitness trials (see the following) and inbreeding depression is more common in predominantly selfing populations of naturally outcrossing species (Stebbins, 1950). Inbreeding depression results in a reduction in offspring fitness and is caused by increased homozygosity in individuals which can effectively 'fix' recessive deleterious mutations in populations or an increase in homozygosity at loci with heterozygote advantage ('overdominance') (Charlesworth and Willis, 2009). With prolonged selfing, these deleterious alleles can be purged from a population through selection (Charlesworth and Charlesworth, 1987; Lande and Schemske, 1985), whereby individuals with the lethal allele may not reach reproductive age and as a consequence, the allele is not passed on and is eventually purged from the population. Thus, over a period of time and if the population does not vanish as a consequence of inbreeding, the effects of inbreeding depression are greatly reduced or even eliminated.

Within the natural *D. jerseyana* populations, many individuals had the same multilocus genotype, and although the progeny array analysis indicated seeds were the product of selfing, trees can sucker from the roots (Eliott, personal observation), so it is possible that identical genotypes could also be clonal. With such a high degree of selfing and loss of genetic diversity, it would be difficult to distinguish between selfed individuals and those that are potentially clonal.

Davidsonia pruriens

The mother trees selected for the progeny array analysis were all identical at each locus examined. The most likely explanation for this is that the trees were grown from seeds collected from one or a few individuals (or from the same population). The trees were highly heterozygous, and consequently, progeny resulting from self-fertilisation (or outcrossing to a parent with the same genotype) would have exhibited the same segregation of alleles.

The progeny array analysis for *D. pruriens* indicated that the species is predominantly clonal. The majority of progeny (89%) were identical to the heterozygous parent, whether from polyembryonic or monoembryonic seed and these were presumably asexually derived through apomixis. Of the remaining progeny, 2% differed from the parental genotype at just one locus where they shared one new allele not found in the mother or potential pollen donor trees. This minor difference may be attributable to a somatic mutation which is not uncommon among connected ramets in clonal species (Gross et al., 2012; Tuskan et al., 1996; van der Merwe et al., 2010), or in apomicts (King and Schaal, 1990) and putative apomicts (Thurlby et al., 2012). This left just 9% of the progeny which were likely to be sexually derived. These sexual progeny were homozygous at up to 71% of the heterozygous parental loci and were likely to be either the result of selfing or outcrossing between parents with the same genotype. The segregation of alleles in the sexually derived progeny was consistent with selfing in a diploid rather than a polyploid organism. In addition, the presence of both asexual and sexual progeny is an indication that both pollination and fertilisation do occur and therefore, *D. pruriens* is likely to be a diploid, facultative apomict, with occasional sexual reproduction.

Apomixis (also known as agamospermy) is defined by Asker and Jerling (1992) as the asexual formation of viable seed and does not include vegetative reproduction. Apomictic seeds may be produced by two main forms of apomixis: gametophytic apomixis (encompassing diplospory and apospory) and adventitious embryony (also referred to as sporophytic apomixis), which typically involves polyembryony (Asker and Jerling, 1992). All forms of apomixis produce seeds derived from the maternal genetic material; however, the probability that a plant will also produce sexually derived progeny, and consequently contribute different levels of genetic diversity to the population, depends on which form of apomixis is involved (Whitton et al., 2008). Apomixis is complex and the following list of references is provided for a more detailed description of the various apomictic processes (Asker and Jerling, 1992; Koltunow 1993; Mogie, 1992; Nogler, 1984; Richards, 1997). The genetic analysis of natural *D. pruriens* populations revealed very low genotypic variation (Eliott et al., unpublished data). Although individuals were highly heterozygous, there was very little variation within populations and where variation occurred, it was most likely attributable to the migration of an individual from a nearby population. This pattern is consistent with apomictic reproduction and there was little evidence that current sexual reproduction contributed to variations in these populations.

Polyembryony in *Davidsonia pruriens*

A proportion of seeds from each of the mother trees in the study were polyembryonic, which means that more than one embryo was present in a single seed (Figure 5.1). The proportion of polyembryonic seeds varied among mother trees with polyembryony observable in as few as 5% and up to 65% of seeds from each mother tree. Polyembryony could only be determined if more than one radicle (root shoot) emerged from a single seed. Dissection of seeds, which is considered a more accurate method for determining polyembryony (Firetti-Leggieri et al., 2013), was not feasible in *D. pruriens* due to the nature of the seeds. *Davidsonia* species seeds are cryptocotylar which means the cotyledons remain within the seed coat and no endosperm is present in mature seeds. This results in embryos that are imbedded in dense, cotyledonous tissue and cannot be easily separated and counted. For this reason, the proportion of polyembryonic seeds observed by germination trials was likely to be an underestimate of the true occurrence, because many embryos are unlikely to develop into seedlings due to the competition for seed resources, limiting embryo maturation (Uma Shaanker and Ganeshaiah, 1997). This may account for the large number of *D. pruriens* progeny from 'monoembryonic' seed that were identical to the parent. These seeds may in fact have been polyembryonic but perhaps only one of several embryos was able to develop. The number of embryos which emerged from a single seed was usually two and rarely three; however, under high-nutrient conditions used for tissue culture, Nand et al. (2004) observed up to seven embryos in a single seed.

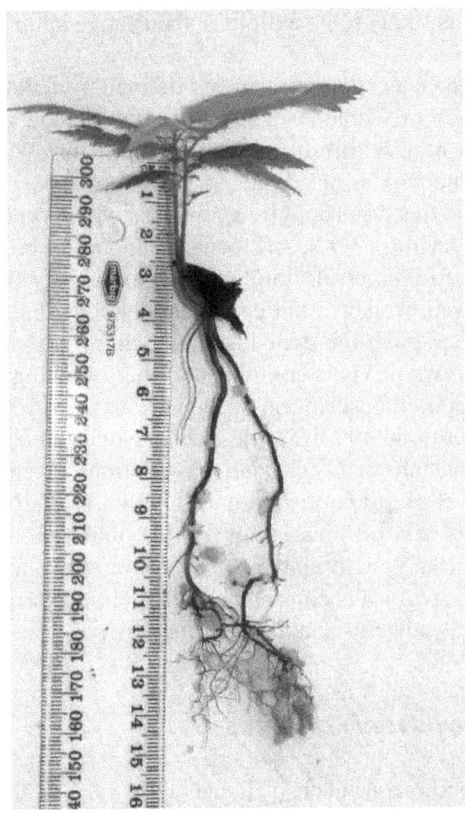

Figure 5.1   A polyembryonic seed in *Davidsonia pruriens*. In this seed, two embryos developed into seedlings but only the root emerged from a third embryo.

D. pruriens is a fleshy fruited, tropical tree. It has polyembryonic seeds with asexual and sexual embryos and is most likely a diploid. These combined features are characteristic of the form of apomixis called adventitious embryony (Richards, 2003). In adventitious embryony, fertilisation initiates a sexual embryo and the formation of endosperm and additional asexual embryos are formed from the somatic cells of the nucellus or the integuments in the ovule (Lakshmanan and Ambegaokar, 1984; Naumova, 1993). Adventitious embryony is considered to be the most common form of apomixis (Carman, 1997; Naumova, 1993) but also the least studied, probably due to its prevalence in tropical trees with a long life cycle (Richards, 2003). Examples of species with adventitious embryony include fruit trees such as *Citrus* and mangoes (*Mangifera* sp.) (Naumova, 1993), mangosteens (*Garcinia* sp.) (Richards, 1997) and forest trees in the Dipterocarpaceae (Singh and Thakur, 2004). Although *D. pruriens* has the features of adventitious embryony, embryological studies are required to confirm this reproductive system in the species.

Progeny Fitness in *Davidsonia jerseyana* and *Davidsonia pruriens*

Germination, seedling growth and survival and seed longevity trials were conducted on seeds collected from six *D. jerseyana* (354 seeds) and five *D. pruriens* (244 seeds) mother trees growing in *ex situ* plantings located in northern NSW. The seedling survival and growth rate were measured over a 12-month period under glass house conditions. There was a significant difference between *D. jerseyana* and *D. pruriens* progeny for all measures of fitness (Table 5.1). In *D. jerseyana*, 79% of the pyrenes contained a potentially viable seed, whereas the majority of *D. pruriens* pyrenes were empty and only 44% contained a seed. In both species, germination of fresh seed was rapid and the majority of seeds germinated within 20 days of sowing. The germination rate was extremely high in *D. jerseyana* (99.6%) but lower in *D. pruriens* (72.4%). There was a similar pattern in seedling survival with more than 99% survival of *D. jerseyana* seedlings compared to 74% in *D. pruriens*. Interestingly, survival of *D. pruriens* seedlings from polyembryonic seed was higher (mean 0.89 ± 0.16) compared to monoembryonic seed (mean 0.70 ± 0.11), indicating a possible advantage of polyembryony due to the increased opportunity each seed has, via numerous embryos, to produce at least one seedling. The seedling growth rate was very even among seedlings and among seedlings from different *D. jerseyana* mother trees which is not unexpected considering the genetic uniformity expected in long-term, highly selfing species. In *D. pruriens*, there was more variation in growth rate among seedlings which may be a consequence of the mix of apomictic and sexually derived progeny and competition between embryos in polyembryonic seed.

Seeds in both *D. jerseyana* and *D. pruriens* are short-lived. For the longevity trial, seeds were stored within the fruit, at ambient temperature, for 3 months prior to sowing. The stored seed germination rate was well below that of fresh seed. Only 5% of the stored *D. jerseyana* seeds germinated, although a higher proportion of *D. pruriens* seeds remained viable with a germination rate of 20%.

Also of interest was the juvenile flowering in *D. jerseyana* seedlings. Juvenile flowering (also known as precocious flowering) was only observed in *D. jerseyana* seedlings, and it only occurred in the progeny from one of the six mother trees (Eliott, unpublished data). The seedlings flowered during September–October which is consistent with the usual flowering time for *D. jerseyana* and was present in 13% of the seedlings at ~36 weeks post germination. Juvenile flowering has also been observed in *D. jerseyana* and small *D. johnsonii* ramets in natural populations (Eliott, personal observation). Spontaneous juvenile flowering in woody species is considered rare (Bolotin, 1975) and many treatments have been employed to induce juvenile flowering to enhance breeding programmes impeded by the long juvenile (non-reproductive) phase in most trees (Meilan, 1997). The juvenile flowering in *D. jerseyana* deserves further investigation as it could benefit growers by providing earlier fruit-bearing trees and accelerate breeding programmes by reducing generation times.

Table 5.1 Progeny Fitness in *Davidsonia jerseyana* and *Davidsonia pruriens* Including the Percentage of Potentially Viable Seed with the Subsequent Germination and Seedling Survival Rate

Species	Fruit with 2 Viable Seeds	Fruit with 1 Viable Seed	Fruit with No Viable Seed	Total Pyrenes with Viable Seed	Germination of Viable Seed	Seedling Survival
D.jerseyana	60.04 ± 16.79 (31.3–76.9)	38.91 ± 17.50 (21.1–68.8)	1.05 ± 1.71 (0–4.0)	79.35 ± 7.93 (65.6–88.6)	99.57 ± 1.06 (97.4–100)	99.34 ± 1.10
D. pruriens	9.87 ± 7.07 (0–18.4)	69.17 ± 12.46 (57.9–87)	20.96 ± 7.24 (13.0–29.2)	44.46 ± 3.51 (39.1–47.7)	72.41 ± 23.07 (36–95)	74.19 ± 9.01
P-value	<0.001	0.010	<0.001	<0.001	0.017	<0.001

Notes: Mean for mother trees in each species ± standard deviation (range). *P*-values from one-way analysis of variance.

Davidsonia johnsonii

No sexual reproduction has been recorded in *D. johnsonii,* and the trees are only known to reproduce clonally through root suckering (Harden and Williams, 2000). The trees flower regularly and generally produce good crops of parthenocarpic (seedless) fruit which may contain pyrenes but viable seeds are not known to develop within them. There have been anecdotal reports of fruit containing seeds but so far, evidence of seed or seed germination has not been documented. There are a number of factors that might be contributing to the sterility in *D. johnsonii.* Harden and Williams (2000) found many flowers to be infertile with styles often reduced or absent (but when present, were mostly shorter than the stamens), the ovary was sometimes vestigial or absent and there was no pollen in the anthers of flowers they examined. This may vary between trees and between populations as viable pollen, which germinated on media and the stigmatic surface, has been obtained from the anthers of some trees (Eliott, unpublished data). In microsatellite analysis, many *D. johnsonii* individuals had three alleles at a single locus, suggesting possible polyploidy (genome duplication) in the species. In this case, the maximum of three alleles at a locus indicates that *D. johnsonii* could be triploid and polyploidy, particularly triploidy, is often associated with sterility (Bretagnolle and Thompson, 1995). The genetic analysis of natural populations supports the lack of fertility in *D. johnsonii* (Eliott et al., unpublished data). The majority of the genetic variation was found among rather than within populations and each population was genetically distinct. There was no evidence of gene flow among populations and most ramets sampled within populations had identical multilocus genotypes, all of which are characteristic of clonal reproduction.

Pollination in *Davidsonia*

Flowers in the *Davidsonia* species present traits which are consistent with a generalised insect pollination syndrome (entomophily) (Faegri and van der Pijl, 1979) (Figure 5.2). The small flowers are actinomorphic (radially symmetrical) and the sepals are fused at the base, forming a fairly narrow opening which conceals the nectaries. According to Knuth et al. (1906), this flower structure is most suited to insects with a long tongue or proboscis such as bees, insects in the Lepidoptera order (moths and butterflies) and long-tongued Diptera (flies and mosquitos) or wasps. The flower opening is dominated by crowded anthers and the pollen adheres to the anther lobes following dehiscence, a feature also considered characteristic of insect pollination (Faegri and van der Pijl, 1979). However, very few potential pollinators have been observed in the flowers of either *D. jerseyana* or *D. pruriens* (Eliott, personal observation).

Many hours have been spent observing *D. jerseyana* flowers (in cultivation and natural populations) for both diurnal and nocturnal pollinators and only a few insects were recorded. Of these, the main visitors were ants, small spiders and wasps. Ants are known pollinators (e.g. Garcia et al., 1995; Luo et al., 2012; Peakall and Beattie, 1991), but their effectiveness as pollinators can be limited, because many species have antibiotic secretions which can inhibit pollen germination and pollen tube growth (Beattie et al., 1984; Dutton and Frederickson, 2012). Spiders and wasps are

Figure 5.2   (See colour insert.) Flowers of *Davidsonia jerseyana* showing the hirsute sepals and ovaries and the anthers before and after dehiscence.

likely to be pollinator predators but during their activities may inadvertently perform pollinating roles themselves (Williams and Adam, 2010). The outer surfaces of the sepals and also the ovaries (in *D. jerseyana* and *D. pruriens*) are covered by irritant hairs which may deter potentially destructive insects and have a filtering effect on potential pollinators (Williams and Adam, 2010), which may explain the low number of flower visitors. In contrast, at peak flowering time in some *D. johnsonii* populations, the inflorescences are abuzz with activity and the flowers clearly attract a large number of insects including bees. The sepals and ovary in *D. johnsonii* flowers are finely pubescent and lack the irritant hairs found in *D. jerseyana* and *D. pruriens* (Harden and Williams, 2000) and this might make the flowers of this species more attractive to pollinators. It is suggested that *D. pruriens* might be wind pollinated (Tony Page, personal communication) as the pollen is very light and has been recorded drifting on air currents and also a lack of potential pollinators has been observed in the species natural range. It is evident that more detailed research is required in order to understand the pollination biology of *Davidsonia* species.

IMPLICATIONS OF THE REPRODUCTIVE SYSTEMS FOR CONSERVATION, CULTIVATION AND DOMESTICATION

The clonal and selfing reproductive systems in the *Davidsonia* species are atypical of long-lived, woody perennials, the majority of which are outcrossing species (Miller and Gross, 2011) with genetically diverse and unstructured populations. In contrast to outcrossing, the reproductive systems in *Davidsonia* species have contributed to low genotypic variation within natural populations and the populations are highly differentiated, because the majority of the variation occurs among populations. Habitat destruction and fragmentation is likely to have further compounded the low genotypic variation by isolating populations, reducing population sizes and increasing the likelihood of population genetic bottlenecks and genetic drift (Rossetto et al., 2004). The most important implication of both the reproductive systems and habitat loss is that the remaining genetic diversity, in all three species, needs to be conserved as a matter of priority. This is not only important for the conservation of the species but also for potential domestication, because genetic diversity is essential for crop improvement (Cooper et al., 2001). In the context of *Davidsonia* species domestication, although the lack of genetic variation that could be exploited for crop improvement is of concern, there are advantages to the reproductive systems in that they provide reproductive isolation which allows growers to easily maintain desirable phenotypes.

Germplasm Collection and Conservation

Germplasm contains the underlying natural genetic variation in a species, and it is important to conserve, characterise and make available as much of this natural variation as possible through germplasm collections. Apart from enhancing *Davidsonia* species conservation efforts, germplasm collections will benefit crop improvement, because any attempt at plant breeding will depend on harnessing and manipulating the available genetic variation to improve genotypes and select for desirable traits (Godwin, 2003). It will also safeguard the existing natural variation against the narrowing of the gene pool which is a frequent outcome of domestication (Godwin, 2003; Olsen and Gross, 2008).

Due to the distribution of genetic variation among populations of the *Davidsonia* species, developing germplasm collections that capture most of the natural diversity would be a relatively easy task compared to collections for outcrossing species. For outcrossers, many individuals within populations would be required, whereas sampling few individuals from *Davidsonia* populations should encompass most of the natural variation within each species. This is particularly the case for the largely clonal *D. pruriens* and clonal *D. Johnsonii*; however, there are rare alleles in the more variable *D. jerseyana* populations, so these would need to be sampled more intensively. In order to maximise the genetic diversity conserved, germplasm collections should aim to sample from as many populations as possible. This is particularly important for clonally propagated species, because individuals are frequently phenotypically similar but may have different genetic origins (Godwin, 2003).

It is also urgent to establish germplasm collections because of the potential for the genetic diversity in natural populations to continue diminishing and because unique genotypes are already likely to have been lost. This is because all three species occur in areas impacted by habitat destruction and fragmentation through the historical clearing of rainforests and more recently, by vegetation clearing for agriculture, housing developments and highway construction. Sampling orchard trees for germplasm collections could also be valuable, because orchards might be harbouring unique genotypes from populations which no longer exist in the wild.

Germplasm collections may consist of seeds, pollen, vegetative material (e.g. tissue culture or shoot-tip cryopreservation) or living plants (Offord and Makinson, 2009). Seeds are the easiest to collect and form the majority of germplasm collections worldwide (Altoveros and Ramanatha Rao, 1998). However, considering the clonality and lack of seeds in *D. johnsonii* and the short-lived seed viability in *D. jerseyana* and *D. pruriens*, alternative approaches are needed. A micropropagation (tissue culture) technique has been developed for *D. jerseyana* and *D. pruriens* (Nand et al., 2004) and using this method or cryopreservation has been recommended for *ex situ* germplasm storage in *Davidsonia* species (Ashmore et al., 2011). Living plant collections are expensive to establish and maintain but would be beneficial for *Davidsonia* species, because they would preserve genetic diversity as well as enable the documentation of phenotypes and traits of interest for domestication.

POTENTIAL FOR DOMESTICATION

The *Davidsonia* species may be considered 'semi-domesticated' according to the definition by Meyer et al., (2012) where a species is under cultivation and subjected to some conscious artificial selection pressures but is not morphologically or genetically distinct from the wild plants. The potential for the domestication of *Davidsonia* species has been acknowledged (Henry, 2010), and it is likely that growers are already selecting for traits of interest. There is anecdotal evidence of a large variation in fruit yield, fruit quality, flowering and cropping consistency (annual, biannual) and time to fruiting age in *Davidsonia* species, and this may be due to genetic differences between populations. For example, in an *ex situ* planting, the fruit-bearing age, fruit yield and fruiting consistency were markedly different between *D. jerseyana* trees sourced from different populations (Eliott, personal observation). There is no doubt that growers are already aware of different phenotypes in their orchards and documenting this information would be valuable to the industry. Traits that could be considered for selecting candidate genotypes for breeding could include fruit yield, fruit flavour and sweetness, flowering and fruiting times, disease resistance and drought tolerance. Traditional breeding through selection would be an achievable goal in the short term, and documenting phenotypes with traits of interest would be the most important first step in this process.

Advantages of the Reproductive System for Domestication

Many of the naturally occurring features of the reproductive systems in *Davidsonia* species are potentially useful for domestication purposes. These include apomixis, polyembryony, putative polyploidy, parthenocarpy, self-compatibility and clonal propagation, all of which are sought-after traits for particular crops and are often aimed for in breeding programmes. It is therefore not surprising that these traits are also among those included in the suite of changes (relative to the wild progenitors) associated with the domestication of perennial crops, known as a 'domestication syndrome' (Miller and Gross, 2011).

The introduction of apomixis into crop plants is considered highly desirable and is gaining increasing attention, because using clonal seed to propagate elite genotypes (e.g. hybrids) has the potential to reduce production costs, reduce breeding time and increase yields (Barcaccia and Albertini, 2013; Ramulu et al., 1999). Among the advantages of apomictic seed is the avoidance of the complications associated with sexual reproduction such as incompatibility barriers, or of vegetative propagation where the transfer of pathogens via propagation material can be problematic (Barcaccia and Albertini, 2013; Savidan, 2000).

Polyembryony is beneficial in allowing the propagation of identical clones and therefore ensuring true-to-type propagation which is not available through sexual reproduction. In mangoes, for example, some cultivars are polyembryonic and for those that are not, achieving polyembryony through breeding is considered a priority, especially for rootstocks (Krishna and Singh, 2007). Apomixis and polyembryony are particularly advantageous for the propagation of elite genotypes; however, they can also hinder breeding programmes aiming for the production of novel genotypes through crossing or hybridisation. This is because the clonal progeny are identical to the maternal parent. In these situations, a method of identifying zygotic (sexual) from the asexually derived progeny is needed and molecular markers are ideal for this purpose and have been employed, for example, in citrus (Yildiz et al., 2013) and mango (Martinez Ochoa et al., 2012).

Polyploidy is prominent among many cultivated crops and is possibly present in *D. johnsonii*, although further investigation is needed to confirm this. The importance of polyploidy in crop improvement is well recognised (reviewed by Paterson, 2005; Udall and Wendel, 2006) and often aimed for inbreeding programmes. This is because polyploidy can generate novel phenotypic variation and adaptive plasticity, both of which are beneficial traits for domestication (Olsen and Wendel, 2013). Other potential benefits of polyploidy include genome 'buffering' against deleterious mutations, increased allelic diversity and increased or fixed heterozygosity (Udall and Wendel, 2006). Among the many polyploid crops, the more prominent examples include sugarcane, potato, cotton, banana, wheat and strawberries.

Parthenocarpy (seedless fruit), as occurs in *D. johnsonii*, is a desirable and commercially valuable trait in species with edible fleshy fruit, because the fruit is easier to consume and process (Varoquaux et al., 2000). In parthenocarpy, the ovary is able to develop into a fruit without the fertilisation of the ovules. Fertilisation of the ovules occurs in some seedless fruit, but the ovules abort soon after fertilisation (stenospermocarpy). The reasons for seedlessness in fruit vary from infertility of the pollen or ovaries, incompatible pollen or abnormal meiosis such as the chromosome imbalance in triploids. Parthenocarpy occurs naturally in many species including bananas, citrus and papaya and is also artificially induced, for example, in seedless watermelons and grapes (Gustafson, 1942; Varoquaux et al., 2000). The reasons for parthenocarpy in *D. johnsonii* are unknown but could be attributed to lack of fertilisation due to pollen and ovary abnormalities (Harden and Williams, 2000) or putative triploidy (Eliott et al., unpublished data).

Self-compatibility is a valuable attribute in cultivated crops, because fruit (or seed) set can be less dependent on pollinators and it also eliminates the need to interplant with suitable cross-compatible mates. Hence, achieving self-compatibility in naturally self-incompatible species has become a major objective in recent breeding programmes for perennial trees such as sweet cherry (Cachi and Wuensch, 2014) and almonds (Socias i Company et al., 2010). Another advantage of self-compatibility is that it allows breeders, through repeated self-fertilising, to produce true-breeding (homozygous) lines. True-breeding lines are valuable for crop improvement, because the selfed progeny have the same phenotypic traits as the parent and also because hybrids can be created by crossing true-breeding lines. The high degree of selfing and consequent homozygosity in *D. jerseyana* has resulted in naturally occurring 'true-breeding lines' which may prove useful for domestication.

Clonal propagation has many obvious benefits, especially for perennial plants with long juvenile phases. Using a variety of clonal methods such as grafting or budding, elite genotypes can be grown true-to-type and can be rapidly disseminated. By avoiding long juvenile phases, clonal propagation also greatly reduces the length of time to crop production and accelerates breeding programmes by reducing evaluation times. For these reasons, more than 75% of perennial crops are propagated clonally (Miller and Gross, 2011) and the natural ability of *Davidsonia* species to reproduce clonally would be advantageous for domestication.

Selection and Hybridisation

The options for crop improvement and eventual domestication of the *Davidsonia* species may be limited. Although *Davidsonia* species have many useful attributes (e.g. parthenocarpy, apomixis and polyembryony), these very attributes could also pose problems for breeding, especially the limited genotypic diversity and unknown potential for sexual recombination due to the largely inbreeding and clonal reproductive systems. The reproductive isolation between the three species may mean that domestication will be most easily achieved within each of the species rather than by hybridisation between species.

The domestication of clonal species is often a single-step process whereby an individual, with exceptional traits, is selected and propagated and subsequent variation to theses clones may only occur through somatic mutations (Zohary, 2004). Selecting elite trees is relatively simple and certainly one of the first steps in *Davidsonia* species crop improvement. This process is already underway for *D. pruriens* in a current project (RIRDC project PRJ-9026), which is evaluating clones from highly productive trees in North Queensland. A similar approach could be taken for selecting high performing *D. johnsonii* clones, *D. jerseyana* homozygous pure-lines and *D. pruriens* trees which are more suited to subtropical growing conditions.

Hybridisation has long been recognised as an important component of plant domestication (Stebbins, 1950) and is extremely valuable in crop plants, because it can provide novel genotypes and hybrid vigour. Hybrid vigour, or heterosis, describes the situation where the phenotypic performance of hybrid offspring is superior to that of their parents. Cross-pollination experiments could be undertaken to see if hybridisation is possible between *Davidsonia* species. Advantages could include, for example, producing novel genotypes by increasing heterozygosity in *D. jerseyana*.

However, caution would need to be applied because gene flow from hybrids to their wild relatives is not uncommon in other species (Ellstrand et al., 1999) and therefore, if hybridisation was successful in *Davidsonia* species, accidental gene flow from hybrids to natural populations could further threaten the endangered species. Fortunately, this is actually unlikely to occur, because the reproductive barriers present between the *Davidsonia* species are likely to preclude gene flow.

Use of Molecular Techniques

The genetic analysis and breeding of long-lived perennial species can be complicated by many factors including long juvenile phases, reproductive systems and ploidy levels. Traditional breeding techniques have many limitations, particularly for long-lived perennial trees; however, breeding programmes can be greatly accelerated with the use of molecular genetic markers. Microsatellite markers, also known as simple sequence repeats (SSRs), are useful for establishing and characterising germplasm collections (Godwin, 2003). In this context, they may be used for identifying genetically distinct populations to prioritise for collection, assessing the genetic diversity captured within the collection, determining the genetic relatedness between accessions and identifying the duplication of genotypes (particularly important for clonal species). The suite of SSR markers developed for the *Davidsonia* genus (Eliott et al., 2013) will enhance a germplasm collection strategy and may also prove useful for breeding programmes.

Genetic markers, such as SSRs, have been used for assessing the genetic diversity in many perennial tree crops (Arias et al., 2012; Jamnadass et al., 2009) and can increase the efficiency and precision of plant breeding programmes through marker-assisted selection (MAS) (Arias et al., 2012; Prentis et al., 2013). In breeding programmes, molecular markers fall into two categories: those which mark the mutation which causes the observed phenotypic variation (perfect markers) or more common are markers that are indirectly associated with the causal mutation (linked markers) (Prentis et al., 2013). To utilise MAS, you need to correlate a genetic maker to a trait of interest and for this, it is necessary to have reliable phenotypic data. This data needs to be obtained from fairly large mapping populations consisting of the progeny of crosses between parental lines that differ in the phenotypic trait of interest. A large number of markers are also necessary, so these approaches require considerable resources. However, with advances in technology such as next-generation sequencing, techniques such as MAS are becoming more feasible. The markers developed for *Davidsonia* species are useful in the immediate future for population genetics, DNA fingerprinting, characterising germplasm collections and identifying individuals from different populations.

CONCLUSIONS

The limited genotypic diversity in all three species of *Davidsonia* is of concern for both *in situ* conservation management and the potential for domestication.

Further research is required to confirm polyploidy in *D. johnsonii* and apomixis and the process leading to polyembryony in *D. pruriens*. A thorough study of the pollination biology in the genus is also needed along with pollination experiments to see if cross-pollination is possible (although with caution, as noted earlier). There are a number of useful traits in the *Davidsonia* species reproductive systems and there appears to be potential for domestication. If domestication were pursued, it would require five main steps: (1) conserving and characterising the existing germplasm, (2) documenting phenotypes with traits of interest, (3) creating or expanding new variation, (4) selecting and fixing desirable genotypes in the progeny and (5) (ideally) correlate phenotypic data to genotypic data to allow for MAS.

REFERENCES

Altoveros, N.C., Ramanatha Rao, V., 1998. Analysis of information on seed germplasm regeneration practices. In: Engels, J.M.M., Ramanatha Rao, V. (eds.), *Regeneration of Seed Crops and Their Wild Relatives. Proceedings of a Consultation Meeting*, 4–7 December 1995. ICRISAT, Hyderabad, India. IPGRI, Rome, Italy, pp. 105–126.

Arias, R.S., Borrone, J.W., Tondo, C.L., Kuhn, D.N., Irish, B.M., Schnell, R.J., 2012. Genomics of tropical fruit tree crops. In: Schnell, R.J., Priyadarshan, P.M. (eds.), *Genomics of Tree Crops*. Springer, New York, pp. 209–239.

Ashmore, S.E., Hamilton, K.N., Offord, C.A., 2011. Conservation technologies for safeguarding and restoring threatened flora: Case studies from Eastern Australia. *In Vitro Cellular and Developmental Biology-Plant* 47, 99–109.

Asker, S.E., Jerling, L., 1992. *Apomixis in Plants*. CRC Press, Boca Raton, FL.

Barcaccia, G., Albertini, E., 2013. Apomixis in plant reproduction: A novel perspective on an old dilemma. *Plant Reproduction* 26:159–179.

Beattie, A.J., Turnbull, C., Knox, R.B., Williams, E.G., 1984. Ant inhibition of pollen function: A possible reason why ant pollination is rare. *American Journal of Botany* 71, 421–426.

Bolotin, M., 1975. Photoperiodic induction of precocious flowering in a woody species *Eucalyptus occidentalis* Endl. *Botanical Gazette* 136, 358–365.

Bretagnolle, F., Thompson, J.D., 1995. Tansley Review No. 78. Gametes with the somatic chromosome number: Mechanisms of their formation and role in the evolution of auto-polyploid plants. *New Phytologist* 129, 1–22.

Cachi, A.M., Wuensch, A., 2014. Characterization of self-compatibility in sweet cherry varieties by crossing experiments and molecular genetic analysis. *Tree Genetics and Genomes* 10, 1205–1212.

Carman, J.G., 1997. Asynchronous expression of duplicate genes in angiosperms may cause apomixis, bispory, tetraspory, and polyembryony. *Biological Journal of the Linnean Society* 61, 51–94.

Charlesworth, D., Charlesworth, B., 1987. Inbreeding depression and its evolutionary consequences. *Annual Review of Ecology and Systematics* 18, 237–268.

Charlesworth, D., Willis, J.H., 2009. Fundamental concepts in genetics: The genetics of inbreeding depression. *Nature Reviews Genetics* 10, 783–796.

Cooper, H.D., Spillane, C., Hodgkin, T., 2001. Broadening the genetic base of crops: An overview. In: Cooper, H.D., Spillane, C., Hodgkin, T. (eds.), *Broadening the Genetic Base of Crop Production*. CABI Publishing, Oxford, U.K., pp. 1–23.

DEC-NSW, 2004a. *Recovery Plan for Davidsonia johnsonii (Smooth Davidsonia)*. Department of Environment and Conservation, Hurstville, New South Wales, Australia.

DEC-NSW, 2004b. *Recovery Plan for the Davidson's Plum (Davidsonia jerseyana)*. Department of Environment and Conservation, Hurstville, New South Wales, Australia.

Dutton, E.M., Frederickson, M.E., 2012. Why ant pollination is rare: New evidence and implications of the antibiotic hypothesis. *Arthropod-Plant Interactions* 6, 561–569.

Eliott, F.G., Connelly, C., Rossetto, M., Shepherd, M., Rice, N., Henry, R.J., 2013. Novel microsatellite markers for the endangered Australian rainforest tree *Davidsonia jerseyana* (Cunoniaceae) and cross-species amplification in the *Davidsonia* genus. *Conservation Genetics Resources* 5, 161–164.

Eliott, F.G., Shepherd, M., Rossetto, M., Bundock, P., Rice, N., Henry, R.J., 2014. Contrasting reproductive systems revealed in the rainforest genus *Davidsonia* (Cunoniaceae): Can polyembryony turn the tables on rarity? *Australian Journal of Botany* 62, 451–464.

Ellstrand, N.C., Prentice, H.C., Hancock, J.F., 1999. Gene flow and introgression from domesticated plants into their wild relatives. *Annual Review of Ecology and Systematics* 30, 539–563.

Faegri, K., van der Pijl, L., 1979. *The Principles of Pollination Ecology*, 3rd revised ed. Pergamon Press, Oxford, U.K.

Firetti-Leggieri, F., Lohmann, L.G., Alcantara, S., da Costa, I.R., Semir, J., 2013. Polyploidy and polyembryony in *Anemopaegma* (Bignonieae, Bignoniaceae). *Plant Reproduction* 26, 43–53.

Garcia, M.B., Antor, R.J., Espadaler, X., 1995. Ant pollination of the palaeoendemic dioecious *Borderea pyrenaica* (Dioscoreaceae). *Plant Systematics and Evolution* 198, 17–27.

Godwin, I., 2003. Plant germplasm collections as sources of useful genes. In: Newbury, H.J. (ed.), *Plant Molecular Breeding*. Blackwell, Oxford, U.K., pp. 134–151.

Gross, C.L., Nelson, P.A., Haddadchi, A., Fatemi, M., 2012. Somatic mutations contribute to genotypic diversity in sterile and fertile populations of the threatened shrub, *Grevillea rhizomatosa* (Proteaceae). *Annals of Botany* 109, 331–342.

Gustafson, F.G., 1942. Parthenocarpy: Natural and artificial. *Botanical Review* 8, 599–654.

Harden, G.J., Williams, J.B., 2000. A revision of *Davidsonia* (Cunoniaceae). *Telopea* 8, 413–428.

Henry, R.J., 2010. *Plant Resources for Food, Fuel, and Conservation*. Earthscan, London, U.K.

Jamnadass, R., Lowe, A., Dawson, I.K., 2009. Molecular markers and the management of tropical trees: The case of indigenous fruits. *Tropical Plant Biology* 2, 1–12.

King, L.M., Schaal, B.A., 1990. Genotypic variation within asexual lineages of *Taraxacum officinale*. *Proceedings of the National Academy of Sciences of the United States of America* 87, 998–1002.

Knuth, P., Ainsworth Davis, J.R., Müller, H., 1906. *Handbook of Flower Pollination: Based upon Hermann Müller's Work: 'The Fertilisation of Flowers by Insects'*. Clarendon Press, Oxford, U.K.

Koltunow, A.M., 1993. Apomixis – Embryo sacs and embryos formed without meiosis or fertilization in ovules. *Plant Cell* 5, 1425–1437.

Krishna, H., Singh, S.K., 2007. Biotechnological advances in mango (*Mangifera indica* L.) and their future implication in crop improvement – A review. *Biotechnology Advances* 25, 223–243.

Lakshmanan, K.K., Ambegaokar, K.B., 1984. Polyembryony. In: Johri, B.M. (ed.), *Embryology of Angiosperms*. Springer-Verlag, Berlin, Germany, pp. 445–474.

Lande, R., Schemske, D.W., 1985. The evolution of self-fertilization and inbreeding depression in plants.1. Genetic models. *Evolution* 39, 24–40.

Luo, C.W., Li, K., Chen, X.M., Huang, Z.Y., 2012. Ants contribute significantly to the pollination of a biodiesel plant. Jatropha curcas. *Environmental Entomology* 41, 1163–1168.

Martinez Ochoa, EdC., Andrade-Rodriguez, M., Rocandio Rodriguez, M., Villegas Monter, A., 2012. Identification of zygotic and nucellar seedlings in polyembryonic mango cultivars. *Pesquisa Agropecuaria Brasileira* 47, 1629–1636.

Meilan, R., 1997. Floral induction in woody angiosperms. *New Forests* 14, 179–202.

Meyer, R.S., DuVal, A.E., Jensen, H.R., 2012. Patterns and processes in crop domestication: An historical review and quantitative analysis of 203 global food crops. *New Phytologist* 196, 29–48.

Miller, A.J., Gross, B.L., 2011. From forest to field: Perennial fruit crop domestication. *American Journal of Botany* 98, 1389–1414.

Mogie, M., 1992. *The Evolution of Asexual Reproduction in Plants*, 1st ed. Chapman & Hall, London, U.K.

Nand, N., Drew, R.A., Ashmore, S., 2004. Micropropagation of two Australian native fruit species, *Davidsonia pruriens* and *Davidsonia jerseyana* G. Harden & J.B. Williams. *Plant Cell Tissue and Organ Culture* 77, 193–201.

Naumova, T.N., 1993. *Apomixis in Angiosperms: Nucellar and Integumentary Embryony*. CRC Press, Boca Raton, FL.

Nogler, G.A., 1984. Gametophytic apomixis. In: Johri, B.M. (ed.), *Embryology of Angiosperms*. Springer-Verlag, Berlin, Germany, pp. 475–518.

Offord, C.A., Makinson, R., 2009. Options and major considerations for germplasm conservation. In: Offord, C.A., Meagher, P.F. (eds.), *Plant Germplasm Conservation in Australia: Strategies and Guidelines for Developing, Managing and Utilising Ex Situ Collections*. Australian Network for Plant Conservation Inc., Canberra, Australian Capital Territory, Australia, pp. 11–34.

Olsen, K.M., Gross, B.L., 2008. Detecting multiple origins of domesticated crops. *Proceedings of the National Academy of Sciences of the United States of America* 105, 13701–13702.

Olsen, K.M., Wendel, J.F., 2013. Crop plants as models for understanding plant adaptation and diversification. *Frontiers in Plant Science* 4, 290.

Page, T., Watkins, M., 2016. Cultivation of Davidson's Plum (*Davidsonia spp*). In: Sultanbawa, Y., Sultanbawa, F. (eds.), *Australian Native Plants: Cultivation and Uses in Alternative Medicine and the Food Industry*. CRC Press/Taylor & Francis Group, Boca Raton, FL.

Paterson, A.H., 2005. Polyploidy, evolutionary opportunity, and crop adaptation. *Genetica* 123, 191–196.

Peakall, R., Beattie, A.J., 1991. The genetic consequences of worker ant pollination in a self-compatible, clonal orchid. *Evolution* 45, 1837–1848.

Prentis, P.J., Gilding, E.K., Pavasovic, A., Frere, C.H., Godwin, I.D., 2013. Molecular markers in plant improvement. In: Henry, R.J. (ed.), *Molecular Markers in Plants*. John Wiley & Sons, Inc., Hoboken, NJ, pp. 67–80.

Ramulu, K.S., Sharma, V.K., Naumova, T.N., Dijkhuis, P., Campagne, M.M.V., 1999. Apomixis for crop improvement. *Protoplasma* 208, 196–205.

Richards, A.J., 1997. *Plant Reproductive Systems*, 2nd ed. Chapman & Hall, London, U.K.

Richards, A.J., 2003. Apomixis in flowering plants: An overview. *Philosophical Transactions of the Royal Society of London Series B: Biological Sciences* 358, 1085–1093.

Rossetto, M., Gross, C.L., Jones, R., Hunter, J., 2004. The impact of clonality on an endangered tree (*Elaeocarpus williamsianus*) in a fragmented rainforest. *Biological Conservation* 117, 33–39.

Savidan, Y., 2000. Apomixis: Genetics and breeding. In: Janick, J. (ed.), *Plant Breeding Reviews.* John Wiley & Sons, Inc., New York, pp. 13–86.

Singh, A.N., Thakur, A. 2004. Polyembryony in *Dipterocarpus retusus. Journal of Tropical Forest Science* 16, 475–476.

Socias i Company, R., Fernández i Martí, À., Kodad, O., Alonso, J.M., 2010. Self-compatibility evaluation in almond: Strategies, achievements, and failures. *HortScience* 45, 1155–1159.

Stebbins, G.L., 1950. *Variation and Evolution in Plants.* Columbia University Press, New York.

Thurlby, K.A.G., Wilson, P.G., Sherwin, W.B., Connelly, C., Rossetto, M., 2012. Reproductive bet-hedging in a rare yet widespread rainforest tree, *Syzygium paniculatum* (Myrtaceae). *Austral Ecology* 37, 936–944.

Tuskan, G.A., Francis, K.E., Russ, S.L., Romme, W.H., Turner, M.G., 1996. RAPD markers reveal diversity within and among clonal and seedling stands of aspen in Yellowstone National Park, USA. *Canadian Journal of Forest Research – Revue Canadienne De Recherche Forestiere* 26, 2088–2098.

Udall, J.A., Wendel, J.F., 2006. Polyploidy and crop improvement. *Crop Science* 46, S3–S14.

Uma Shaanker, R., Ganeshaiah, K.N., 1997. Conflict between parent and offspring in plants: Predictions, processes and evolutionary consequences. *Current Science* 72, 932–939.

van der Merwe, M., Spain, C.S., Rossetto, M., 2010. Enhancing the survival and expansion potential of a founder population through clonality. *New Phytologist* 188, 868–878.

Varoquaux, F., Blanvillain, R., Delseny, M., Gallois, P., 2000. Less is better: New approaches for seedless fruit production. *Trends in Biotechnology* 18, 233–242.

Watson, G.C., 1987. A comparative study of the New South Wales members of the family Davidsoniaceae Bange with a view to their conservation and development. Diploma in Natural Resources thesis, The University of New England, Armidale, New South Wales, Australia.

Whitton, J., Sears, C.J., Baack, E.J., Otto, S.P., 2008. The dynamic nature of apomixis in the angiosperms. *International Journal of Plant Sciences* 169, 169–182.

Williams, G., Adam, P., 2010. *The Flowering of Australia's Rainforests: A Plant and Pollination Miscellany.* CSIRO Publishing, Collingwood, Victoria, Australia.

Yildiz, E., Kaplankiran, M., Demirkeser, T.H., Uzun, A., Toplu, C., 2013. Identification of zygotic and nucellar individuals produced from several Citrus crosses using SSRs markers. *Notulae Botanicae Horti Agrobotanici Cluj-Napoca* 41, 478–484.

Zohary, D., 2004. Unconscious selection and the evolution of domesticated plants. *Economic Botany* 58, 5–10.

CHAPTER 6

Cultivation of Desert Limes (*Citrus glauca*)

Jock Douglas

CONTENTS

INTRODUCTION

Botany and Distribution

Citrus glauca is the botanical name for the most widely occurring Australian native *Citrus*. Common names are many – desert lime, bush lime, limebush and wild lime – but the name desert lime is now the accepted name for this interesting *Citrus* species. Desert limes grow wild intermittently through a large section of inland Queensland and New South Wales (NSW). The northern limit is a line running from near the coast north of Rockhampton and through central Queensland to

Winton in the far west of Queensland. From there, desert lime occurrence extends south through Roma and Goondiwindi districts and down through mid-western NSW to the Dubbo area. The preferred soil type appears to be the factor that influences occurrence. Clay-based soil is a common requirement for the species, and it is often found where there is a change in soil type in the landscape. There are some minor occurrences in the central South Australia, particularly just north of Port Augusta. It is notable that in this area the desert lime grows in deep sands. Fruit in the northern range through Queensland is typically larger than fruit on trees in the southern occurrence in South Australia.

This species of native *Citrus* has evolved in harsh climatic and soil conditions, resulting in some remarkable characteristics. Variations in the species have been noted with adaptation for locality. For instance, the trees along the frost prone and exposed Ward River south of Tambo in Western Queensland have fine leaves compared to leaves on trees directly east in more tropical central Queensland. The desert lime is very drought tolerant and able to withstand extremes. In a grafted commercial desert lime stand near Roma (Queensland State), temperatures of 45°C and (–)8°C have been recorded with no noticeable ill effect. The time from flowering to fruiting is the shortest among any *Citrus* species: it is 10–12 weeks (Sykes, 1997). The species flowers mainly in spring and fruit ripens in summer (Figures 6.1 and 6.2).

Desert lime emerges as a thorny bush with few leaves, giving suitable protection from grazing native animals. Thorniness is an interesting characteristic in that seedling trees have sharp thorns, but once the tree gains height and branches are above the browse height of a large kangaroo, the tree becomes thornless. This is not

Figure 6.1   Botanical illustration of *Citrus glauca* – three forms. (Illustrated by David Mackay. Flora of NSW, Copyright Royal Botanic Gardens and Domain Trust.)

Figure 6.2   (See colour insert.) Photo of a desert lime (*Citrus glauca*) tree. (Photo by Jock Douglas. Copyright Australian Desert Limes.)

simply an age factor as large trees have been observed to be thornless except for low hanging branches which retain thorns. Again there is a variation in characteristics, with thorniness in South Australian trees being retained throughout the height of the tree regardless of age.

The desert lime is the only known member of the Citrus family which is a xerophyte, being adapted to withstand drought by dropping leaves. Seedlings develop a large root system before making vigorous aerial growth, and under extreme drought, the leaves fall and the leafless twigs carry on photosynthesis. In the wild, mature trees vary in height from around 4 to 15 m, depending on local conditions and genetics. Trees typically flower in August–September and the fruit is ripe in November–December. The fruit is about the size of a smallish grape, has thin and virtually tasteless skin and is yellow green when ripe. The fruit is juicy, pleasantly acid and often seedless. Large native trees have been known to produce over 60 kg of fruit in very favourable seasons (Figure 6.3).

Figure 6.3 Photo of a typical desert lime commercial plantation layout and row plantings.
(Photo by Jock Douglas. Copyright Australian Desert Limes.)

TRADITIONAL USE AND ECONOMIC POTENTIAL

Knowledge of the food plants of the Aborigines is sketchy, relying as it does mainly on early European accounts and the oral traditions of the community as a whole, which includes Aboriginal, Islander and pioneer settler sources. Certainly desert limes were a known food source for indigenous people. Little is recorded, but today's indigenous descendants in the Maranoa and Fitzroy river catchments of Queensland regard these wild limes as having been an important food and much favoured when in season. Leichhardt, an explorer, recorded making lime-based dishes in central Queensland during his successful journey of exploration in 1844 to the Arnhem Land and Port Essington. The central Queensland town of Taroom has the desert lime tree association, being named from the local Waka word *tarum* meaning *wild lime*. Pioneering families quickly recognised the flavour and health attributes of desert limes. European settlement records show that it was not only eaten from the tree as a tart but refreshing fruit, but was also harvested for making jams, chutneys and cordial.

The economic potential of desert limes is excellent. An assessment of the positives and negatives is given to show how this conclusion has been reached.

Current and Potential Benefits

- The trees are relatively easy to grow and can be grown over a wide climatic zone.
- It is a native Citrus fruit with a distinctive appealing flavour, making it suited to a wide range of food and beverage products.

- Desert lime fruit has health-enhancing components, notably high level of natural folate, vitamin C and antioxidants (Konczak et al., 2009; Maen and Cock, 2015; Zhao and Agboola, 2007).
- The background of desert limes is an interesting 'food story', which is important for getting differentiated products into high value markets.
- Desert lime fruit retains its structure and flavour well when frozen. This allows for storage and convenience of supply throughout the year.
- Trees are attractive as a home garden plant and exploiting this characteristic is an important strategy to broaden consumer awareness.
- It has a wide range of uses which so far include the following: the fruit is used by cooks and chefs as a garnish for fish and chicken servings; used in syrup and as an ingredient in jams, chutney, aioli, apple sauce, relish, paste, cordial, cider, slush drinks, ice cream and yoghurt; dried into a powder for herb and spice mixes and in yoghurt; and made into a puree with aseptic packaging for manufacturing.
- Seasonal supply of chilled desert lime fruit in punnets to domestic consumers has considerable potential but has not been trialled or developed.

Current and Potential Drawbacks

- Too little is known by consumers about desert limes and changing this is a major challenge that must be met as production increases.
- The market for desert limes could be affected by oversupply; therefore, growth in production would need to be matched by developing the growth in demand.
- The agronomy must be carefully managed, and although this may appear relatively straightforward, there is limited experience in this new area.
- Desert lime fruit production is labour intensive, and while innovation in mechanisation is lowering this cost, further development is needed.
- While storage of whole frozen limes gives the capacity for year-round supply, frozen storage is quite costly.

AGROCLIMATIC REQUIREMENTS

The climatic range of desert limes is extensive. Commercial plantings of grafted desert limes have been established near Roma in south-west Queensland, Gympie in south-east Queensland, Glenrowan in north-east Victoria, and southern part of South Australia and Western Australia. Home garden plantings have been successfully established in all states including Tasmania. Generally, the desert lime will thrive where commercial *Citrus* species thrives, but its range is wider because of its heat and frost tolerance. The desert lime requires considerably less water than non-native Citrus, presumably because its fine leafed foliage will have lower transpiration rates. This has been observed where orange and mandarin trees have been planted between rows of desert limes, and during a drought, the orange and mandarin trees showed early signs of moisture stress while the (grafted) desert limes were unaffected.

AGRONOMY

Agronomy for desert limes is quite similar to other *Citrus* species, but the particular nature of the native plant needs to be considered. Following are some points to be noted based on the experience of the author:

- Grafted desert lime trees are shallow rooted, with 80 cm maximum root depth noted on 10-year-old trees.
- Soil condition is a vital element and management should aim for a high level of organic matter with a balance of mineral elements including trace elements suited to *Citrus*. Soil tests, leaf tests and professional analysis advice are recommended.
- Trees respond well to fertiliser both for growth and for fruit production. Foliar applications of nutrients have proved to be very successful.
- Young grafted trees do not compete well with older trees, which should be a consideration for both garden and commercial plantings.
- The need for pruning is minimal, aiming for maximum foliage but with shape and ground clearance retained.
- Insect attack is similar to other *Citrus* species and needs to be managed. Similarly, pathogens such as *Phytophthora* may require management if this pathogen is present in the soil. There is one important exception and this is addressed in section 'Biotic Constraints'.
- Watering regimes should aim to keep soil moisture at an optimum level, bearing in mind that *Citrus* trees of any sort prefer well-drained soil.
- A short break in watering is recommended in July. Frost can also provide this sought-after reduction of sap flow. The short stress period influences the spring shoots on the trees which may be either blossoms or leaves. Stress at this time promotes flowering.
- The soil preference is a sandy loam; however, grafted desert limes will also grow well on clay-based soils when these are mounded.
- The rootstock should be selected to suit the soil type. Troyer is a preferred and proven rootstock for most soils although Trifoliata is recommended as the rootstock for heavy clay soils.

The following recommendations are based on the experience of the author:

- Row spacings recommended are 4 m wide and tree spacings are 4 m. However, it should be quite possible to plant 'hedge rows' of trees with 2.5 m spacings and row widths suited for machinery to be used.
- Underground soakage lines (subsurface drip irrigation) are recommended for watering, being placed about 100 mm underground and at a distance of 1 m on either side of the tree lines.
- Fruit will increase in size until fully mature under the right conditions. Large, juicy fruit is produced with optimal soil moisture and water applied to achieve this, right through from fruit set to ripening.
- Yield expectations will depend on securing proven grafted plant selections, but under ideal growing conditions, trees should produce 1 kg in year 3 from planting, 3 kg in year 4 and then have a 1 kg per year increase per year through to year 12. This would see about 11 kg of limes produced in year 12 from planting. It is believed that the grafted trees will keep growing beyond 12 years (Figure 6.4). Fruit yield is directly related to the extent of the tree canopy.

Figure 6.4 (See colour insert.) Photo of ripe desert lime (*Citrus glauca*) fruits. (Photo by Yasmina Sultanbawa. Copyright University of Queensland.)

GENETIC IMPROVEMENT AND PROPAGATION

As a species, *C. glauca* or desert lime is highly variable in the wild; consequently, careful selection of proven trees as a propagation source for commercial production is essential. Features to be considered in 'mother' tree selection are as follows:

- An outstanding fruiting capacity observed over at least 5 years through a variety of seasonal conditions
- Has large fruit compared with that from other trees in the locality
- Has good canopy shape, selecting for a low-set and rounded rather than a tall narrow canopy
- Vigorous healthy tree, displaying good growth capacity and ability to resist pests and diseases

Genetic improvement is possible through careful selection in the wild. The desert lime is relatively easy to hybridise and crossing selections with preferred characteristics is possible, although progress would be slow.

Plants can be propagated from seed. This has been trialled with seedlings planted in irrigated lines along with grafted plants on seedling rootstock. The seedlings propagated from seed retained a massive amount of thorns. At 8 years of age, they were approximately 2.5 m high and had not started fruiting.

Grafting is the only recommended propagation method for commercial desert lime production. Budding has been tried by many who are well experienced in this technique but without success. A wedge graft is used, firm taping applied and plastic bags tied over newly grafted plants to prevent desiccation and maintain humidity. Bags are retained until shoots appear on the scions.

Grafting material is collected from proven mature trees. The lack of thorns in the grafted trees is a major benefit. The desert lime has potential as a rootstock with its capacity to tolerate salt and its hardiness. Desert limes are compatible with a wide range of rootstocks and those successfully grafted include Troyer, Trifoliata, Benton, Cleopatra and Flying Dragon. No rootstock trialled has proved incompatible. Troyer is generally preferred, while Flying Dragon used recently is producing very vigorous seedlings.

Observation of natural propagation and growth has led to the following conclusions: Birds, which eat the fruit and (principally) emus, spread the seed. Under very favourable conditions, germination and seedling emergence occurs. The seedling spreads laterally where conditions permit, developing into an intense thorny clump. A clump of desert limes with close to a thousand seedlings has been examined to find that all the roots were connected. The clump was basically one organism. From these seedling clumps, a small number of trees, perhaps three or four, will emerge and mature. Meanwhile the thorny seedlings in the clump die off. During this process, which may take well over 50 years to occur, the clump has been protected from fire as virtually no grass will be growing within or close to it. Fruiting is limited until trees start to emerge from seedling clumps. The few trees that remain and mature become favoured shade camps for native animals – and now domestic animals. The manure provided is a natural fertiliser source which this native species needs for growth and fruit production.

The nature of desert limes clumps with their thorniness and bare ground–colonising habit is often seen as incompatible with grassed paddocks and the grazing of cattle and sheep. Modern woody weed management techniques such as blade ploughing make it relatively simple to destroy clumps. Given the length of time taken for the few trees to emerge from clumps, it is likely that the natural regeneration of desert limes will decline over its range. This decline could be exacerbated if recurrence of drought increases through climate change as predicted. Insect attack through borers is prevalent at any time in the wild and is normally withstood but can cause tree deaths in severe drought, as has recently been noted.

BIOTIC CONSTRAINTS

Birds are not likely to be a problem in commercial production, although in a drought year at Roma, there was some slight red-winged parrot and rosella damage, but this did not recur the following year when the season improved and the parrots had more preferred food sources available. No other birds or fruit bats have been seen eating fruit. Given the opportunity, emus can cause considerable damage.

Seedling leaves will be readily eaten by rabbits, hare and wallabies, causing severe damage and plants must be protected from these animals.

In the wild and in desert lime stands throughout Queensland and NSW, a gall fly is endemic. The gall occurs on the leaves and stems without any great detrimental effect. However, when fruit is developing, the gall fly lays its eggs within maturing fruit, which quickly develop as galls containing pupae in lime fruit. This is not evident until cut open. The fly emerges through a small hole it makes in the skin, evident as a tiny black mark. Gall fly can be readily controlled by two treatments of dimethoate or similar as a foliar spray. The first foliar spray should be done in early spring prior to flowering; the second soon after this at early fruit set. There is natural control of this insect through a parasitising fly and care should be taken in timing of insecticide applications. For intended commercial desert lime production, the location becomes very important. It is possible to completely avoid this insect pest by not growing desert limes in the same locality as wild trees.

Desert lime trees are prone to attack from borers but cope well with this unless stressed, as the sap flow of the tree normally plugs the borer hole. Insecticide treatment for borers is commercially available if needed.

Spined Citrus bugs are attracted to desert lime trees as the fruit ripens. These bugs sting the fruit but damage is limited. The bugs have an unpleasant pungent smell and can become lodged in amongst fruit while harvesting. Scale insects are common to all *Citrus* and can occur in desert limes as well. This insect is easily managed with foliar application of white oil. The Citrus butterfly can lay eggs on young trees with resulting caterpillars causing defoliation. This can cause a setback to growth but rarely kills the tree.

Fungal infestations such as sooty mould as well as *Phytophthora* can be a problem, especially in extended wet periods. Management techniques for control are well known and available. Mounding to promote drainage is a useful preventative strategy against root and collar diseases.

HARVESTING AND PROCESSING

Desert lime fruit ripens rapidly with fruit changing colour from green to light green to yellowish green and finally light yellow. When ripe, the fruit falls readily. Ripe fruit on the ground quickly deteriorates; therefore, only fruit picked from trees is of the right quality. The usual duration for harvesting a stand of commercial desert limes is 3 weeks. This may extend slightly if cool, cloudy weather prevails. Generally, harvest time coincides with the approaching height of summer when hot weather prevails. Ripening of fruit varies from year to year and harvesting should start immediately after the fruit begins turning from light green to yellowish green. Fruit continues ripening for several days after being picked.

Hand picking is general practice. This can be assisted with D-shaped catching baskets placed under trees and fruit stripped from twigs and branches into these. Baskets are emptied into slotted crates on the back of an accompanying vehicle. Desert limes will heat rapidly when in a mass and handling must be designed to avoid this. In hot conditions, the morning pick of lime crates is removed from the paddock at midday and stacked in a cold room to cool. Likewise, the afternoon pick of limes is placed in a cold room. Therefore, the pick for 1 day is cooled and cleaned and packed the following day. Two teams of people are involved for a continuous operation: one in the field picking, the other in the shed cleaning and packing. The application of olive-picking technology and equipment is proving to be a useful innovation. Shade cloth is rolled out under trees and a handheld electromechanical olive harvester is used.

Typically, considerable leaf matter is present with fruit when picked, which has to be removed. A reverse running belt cleaner has been devised and also a slow rotating cylinder. The final cleaning is done with limes running over a brush-cleaning unit with air suction at the last stage to remove any remaining loose leaf matter. There is scope for innovation and further development of this process.

Packaging and Storage

Limes coming off the cleaning line are then (if necessary) graded into sizes with a sizing belt machine. Limes are then ready for weighing into plastic bags and placement in cartons. These are then put back into refrigeration and taken from there to the freezer storage at $-21°C$.

Marketing

The principles of successful marketing vary little across industries and businesses. What needs to be borne in mind is that desert limes are little known in the food industry – the major target audience for marketing. Hence, the market largely needs to be created, with the desert lime positioned as a differentiating factor that adds value. The defining features of desert limes need to be well known and described. Success in marketing will come from finding new and expanded uses for desert limes in food, beverage and nutraceutical products, rather than waiting to be found.

Recently (December 2014), there is a shortage of desert limes due to the major supplier closing down because of drought and lost water access. The whole of the desert lime plantation near Roma has been lost, and this had been the largest supplier to Australian and overseas markets. Stored frozen limes are in short and diminishing supply. A shortage is likely to continue, possibly for 5 or more years. Nonetheless, information on marketing of desert limes provided is based on what has been learned by the Douglas family and its Australian desert limes business prior to the disastrous 2014 situation at Roma.

These tenets were adopted by Australian desert limes (ADL) and are provided as a guide:

- Plan well ahead. Desert lime production and marketing is a long-term exercise.
- Secure the supply base. Manufacturers and customers generally need to know that the supply will be constant.
- Decide that excellence is essential – in product quality, packaging and supply service.
- Have control or at least have influence right through the value chain.
- Know the story – discover points of difference and market them.
- Develop cordial relationships with customers, suppliers and support people. Having good relationships is the key to good business and underpins the enjoyment of effort.
- Extend networks. Help people and they will be helpful – with information, ideas and support.

- Continually expand skills and knowledge.
- Manage ourselves harmoniously as a family business to sustain enthusiasm and creativity.

Pricing and Costs

The economics of desert limes production is not easily defined because of the variables. However, a useful source of information is available in the *RIRDC New Crop Industries Handbook* (pp. 357–367) in the article by Hugh Macintosh of IQ Agribusiness.

The 2014 wholesale price for bulk frozen limes has been consistently at or above $15 per kg. Production cost of picking, cleaning, packaging and placing in storage is estimated at $2.50 per kg. Recommended agronomy, harvesting and handling should be noted for estimating overall production costs.

CONCLUSIONS AND RECOMMENDATIONS

Desert limes have good potential as a commercial crop. Sites for production should best be away from wild desert lime trees, should have a suitable soil type and a secured supply of water for irrigation. Seedlings with proven genetics are essential for an assured high level of production.

The natural occurrence of desert limes is shrinking in its grazed grasslands habitat. Two factors are involved: While mature trees are valued as attractive shade providers, the seedling clumps which produce these are seen as woody weed encroachments and often mechanically removed. Additionally, the longevity of mature trees is diminishing, with dry spells and drought widespread in the last 10 years. These conditions are predicted to become more prevalent with climate change.

The fruiting capacity of trees and their growth habit has been noted as highly variable, factors which will be important for conservation of the species and for selection of commercial cloning material. Little is known about this variability or the factors impacting on the capacity of *C. glauca* to naturally regenerate. What is known is that while the species is well adapted to semi-arid landscapes, its presence is highly localised and often remote. Identification of characteristics important for commercial selection while the species is still widespread should be undertaken.

A desert lime study should be undertaken to meet the following:

1. Define the characteristics and map the occurrence of *C. glauca* across the full range of habitats in which it naturally grows. The study would aim to add substantially to the knowledge of this species for its selection and use in commercial production and for its conservation as a valued native tree species.
2. Identify characteristics important for commercial selection. These include features such as canopy size, shape, thorniness, fruiting capacity, the size of fruit and yield.
3. Identify associated insects and fungi and the beneficial or detrimental effects these may have on tree growth, fruit production and fruit quality.

REFERENCES

Desert Lime. http://www.anfil.org.au/key-native-species/flavour-of-the-month-may/.

Konczak, I., Zabaras, D., Dunstan, M., Aguas, P., Roulfe, P., Pavan, A., 2009. *Health Benefits of Australian Native Foods: An Evaluation of Health-Enhancing Compounds.* Rural Industries Research and Development Corporation, RIRDC Publication No.09/133, Canberra, Australian Capital Territory, Australia.

Maen, A., Cock, I.E., 2015. Inhibitory activity of high antioxidant Australian native fruits against the bacterial triggers of selected autoimmune diseases. *Pharmacognosy Communications* 5, 48–58.

RIRDC, 2014. Desert Lime. https://rirdc.infoservices.com.au/items/14-113.

Sykes, S., 1997. Australian native limes (*Eremocitrus and Microcitrus*) – A citrus breeder's viewpoint. *Australian Bushfoods Magazine* 3, 12–15.

Zhao, J., Agboola, S., 2007. *Functional Properties of Australian Bushfoods.* Rural Industries Research and Development Corporation, RIRDC Publication No. 07/030, Canberra, Australian Capital Territory, Australia.

CHAPTER 7

Cultivation of Australian Finger Lime (*Citrus australasica*)

Sheryl Rennie

CONTENTS

INTRODUCTION

Australian native finger lime is emerging as a unique ingredient in the culinary industry. There is considerable potential to develop the plant into a significant horticultural crop in Australia. The fruit has already found a variety of uses, from fresh fruit to sauces, marinades and also desserts. Preliminary work has been done towards commercialisation of this crop, with a few growers already adopting a scientific approach towards cultivation and production. However, significantly more work needs to be done to meet the anticipated demand.

BOTANY

Australian native finger lime is a member of the family Rutaceae and is native to the rainforests of northern New South Wales (NSW) and south-east Queensland. *Microcitrus australasica*, subsequently reclassified as *Citrus australasica*

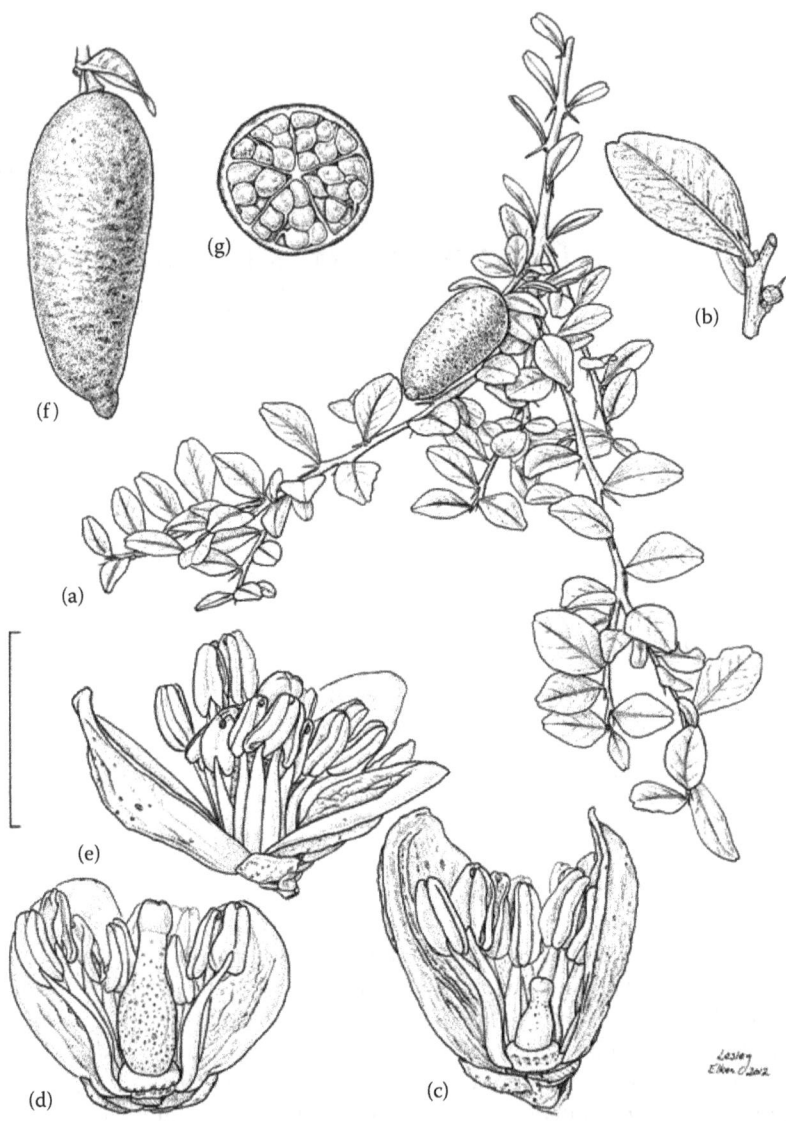

Figure 7.1   Botanical illustration of *Citrus australasica*. (a) Fruiting branch, (b) leaf detail showing spines, (c) male flower, (d) bisexual flower (with petal and stamens removed),
(e) bisexual flower, (f) mature fruit, (g) cross section of fruit. Scale bar: (a) 4 cm,
(b and g) 2 cm. (c, d and e) 0.5 cm. (f) 2.5 cm. (From *Flora of Australia, Volume
26: Meliaceae, Rutaceae and Zygophyllaceae*, CSIRO Publishing/Australian
Biological Resources Study (ABRS), Canberra, Australia, 2013; Illustrated by
Lesley Elkan. Copyright Royal Botanic Gardens and Domain Trust.)

Figure 7.2   (See colour insert.) Photo of finger lime (*Citrus australasica*) tree with finger limes. (Photo by David Hancock. Copyright Skyscans.)

(Mabberley, 1998), is one of the six species which are uniquely Australian, geographically. Specimens have also been found as far south as Bellingen in NSW and as far north as Melany in Queensland. As an understorey tree in their natural habitats, they can reach heights of 10 m and have been estimated to be up to 100 years old (Figures 7.1 and 7.2).

Branches have thorns at leaf axils. Leaves are entire with inconspicuous wings. Flowers are small; white; glabrous, with a notched tip; and crenate towards the upper apex. Fruits are cylindrical, 4–12 cm long, and when opened, the pulp of the fruit separates into pearl-like structures resembling caviar. The colours of the fruit can range from different shades of greens and purples (Figure 7.3).

Figure 7.3   (See colour insert.) Photo of finger limes (*Citrus australasica*) in a range of skin colours. (Photo by David Hancock. Copyright Skyscans.)

Finger limes show considerable genetic diversity, as is evident in the wide palette of colours, size and shape of fruit and pulp, variation in tree size and shape and degree of seediness (from almost seedless to very seedy), to cite a few common traits.

TRADITIONAL USE AND ECONOMIC POTENTIAL

The growing region of the finger lime in NSW and Queensland was drastically cleared for its mature timber to enable dairy farming. Early settlers used the fruit, but little is known of indigenous use and this knowledge may even be lost. However, the wood from finger lime trees was used by both indigenous communities and early settlers. Hammer and axe handles, walking sticks and other tools were made from the wood because of its strength.

Demand for finger limes has grown over the last decade, with export-grade fruit being sold fresh to the European Union, Asia, United Kingdom, Canada and Japan. The finger lime has been compared to and promoted as 'citrus caviar' due to the caviar-like appearance of the pulp. Asian seafood dishes were particularly enhanced by using finger lime. The native food industry is one of the largest buyers of finger limes in Australia. However, a research and development (R&D) effort must be developed to utilise lower-grade fruit to make jams, cordials, pickles and chutneys, to increase profits for growers and processors. There has also been interest expressed by the pharmaceutical and cosmetic industries to use finger lime as an ingredient.

The volatile constituents of the peel solvent extract of the Australian finger lime *C. australasica* were analysed by GC–MS. While limonene is a common component in *Citrus* species, isomenthone, which is rare in *Citrus* species, was a major component. Four of the six novel terpenyl esters identified have not been previously identified in a natural product (Delort and Jaquier, 2009).

AGROCLIMATIC REQUIREMENTS

Finger limes are endemic to northern NSW and south-east Queensland. However, commercial plantations have been established in southern NSW, South Australia and North Queensland. However, tree growth has not reached the vigour and health observed in the tree's natural habitat (Figure 7.4).

Finger limes can tolerate a wide range of soil types, provided they are well drained. A soil pH around 5.0–6.5 is considered ideal. Regions with heavy frosts, strong winds and intense sun should be avoided.

CROP MANAGEMENT

Best management practices need to be observed when growing finger limes, because although they are a native plant, they are not easy to grow. Commercial

Figure 7.4   Photo of commercial plantation of finger lime (*Citrus australasica*) trees. (Copyright R & R Deaker/Australian Fingerlime Caviar.)

crops are planted in rows 4–5 m apart, with trees 2.5–3 m apart within rows, depending on variety or cultivar, usually as hedge rows at a density of 600–800 trees/ha. Each cultivar has a different growth habit, management and harvesting requirements. Light pruning has also been found to be beneficial. Being a native tree, they do not have a high fertiliser requirement and are at risk due to over-fertilising. Windbreaks are essential as fruit are damaged by high winds.

Fruiting commences after 2–3 years and peaks after the 5th–6th years. The fruit takes around 5 months after flowering to mature. The fruiting times are between December and May, with a peak between March and May. There is variation in fruiting times among the different cultivars; therefore, it is necessary to plant a variety of cultivars to ensure continuous availability of fruit over a 5-month period. Each cultivar will bear fruit for 5–8 weeks a year. Fruits must also mature on the tree and do not ripen after picking. Further, fruits do not mature uniformly and can be easily damaged, which makes handpicking imperative. Trees maintain fruiting capacity for many years.

First-grade fruit is supplied to the export and domestic markets. Second-grade fruit is frozen to supply off-season demand, while the remaining fruits are brought to a market with processors of jams, chutneys and condiments.

Yields of 20 kg per tree can be obtained from 5–6-year old trees, but yield is largely dependent on cultivar and prevailing weather conditions, mainly the availability of water. Under ideal conditions, 50% of fruit would be first-grade quality and 30% second grade.

GENETIC IMPROVEMENT AND PROPAGATION

Over 65 varieties of finger lime have been identified in the wild, some of which have been selected for trials. Many varieties do not breed true; therefore, selection for commercial plantings is a long and laborious process. Today's export cultivars, some of which are registered with the Australian Citrus Recording Association (ACRA), have been selected for low seed count, taste, colour, skin quality and shelf life under chilled transportation.

Breeding programmes have also been initiated. The blood lime, a cross between a mandarin and a finger lime, has blood-red rind, flesh and juice (Sykes, 2002). 'Australian Blood' and 'Australian Sunrise' were hybrids developed by the CSIRO and granted Plant Breeders' Rights (The Registrar, Plant Breeders' Rights, 1997).

Some cultivars available for retail within the industry are Rainforest Pearl, Rainforest Jade, Rainforest Diamond, Rainforest Ruby, Rainforest Garnet, Byron Sunrise, Alstonville and Emerald.

Cuttings have a low strike rate, usually below 50%. Propagation by grafting scions from selected mother plants onto rootstocks (commonly *Citrus trifoliata* or Troyer citrange rootstocks) is the chief means of obtaining commercial planting material. The *Citrus trifoliata* rootstock confers some resistance against infection by *Phytophthora* root rot.

BIOTIC CONSTRAINTS

Finger limes are susceptible to common diseases affecting citrus in general. Dieback of branches and melanose can be a major problem. Melanose is caused by the fungus *Diaporthe citri*, resulting in dark-brown to black spots on the leaves, branches and fruit. Common pests include scale insects, aphids, mealybugs, spined citrus bug, bronze orange bug, caterpillars, snails, katydids and grasshoppers. In 2007, the NSW Department of Primary Industries examined the host status of finger limes to Queensland fruit fly and concluded that it was a non-host. Oleocellosis, where rind damage is caused by release of oil from the damaged oil glands due to physical damage, is also common.

CONCLUSIONS AND RECOMMENDATIONS

Finger lime has a considerable potential as a horticultural tree crop. Considerable R&D work needs to be done to produce elite planting material, to reduce variability in yield and phenology.

BIBLIOGRAPHY

Australian Native Plants Society (Australia). http://anpsa.org.au/c-aust.html.

Australian Plants online – March 2001. http://anpsa.org.au/APOL21/mar01-7.html#photo1.

Delort, E., Jaquier, A., 2009. Novel terpenyl esters from Australian finger lime (*Citrus australasica*) peel extract. *Flavour and Fragrance Journal* 24, 123–132.

Finger Lime. Publication No. 14/114, Rural Industries Research and Development Corporation, (www.rirdc.gov.au.) and Australian Native Food Industry Limited (www.anfil.org.au).

Flora of Australia, Volume 26: Meliaceae, Rutaceae and Zygophyllaceae, 2013. CSIRO Publishing/Australian Biological Resources Study (ABRS), Canberra, Australia.

Hardy, S., Wilk, P., Viola, J., Rennie, S., 2010. Growing Australian native finger limes. Primefact 979. Industry and Investment, Orange, New South Wales, Australia.

Mabberley, D.J., 1998. Australian Citreae with notes on other Aurantioideae (Rutaceae). *Telopea* 7, 333–344.

Native Australian citrus. http://citruspages.free.fr/australian.html#finger.

Ryder, M., Latham, Y., Hawke, B., 2008. *Cultivation and Harvest Quality of Native Food Crops*. Rural Industries Research and Development Corporation, RIRDC Publication No. 08/019, Canberra, Australian Capital Territory, Australia.

Sykes, S.R., 2002. 'Australian Outback', 'Australian Blood' and 'Australian Sunrise'. *Plant Varieties Journal* 15, 18–21.

The Registrar, Plant Breeders Rights Office, 1997. *Plant Varieties Journal* 10(2), 9.

CHAPTER 8

Production of *Terminalia ferdinandiana* Excell. ('Kakadu Plum') in Northern Australia

Julian Gorman, Kim Courtenay and Chris Brady

CONTENTS

INTRODUCTION

The commercialisation of *Terminalia ferdinandiana* presents an outstanding economic opportunity for vast areas of currently relatively economically unproductive and unpopulated tropical savannahs. This economic base could help support Indigenous people to maintain or retain connection with their traditional country, benefiting both people and the environment. This chapter seeks to outline the opportunity resulting from the exceptional properties of this species; provide a brief summary of the story of commercialisation thus far; explore some of the potential co-benefits that could be achieved in areas such as Indigenous health, employment and training and environmental management; outline some lessons learned from wild harvest and enrichment planting; and outline some of the challenges that need to be met.

T. ferdinandiana is endemic to the monsoon tropics of northern Australia and has a long history of medicinal/nutritional use by Indigenous Australians. The fruit and other parts of the tree have exceptional chemical properties including extraordinary high content of vitamin C and antioxidants. The nutraceutical, pharmaceutical, antimicrobial and preservative potentials of the fruit are driving market demand. Despite the demand, there is relatively limited supply as a result of an uncoordinated approach to wild harvest and little cultivation.

The majority of the natural occurrence of *T. ferdinandiana* falls on Indigenous tenured land. Efforts are being made to involve Aboriginal people in economic opportunities derived from the commercialisation of *T. ferdinandiana*. Meeting market demand in the short term requires a coordinated wild harvest of fruit from multiple ranges. Given the species distribution, density and yields, there is no reason why this cannot be done in an ecologically sustainable manner. To ensure Aboriginal involvement and benefit, wild harvest needs to incorporate the cultural and social protocols of the different Aboriginal groups on their Clan Estates. In the longer term, the large market for this fruit is likely to drive more conventional cultivation, requiring agronomic research and development. There is likely to be a variety of cultivation practices acceptable to Aboriginal people such as mixed and enrichment plantings, and it is critical that Aboriginal people are given the support and advice to be able to participate in the growth of this industry in a way that suits their aspirations, knowledge and cultural protocols.

BOTANY

Terminalia ferdinandiana Excell. is part of the Combretaceae family which contains 20 genera and 500 species that are widespread in tropical and subtropical regions of the world (Dunlop et al., 1995). The genus *Terminalia* consists of about 200 species, of which 29 species or subspecies are native to Australia (Dunlop et al., 1995) with 14 species in the Kimberley of Western Australia (WA), 12 in Northern Territory (NT) and 16 in North Queensland (Pedley, 1995). Although there is some taxonomic uncertainty, *T. ferdinandiana* Excell. is endemic to Northern Australia and ranges from just south of Broome (WA) through both the coastal and inland areas of Northern WA into the NT, where it occurs across the wet/dry tropics from the west, across Arnhem Land to the Gulf of Carpentatria (Dunlop et al., 1995; Pedley, 1995).

T. ferdinandiana Excell. is a small to moderately sized semi-deciduous tree (Pedley, 1995) which was originally described as *T. edulis* by Ferdinand von Mueller in 1860 (Williams, 2011). It is closely similar to *T. carpentariae*, *T. hadleyana* and *T. latipes*, which is sometimes considered as a subspecies of *T. l. psilocarpa*

Figure 8.1 Photo of mature Kakadu plum (*Terminalia ferdinandiana*) tree. (Photo by Kim Courtenay. Copyright Kimberley Training Institute.)

(Byrnes, 1977; Pedley, 1995; Wheeler, 1992). Sometimes it is also combined with *T. prostrata* (Dunlop et al., 1995). The exact taxonomy of *T. ferdinandiana* Excell. is uncertain as there are a number of natural hybrids (Cunningham et al., 2009; Kenneally et al., 1996) (Figure 8.1).

Across its range, there is some variation in flowering and fruiting, but generally, the *T. ferdinandiana* flowers at the end of the dry season (September–November) and fruits from the middle of the wet season (January–June) to the early part of the dry season. It produces smooth fleshy ovoid drupes; the fruits can be highly variable in shape but are yellow-green when ripe (Brock, 2001) (Figure 8.2).

The species is known by a variety of common and language names, including Kakadu plum, bush plum, billygoat plum, green plum, salty plum, Gabiny (Yawuru people from the Broome area), Gubinge (Bardi people north of Broome), Kabinyn (Nyul Nyul people of the Damier Peninsular), Madoor (Bardi people near One Arm Point – Dampier Peninsula), Manmohban (Kune language in Maningrida area), Kerewey (Matngala people near Daly River), Mi-marral (Wadeye Area) and Murunga (East Arnhem Land). To avoid any bias in use of name, we will use its Latin name, *T. ferdinandiana*, in this chapter.

Distribution and Ecology

T. ferdinandiana is a major understory component of *Eucalyptus tetradonta* and *Eucalyptus miniata* woodland and open forest vegetation in the NT. This vegetation type is widespread across the northern part of the NT (Woods, 1995), making

Figure 8.2 (See colour insert.) Photo of Kakadu plum (*Terminalia ferdinandiana*) fruit.
(Photo by Kim Courtenay. Copyright Kimberley Training Institute.)

T. ferdinandiana very common. The few studies that have measured this species density show that there is high variability across its range (Whitehead et al., 2006; Woods, 1995), likely to be related to soil types, drainage and climatic conditions (rainfall). The density of *T. ferdinandiana* trees of fruit-bearing age (more than 2 m in height) has been recorded to be in excess of 500 trees/ha, with the highest densities on or near the coast (Woods, 1995). The coastal strip in the Darwin region was found to have mature trees (>2 m in height) at 272 ± 169 trees per ha (Whitehead et al., 2006). At intervals from the coast central Arnhem land to 50 km inland densities averaged 14.4 ± 24.3 trees/ha with the highest density being along the narrow coastal strip (82 trees/ha) and on clay soils 40 km inland (31 trees/ha) (Gorman et al., 2006; Whitehead et al., 2006). In certain areas, densities are extremely high and could be thought of as natural plantations.

TRADITIONAL USE AND ECONOMIC POTENTIAL

Of the 200–250 species of Terminalia worldwide, at least 50 species are used as food by people with the fruits, nuts and gum eaten (French, 2013). The kernels of the Indian almond (*T. catappa*) and *T. grandiflora* have a protein content of 20% and are eaten in a number of tropical countries (Brand Miller et al., 1993; Dunlop et al., 1995). The flesh of *T. catappa, T. carpentariae* and *T. macrocarpa* have been eaten by Aboriginal people (Byrnes, 1977; Coode, 1973; Cunningham et al., 2009; Pedley, 1990). The fruit of *T. ferdinandiana* was consumed by Aboriginal people on hunting trips for quick energy and refreshment (Brock, 2005). Central Arnhem Land tribes regarded it more as a medicine than a food (Isaacs, 1987). In addition to the fruit being eaten for medicinal purposes (to treat colds and congestion) (Lindsay et al., 2001; Puruntatameri et al., 2001; Raymond et al., 1999), the sap was roasted and the bark was boiled in water and used to treat skin conditions and sores, or drunk as a tea for colds and flu (Lindsay et al., 2001). Traditionally, both the fruit and seed of *T. ferdinandiana* were eaten raw.

Market demand often determines the approach taken in commercialising a product. In the case of *T. ferdinandiana*, the market has fluctuated greatly since the mid-1980s when its potential for pharmaceutical, medicinal and nutraceutical values became apparent and was supported by scientific data. It is currently used as a food ingredient, in beauty products and increasingly for its functional properties, significant antioxidant capacities and high levels of vitamin C, vitamin E, lutein, folate and certain minerals (Cunningham et al., 2009).

Commercial attention was drawn to this species as a source of vitamin C over 20 years ago (Brand et al., 1982), but commercial wild harvest only began in 1996. At this time, the exceptionally high percentage of vitamin C of the fruit quickly attracted the attention of a large multi-national company as well as several smaller Australian companies and resulted in investment in research from these private companies. In recent years, research has been inconsistent and ad hoc with investment from state and federal governments. Growth of the industry has been slow. Recent research, relating to the antioxidant and antimicrobial qualities of the fruit, has once again drawn the attention of private enterprise and has greatly increased demand and the need for further research and development of models of harvest and supply.

In the 1990s, there was commercial interest in *T. ferdinandiana* from a private company called Access Business Group International LLC (ABG) (a business of the Alticor Group and sister company of Amway) in partnership with Cognis Australia Pty Ltd. This resulted in agronomic research and a plantation of approximately 6000 trees being established just east of Darwin (NT) at a place called Wildman River (Cunningham et al., 2009; Robinson, 2010). In May 2004, ABG and Cognis believed they had identified chemotypes with a high production potential and were reported to have exported tissue culture to Amway's facilities in Brazil; this landed them in a big controversy. The controversy surrounding the export of tissue culture was the demise of the ABG and Cognis Australia Research Group (see Cunningham et al., 2009), and it could be construed that this controversy has stalled the further development of this product in this company's product range.

As of 2014, the University of Western Australian is co-ordinating a research development project funded by the Rural Industry Research Development Corporation to collect genetic material suitable for cultivation from across the species range and to grow this material in a Government arboretum and in time develop suitable cultivars. They are also to look at variation of fruit chemical composition over the species range and processes and procedures of best practice in fruit harvest.

In 2013, demand for *T. ferdinandiana* increased greatly when new research showed antimicrobial properties of the fruit, which may replace some of the chemical preservatives used by the seafood industry (ABC, 2013; Williams et al., 2014). *T. ferdinandiana* was also awarded the Editor's Choice Award for 'Best Novel Food

Ingredient' at the Natural Products Expo West/Engredea Show in Anaheim CA in 2014 attracting exposure and demand.

This section demonstrates that there had already been substantial research and development on the commercialisation of *T. ferdinandiana* stemming back a number of decades. Growth of this industry will depend on continued collaboration across Northern Australia in a very integrated and inter-disciplinary fashion.

Chemical Composition

Aboriginal people have long recognised that the fruit of *T. ferdinandiana* has medicinal and nutraceutical benefits, and later this has been quantified by laboratory chemical analysis. Over 30 years ago, Brand et al. (1982) identified the commercial potential of the *T. ferdinandiana* fruit as a source of natural ascorbic acid (vitamin C). The wet weight of vitamin C in *T. ferdinandiana* fruit was found to average 3.5%, and it was later recorded as high as 5.5% (Woods, 1995). This is extremely high relative to other fruits including that of the Barbados cherry, *Malpighia glabra*, native of Brazil, and was once claimed to have the highest vitamin C content among all the fruits in the world, at 1.7% wet weight (Clein, 1956; Johnson, 2003). This finding was supported by a more recent study by Konczak et al. (2010).

Another important component of *T. ferdinandiana* fruit is the phenolic compounds, responsible for high antioxidant capacity (Konczak et al., 2010, 2014; Ohno et al., 1999; Williams et al., 2014). Phenolic-rich fruit extract has recently been found to have pronounced anti-inflammatory and chemopreventive properties (Tan et al., 2011), further supporting many traditional uses of *T. ferdinandiana* as a medicine (Konczak et al., 2010). In a comparison with 12 native Australian fruits, *T. ferdinandiana* was found to be the richest source of ascorbic acid (Netzel et al., 2007) and more recent studies have found prolific ellagic acid levels in *T. ferdinandiana* (Williams et al., 2014) and the ability to inhibit microbial growth on prawns.

There appears to be a considerable variability of chemical properties of fruits across the species range (Konczak, 2014) along with the other phenotypic properties such as fruit size, ratio of kernel to flesh, fruit yields and timing of fruiting, tree size and shape. This is likely due to a variety of factors including genetics but also biophysical factors such as climate and soil conditions. The variety may be confounded by inconsistency in the fruit collected for analysis as ripening stage as well as storage condition would impact greatly the chemical composition (Konczak et al., 2010).

Along with this variation, a further consideration in cultivation of *T. ferdinandiana* is the degree of impact of horticultural influences (irrigation, fertiliser, soil type, pollination) on the growth and the chemical composition of the fruit. Preliminary findings (Konczak et al., 2013) indicate that there was little variation in the nutritional levels of fruit sampled from cultivated trees (watered with drip irrigation only) compared to those from wild trees in the same general region. There is a need for further research and quantification of this finding.

CROP MANAGEMENT

In the 1990s, there was commercial interest in *T. ferdinandiana* from a private company called Access Business Group International LLC (ABG) (a business of the Alticor Group and sister company of Amway) in partnership with Cognis Australia Pty Ltd. This resulted in agronomic research and a plantation of approximately 6000 trees being established just east of Darwin (NT) at a place called Wildman River (Cunningham et al., 2009; Robinson, 2010).

There has been little research on fruit yields of *T. ferdinandiana* but Woods (1995) conservatively estimated that the yields of tree in semi-domesticated state (parks) ranged from 15 to 24 kg per tree per season and cited having observed mature trees yielding up to 40 kg in one season. There is likely to be greater variability in the wild than that found by Woods (1995), but using 15 kg per tree, this is the equivalent of 1500 kg of *T. ferdinandiana* fruit per hectare at 100 trees/ha and given the vast distribution, the potential for wild harvest across Northern Australia is considerable. Yields are generally found to be higher in semi-horticultural settings (Woods, 1995), with yield and age of maturity of uncultivated wild trees found to vary considerably. The influence of irrigation, fertiliser and other benefits of cultivation on the yield and chemical composition of fruit are yet to be measured and quantified.

Supply of Fruit for Existing Markets

Estimated demand for *T. ferdinandiana* for its natural antimicrobial values is estimated to be in excess of 50 t of fresh fruit in 2015/2016, and there is likely to be demand from other markets. This raises the issue of short- and long-term solutions in meeting these market demand issues. Wild harvest may meet the demand in the short term whilst cultivation is likely to be important in the longer term. The potential industry is an exciting opportunity for Northern Australia and if developed strategically provides an opportunity for development in a way that provides an economic opportunity for Indigenous people with benefits for the environment in terms of carbon storage and increased biodiversity. Currently, the extremely low density of people living on country outside of city centres is a great challenge for land management. There are few livelihood options in these areas, contributing to the low socio-economic status of Indigenous people that are living there. *T. ferdinandiana* offers an opportunity for income which in turn has the potential to improve livelihoods and contribute to people remaining within their country.

Wild Harvest of *T. ferdinandiana* in the Northern Territory

Market demand has prompted some commercial wild harvest of fruit in the NT over the years with varying degrees of success. In 1999, the Australian Government funded a research project titled 'Feasibility of small scale commercial plants harvest by Indigenous communities' through the Key Centre for Tropical Wildlife, Charles Darwin University. Part of this research project considered the feasibility of wild

Figure 8.3   (See colour insert.) Photo of Indigenous rangers sorting wild harvested Kakadu
plum (*Terminalia ferdinandiana*) fruit. (Photo by Julian Gorman. Copyright Charles
Darwin University.)

harvest of *T. ferdinandiana* and found potential for remote communities to generate income with fairly minor initial capital outlay or training (Gorman et al., 2006;
Whitehead et al., 2006). The resulting research and network was useful when Coradji
Pty Ltd began purchasing fruit from the NT. A number of Indigenous Ranger groups
across the NT including the Djelk Rangers (Maningrida), Thamarrurr Rangers
(Wadeye), Wudicupildiyerr Rangers, Adjumarllarl Rangers (West Arnhem Land)
and Asyrikarrak Kirim Rangers (Peppimenarti) began wild harvesting *T. ferdinandiana* with support of the Northern Land Council. The wild harvesting was at a relatively small scale between 2005 and 2010, and never expanded greatly for a number
of reasons (Figure 8.3).

Given the demand for *T. ferdinandiana* fruit and the large natural stands of this
species across Northern Australia, there is an opportunity for a substantial wild harvest, especially in the period before substantial cultivation. As much of this natural
resource occurs in Aboriginal land (especially in the Northern Territory), this provides an economic opportunity for Aboriginal people.

CASE STUDIES

We will now describe case studies of communities who have been working
towards the development of the industry: one from Wadeye in the Northern Territory
who have been involved in wild harvest and one from Broome in the Kimberly, West
Australia, where the Kimberly Training Institute has played a leading role in cultivation of *T. ferdinandiana*.

Wadeye Case Study

Wadeye is used here as a case study as it has a substantial natural abundance of *T. ferdinandiana*. The community has shown enthusiasm to be a part of the industry and during the past decade has been working through and finding solutions to many of the issues that have restricted harvest. Over 300 people have been involved in the *T. ferdinandiana* harvest in Wadeye annually in recent seasons.

Wadeye, based on the former catholic mission of Port Keats, is a remote community of approximately 3000 people (the largest Indigenous community in the Northern Territory) on the western coast of the Northern Territory. Wadeye is on the tradition lands of the Diminin people but is now home to more than 20 clan groups from the surrounding Thamarrurr region. The Thamarrurr region is approximately 20,000 km^2 and includes two substantial river systems, the Moil and the Fitzmaurice. Within the region, *T. ferdinandiana* is a very common mid-storey tree; and there are substantial areas where it is the dominant species, forming what could be considered natural plantations.

The Northern Land Council supported numerous ranger groups to pick *T. ferdinandiana* and sell them to Coradji Pty Ltd. This included the Thamarrurr Rangers, based in Wadeye. Between 2007 and 2010, they harvested a maximum of 400 kg of fruit per year. The fruit had to be sorted, packed, frozen and delivered to Darwin where they received $20 per kg, receiving payment for the Ranger Programme. It was hoped that this opportunity to generate income would be utilised by other local people, and this would bring the surrounding outstations (settlements on Clan Estates) within the ambit of the Ranger Programme.

In 2011, the Thamarrurr Rangers established an Enterprise Centre with the explicit aim of promoting the use of natural resources to provide economic benefits to the community. The harvest of *T. ferdinandiana* was one of the objectives of the Enterprise Centre. The Enterprise Centre funded a person who could drive the harvest. All successful business require a person or group of people to promote and co-ordinate activities.

In order to expand the harvest, community members were encouraged to harvest from their own 'country' and bring the fruit daily to a central collection point where it was graded, weighed and frozen. Pickers were paid an immediate upfront cash payment of $10 per kg. This model meant Aboriginal people were their own boss in their own 'country', something rarely seen in employment of Indigenous people in Northern Territory communities. Upfront payments suited Aboriginal people. These two factors were key to engaging people in the harvest. The harvest in 2011 exceeded 2500 kg and was only limited by money available to pay pickers and freezer space.

Cultural Considerations

Like other landowners, traditional Aboriginal land ownership provides rights to the resources of the land and obligations to protect it. Therefore, all harvesters require the permission of Traditional Owners (TOs). In order to ensure that this occurs, all harvesters are required to identify where they had harvested and if that was their family estate or the name and relevant details of the TO that gave permission. Having local people involved and in control at the collection point is essential to manage this issue. Damage to trees occurred, such as broken branches and trees pushed over. This was of concern to many TOs. A community education campaign was established, with advertisements on the local television network and notices placed around town. A rule was established that fruit would not be purchased from people known to break trees.

Market Requirements

The market demanded fruit of high quality and had requirements on size, maturity cleanliness, etc. A community education programme was established which included some community members visiting the Coradji processing plant in Sydney, to see the quality of fruit required and the end products. These people were involved in educating others in the community. Work to improve the storage and transportation of product continues. A blast freezer and purpose-built bulk freezer space is currently being established in Wadeye. The University of Queensland has been assisting in establishing protocols for the handling and storage of fruit.

Legal Requirements

The harvest takes place under a 'Permit to Take' issued by Parks and Wildlife Commission of the NT. The permit conditions require the harvest is sustainable and has permission of landowners. The issue of ecological sustainability is essential in broader terms and as such a research programme is currently being initiated to provide information of the size of a sustainable harvest and information of the management of the harvest (such as impact on trees, mitigating the effects of fire and weeds, etc.) and potential impacts on other species. Currently, the harvest is estimated to constitute less than 0.1% of the available fruit.

In the NT, any commercial harvest of a native product on Aboriginal Land need permission from the TOs in accordance with Section 19 of the *Aboriginal land Rights Act* (NT) 1976. Currently, an agreement is in place with TOs, organised through the local body TRACC whereby TOs have agreed to the harvest and receive a $1 per kg royalty. TOs have agreed that this payment can be made into an account for general community benefit. A formal Land Use Agreement (LUA) may be required in the future. The LUA process involves the NLC negotiating with all TOs of an area including on royalties, which are paid to the NLC and then paid to TOs. Although suitable for large complex projects (such as mines), this process can be inhibiting for small-scale enterprises. Reform and streamlining of this process with regard to sustainable harvest of natural resources should be considered.

With the now substantial harvest, the Enterprise Centre began to investigate ways to market fruit that would ensure the harvest could continue in the long term and provided maximum benefit to the community. This eventually led to the local womens' centre, the Palngun Wurnangat Association (PWA), purchasing Coradji Pty Ltd. To attract investment into the industry requires a consistent, guaranteed supply. This is very difficult to achieve from a single community, where there are risks such as cyclones and other climatic factors, social disruptions or equipment failures. One way to achieve this constituent supply, as the industry is dependent upon wild harvest, is to source fruit from a range of communities. To this end, establishment of a central processing hub is being pursued.

Cultivation and Wild Harvest in Kimberley, Western Australia

The Kimberley region lies on the western side of the Australian tropical savannah and over the last 30 years has been the location with a number of training, research and commercial enterprise initiatives related to *T. ferdinandiana*. These activities have involved Aboriginal groups, state and federal governments and private enterprise. The Kimberly Training Institute (KTI – formally known as the Kimberley College of TAFE) has played a leading role in developing cultivation of *T. ferdinandiana* which is known as 'Gubinge' in this region from the language name from the Kimberley Nyul Nyul Clan.

Wild harvesting has supplied the industry through its early stages and is certain to continue playing a major role, but cultivation is emerging as a new dimension with multiple co-benefits. Cultivating *T. ferdinandiana* in communities introduces a range of horticultural skills that can be applied to the cultivation of other high value native and exotic species through the promotion of nursery and propagation skills; the establishment of community food gardens; and the planting of wood lots, shade trees and ground covers (lawns) for dust prevention.

T. ferdinandiana has quite different phenotypic qualities across its range including fruit structure and composition (Konczak et al., 2014) as well as time of flowering, fruit set and harvest. KTI horticulture lecturer Kim Courtenay, a long-term Broome local, saw the opportunity to assist in the development of the emerging industry and started incorporating cultivation of *T. ferdinandiana* into his practical training delivery. Through seedling germination trials over the following years, propagation rates were increased from around 5%–90%.

As part of the Pathways to Better Business programme, seedlings raised through the propagation trials were used to assist three local Aboriginal communities establish plantations through practical training delivery. These plantations were watered by automatic drip irrigation systems installed as part of the practical training delivery and were soon thriving, demonstrating that *T. ferdinandiana* could be cultivated on a commercial scale.

The concept of 'enrichment planting' was born and in the following year (2007), this technique was absorbed by the Balu Buru Centre at the Broome 12 mile. The first trials were established through practical training programmes involving predominantly Aboriginal students with a high percentage of minimum security prisoners from Broome Prison. The 'enrichment planting' aimed at planting seedlings in existing *T. ferdinandiana* natural plantations.

Savannah Enrichment

A fundamental principle of the 'savannah enrichment methodology' is to combine traditional burning practices with modern techniques in agronomy to systematically transform savannah plant communities dominated by fire-tolerant species into woodlands with enriched biodiversity and the capacity to produce commercial quantities of prized bush foods. This technique, still in the research and development stage, combines traditional Aboriginal burning practices with specialised propagation methods and drip irrigation to reverse the consequences of regular hot fires and establishes long-lived trees which store carbon. 'Savannah enrichment' is currently being promoted as a new methodology under the Federal Government's Carbon Farming Initiative. There is potential that this strategy could transform vast areas of mismanaged country into a great carbon sinks, providing an additional income stream, supporting local livelihoods and long-term sustainable land management.

The Balu Buru Tropical Savannah Enrichment Planting Project is being implemented on a 19.2 ha site adjoining horticultural farms in an area known as the Broome 12 mile. Since the wet season of 2007/2008, KTI has planted areas of around one hectare each year, exploring the savannah enrichment concept through ongoing practical training activities. *T. ferdinandiana* and several other Indigenous species with commercial potential, including native ebony wood, pindan walnut and desert yam, have been trailed. The core aim is to develop and refine new concepts pertaining to tropical savannah agronomy; these can be used/applied to develop the remote communities into sustainable and culturally appropriate commercial enterprises.

In December 2014, the Australian Agroforestry Foundation (AAF) in partnership with the KTI ran the first Masters Tree Growing Workshop. The course attracted people from Kimberley that were interested in growing *T. ferdinandiana* and explored the emerging market potential that this species had while outlining some of the specialised cultivation techniques developed through KTI training programmes over the last 15 years. KTI and the core team of trainers associated with the AAF will continue to support and advise people interested in growing *T. ferdinandiana* to meet the market demand in the future. The course recognised the role of cultivation in the future.

CONCLUSIONS AND RECOMMENDATIONS

The situation as of 2015 is that there is a big demand for fruit of *T. ferdinandiana* but a very limited supply. There are a few small plantations of *T. ferdinandiana* trees in the Kimberley, WA, which have potential to produce fruit in the longer term but even in full production would be very short of supplying the current market demand. The only possible way to come close to meeting demand in the short term is through a coordinated wild harvest across a number of regions. In the longer term, there need to be agronomy and more cultivation of this species in ways which can involve Aboriginal communities.

With the underlying aim being to refine and present commercially viable models, there is still work to be done on

- The benefits of varying nutrition and watering regimes
- The most efficient ways of setting up an Enrichment planted area in terms of site preparation
- The potential for the use of controlled burns in this process
- The control and management of acacia species, particularly *Acacia eriopoda* which often dominate regularly burnt areas of savannah, encouraging the ongoing cycle of hot fires

A central hub could initially be set up in Darwin and, depending on the number of regional collection points, could perhaps expand to Broome and other centres. At these central hubs, fruit has to be stored and processed to a standard required by the buyers and stipulated in the contract. Product development, labelling and advertising are the other issues which would be dealt with in such a hub. Most importantly, regional hubs need to have a percentage ownership of the central hub, ensuring a flow through of profits and continuity along the value chain.

There are many research and development gaps that need to be filled as well as strategies to assist in the development of Northern Australia in this industry as a whole. Some of the recommendations include

- Studies measuring density and yields of trees in areas wanting to commercially wild harvest to measure impacts of harvest and ensure sustainability
- Further refining of models of wild harvest suitable to different regions for Indigenous involvement
- Agronomy and cultivation trials on Aboriginal land
- Training and capacity building in cultivation best practice taking into account Aboriginal aspirations (Master Tree Growers)
- Processing, marketing and product development

REFERENCES

ABC, 2013. Bush plum proving fruitful for seafood industry. Accessed 19 September 2013. http://www.abc.net.au/news/2013-09-19/kakadu-plums-improving-prawns/4968046.

Brand, J.C., Cherikoff, V., Lee, A., 1982. An outstanding food source of vitamin C. *Lancet* 2(8303), 873.

Brand Miller, J., James, K.W., Maggiore, P., 1993. *Tables of Compositions of Australian Aboriginal Foods.* Aboriginal Studies Press, Canberra, Australian Capital Territory, Australia.

Brock, J., 2001. *Top End Native Plants: A Comprehensive Guide to the Trees and Shrubs of the Top End of the Northern Territory.* Reed New Holland, Sydney, New South Wales, Australia.

Byrnes, N.B., 1977. *A Revision of the Combretaceae in Australia.* Queensland Herbarium, Brisbane, Queensland, Australia.

Burgess, C., Johnston, F., Berry, H., McDonnell, J., Yibarbuk, D., Gunabarra, C., Mileran, A., Bailie, R., 2009. Healthy country, healthy people: The relationship between Indigenous health status and 'caring for country'. *Medical Journal of Australia* 190, 567–572.

Burgess, C.P., Johnston, F.H., Bowman, D.M., Whitehead, P.J., 2005. Healthy country: Healthy people? Exploring the health benefits of Indigenous Natural Resource Management. *Australian New Zealand Journal of Public Health* 29, 117–122.

Clein, N.W., 1956. Acerola juice, The richest known source of vitamin C; A clinical study in infants. *Journal of Pediatrics* 48, 140–145.

Coode, M.J.E., 1973. Notes on *Terminalia L.* (Combretaceae) in Papuasia. *Contributions from Herbarium Australiense* 2, 1–33.

Cunningham, A.B., Courtenay, K., Gorman, J.T., Garnett, S., 2009. Eco-enterprises and Kakadu Plum (*Terminalia ferdinandiana*): 'Best laid plans' and Australian policy lessons. *Journal of Economic Botany* 63, 16–23.

Dunlop, C.R., Leach, G.J., Cowie, I.D., 1995. *Flora of the Darwin Region*, Vol. 2. Northern Territory Botanical Bulletin No. 20. Conservation Commission of the Northern Territory, Darwin, Northern Territory, Australia.

French, B., 2013. Edible *Terminalia* fruit and nuts. Food Plants International. www.foodplantsinternational.com.

Gorman, J., Whitehead, P.J.,. Griffiths, A.D., 2006. An analysis of the use of plant products for commerce in remote aboriginal communities of northern Australia. *Journal of Economic Botany* 60, 362–373.

Isaacs, J., 1987. *Bush Food, Aboriginal Food and Herbal Medicine*. Ure Smith Press, Sydney, New South Wales, Australia.

Johnson, P.D., 2003. Acerola (Malpighia glabra L., M. punicifolia L., M. emarginata D.C.): Agriculture, production and nutrition. *World Review of Nutrition and Dietetics* 91, 67–75.

Kenneally, K.F., Choules Edinger, D., Willing, T. 1996. *Broome and Beyond: Plants and People of the Dampier Peninsula, Kimberley, Western Australia*. Department of Conservation and Land Management, Perth, Western Australia, Australia.

Konczak, I., Maillot, F., Dalar, A., 2014. Phytochemical divergence in 45 accessions of *Terminalia ferdinandiana* (Kakadu plum). *Food Chemistry* 151, 248–256.

Konczak, I., Zabaras, D., Dunstan, M., Aguas, P., 2010. Antioxidant capacity and hydrophilic phytochemicals in commercially grown native Australian fruits. *Food Chemistry* 123, 1048–1054

Lindsay, B.Y., Waliwararra, K., Milijat, F., Kuwarda, H., Pirak, R., Muyung, A., Pambany, E., Marryridj, J., Marrfurra, P., Wightman, G., 2001. *MalakMalak and Matngala Plants and Animals – Aboriginal Flora and Fauna Knowledge from the Daly River Area, Northern Australia*. Conservation Commission of the Northern Territory, Darwin, Northern Territory, Australia.

Netzel, M., Netzel, G., Tian, Q., Schwartz, S., Konczak, I., 2007. Native Australian fruits – A novel source of antioxidants for food. *Innovative Food Science and Emerging Technologies* 8, 339–346.

Ohno, Y., Fukuda, K., Takemura, G. et al., 1999. Induction of apoptosis by gallic acid in lung cancer cells. *Anticancer Drugs* 10, 845–851.

Pedley, L., 1995. Combretaceae. *Flora of Australia* 18, 255–293.

Puruntatameri, J., Puruntatameri, R., Pangiraminni, A. et al., 2001. *Tiwi Plants and Animals: Aboriginal Flora and Fauna Knowledge from Bathurst and Melville Islands, Northern Australia*. Parks and Wildlife Commission of the Northern Territory and Tiwi Land Council, Darwin, Northern Territory, Australia.

Raymond, E., Blutja, J., Gingina, L. et al., 1999. *Wardaman Ethnobiology*. Government Printer of the Northern Territory, Darwin, Northern Territory, Australia.

Robinson, D.F., 2010. Traditional Knowledge and Biological Product Derivative Patents: Benefit-Sharing and Patent Issues Relating to Camu Camu, Kakadu plum and Acai Plant Extracts. Traditional Knowledge Bulletin – Tropical Issues Series, April 2010, United Nations University, Tokyo, Japan.

Tan, A.C., Konczak, I., Ramzan, I., Zabaras, D., Sze, M.Y., 2011. Potential antioxidant, anti-inflammatory, and proapoptotic anticancer activities of kakadu plum and illawarra plum polyphenolic fractions. *Nutrition Cancer* 63, 1074–1084.

Williams, C., 2011. *Medicinal Plants in Australia*, Vol. 2: Gums, Resins, Tannin and Essential Oils. Rosenberg Publishing, Kenthurst, New South Wales, Australia.

Williams, D.J., Edwards, D., Pun, S., Chaliha, M., Sultanbawa,Y., 2014. Profiling ellagic acid content: The importance of form and ascorbic acid levels. *Food Research International* 66, 100–106.

Woods, B.E., 1995. *A Study of the Intra–Specific Variations and Commercial Potential of Terminalia ferdinandiana Excell. (The Kakadu Plum)*. School of Chemical Science, Northern Territory University, Darwin, Northern Territory, Australia.

Wheeler, J.R., 1992. Family 81 Combretaceae. In: Wheeler, J.R. (ed.), Flora of the Kimberley Region. Department of Conservation and Land Management, Como, Western Australia, Australia, pp. 551–559.

Whitehead, P.J., Gorman, J., Griffiths, A.D., Wightman, G., Massarella, H., Altman, J., 2006. Small Scale Commercial Plant Harvests by Indigenous Communities: A report for the RIRDC/Land and Water Australia/FWPRDC/MDBC Joint Agroforestry Program. RIRDC, Barton, Australian Capital Territory, Australia.

CHAPTER 9

Cultivation of Lemon Aspen (*Acronychia acidula*)

Rus Glover

CONTENTS

INTRODUCTION

The name 'lemon aspen' has commonly been used in the past to refer to two species: *Acronychia acidula* and *Acronychia oblongifolia*. The latter is also known as 'southern white aspen'. The former, *Acronychia acidula*, has been described as the 'true' lemon aspen, and is accepted as such in the industry.

BOTANY

Lemon aspen is a small- to medium-sized tree that can reach up to 15 m. It is endemic to Queensland, Australia, and thrives in northeast Queensland and south-wards as far as coastal central Queensland. The tree can be found from near-sea level to an altitude of 1150 m. It grows on a variety of soil types in well-developed upland

and mountain rainforest, particularly on the Atherton Tablelands where it is known locally as 'pigeon berry'.

Leaf blades are glossy green and about 10–20 × 5–11 cm in size. The underside of the leaf blade is only slightly paler than the upper surface. Crushed leaves often emit an odour like mango skin.

Flowers are white to cream and about 9.5 mm long. Stamens eight, four long and four short, in one whorl, long and short stamens alternating. The disk and style are yellow. The ovary and stigma are green (Figures 9.1 and 9.2).

Figure 9.1 Botanical illustration of *Acronychia* species – flowering branch. (Illustrated by David Mackay. Flora of NSW, Copyright Royal Botanic Gardens and Domain Trust.)

Figure 9.2 **(See colour insert.)** Photo of lemon aspen (*Acronychia acidula*) flowers. (Photo by Rus Glover. Copyright Woolgoolga Rainforest Products.)

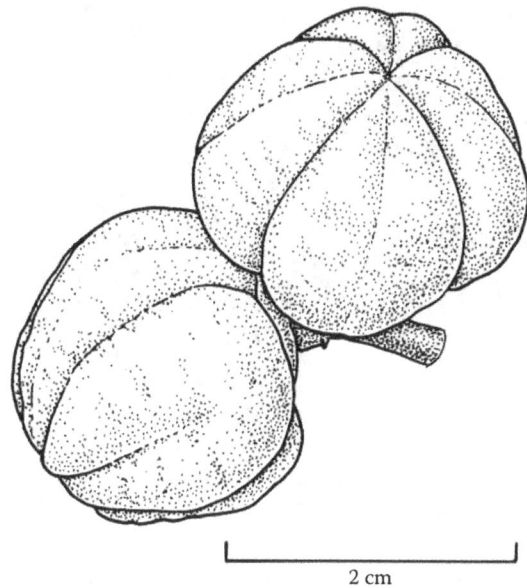

2 cm

Figure 9.3   Botanical illustration of *Acronychia* species – fruit. (Illustrated by Lesley Elkan. Copyright Royal Botanic Gardens and Domain Trust.)

Fruit is a pale lemon colour, globular and about 20 mm in diameter. The core contains small black seeds in most cases. Fruit is generally produced from January to mid-April, although the season is later in the south (Figures 9.3 and 9.4).

TRADITIONAL USE AND ECONOMIC POTENTIAL

Indigenous communities in northeastern Queensland regularly consumed lemon aspen, although details are scarce. It was one of the first fruits to be used by early settlers, usually as a component of a drink.

Lemon aspen is generally traded frozen, whole, as a puree and as a juice. It is primarily used in its processed form, although it can be eaten fresh. Its strong lemon and grapefruit taste lends itself to use as fruit flavouring agent in both sweet and savoury products.

Total production of lemon aspen in 2011 was approximately 5 t and production has increased since. Demand for lemon aspen currently exceeds supply, leaving many in the industry positive about its potential.

There are commercial plantations in north, central and south Queensland and several in northern New South Wales (NSW). Most are components of multi-species plantings. Trial plantings are taking place on the south coast of NSW and Victoria.

Figure 9.4   (See colour insert.) Photo of Lemon aspen (*Acronychia acidula*) fruits. (Photo by Rus Glover. Copyright Woolgoolga Rainforest Products.)

AGROCLIMATIC REQUIREMENTS

While its natural range is tropical (both lowland and highland), lemon aspen will tolerate quite low temperatures and mild frost, particularly after establishment. The tree grows well in different types of soil. Sandy soils benefit from added organic matter and clay-based soils need to be well drained, while extremely dry soils should be avoided. pH range is best at 4.5–5.5. Plants grow well with summer/autumn rainfall pattern and benefit from irrigation when establishing. No information is available concerning the effects of irrigation on commercial plantings and yield.

AGRONOMY

Plantation layout needs to take several variables into account:

- Slope and field orientation. Northerly aspects and slopes maximise sunlight, especially during winter, particularly on the east coast of NSW/Queensland. Maximising sunlight should be a priority in any locality.
- Ideally, orientation of rows should be north/south to provide equal amount of sunlight to all plants all-year round. This is not always possible due to slope and/or land shape. A balance needs to be found between being able to effectively mow/slash and access inter-rows with machinery, minimise erosion and maximise sunlight.
- Inter-row spacing needs to allow for tractor access for slashing, tree maintenance and harvesting.

Little is known about the nutrient requirements of lemon aspen apart from the pH and nutrient status of soils where endemic populations occur. As experience with plantation crops accumulates, firmer guidelines and objective assessment techniques, such as leaf analyses, will come into common use. Until then, some general principles should be kept in mind when designing a fertiliser programme:

- Many Australian native plants are intolerant to high phosphorus contents, so a relatively low phosphorus fertiliser, suitable for natives, should be used.
- In the first season following establishment, the aim should be to maximise growth to develop a good plant structure and produce several flushing's of next season's fruiting wood. Lemon aspen needs to be pruned within the first year to provide a 'shrubby' shape consisting of several main trunks and can then be lightly shaped in subsequent years to allow light to penetrate the canopy. Frequent but relatively small fertiliser applications, particularly of nitrogen, potassium, phosphorus and trace elements, are desirable.
- After harvest, vegetative growth, which will form next year's fruiting wood, should be encouraged. A complete micronutrient foliar spray may be advisable in spring into summer as flowers are being initiated.
- Experience has shown that trees benefit from applications of good-quality compost after fruiting has been completed in March/April and in early spring.

The majority of fruit is produced on or near the ends of growth that has matured during the year. Plants are susceptible to wind, particularly during the flowering and fruiting period, and natural windbreaks can provide protection.

GENETIC IMPROVEMENT AND PROPAGATION

There are no commonly known cultivars and lemon aspen is usually grown from seed. Nurseries have been selecting seed from plants that produce fruit heavily in the area that they are growing in. The plant can be clonally propagated using cuttings and have also been air layered (marcotting) successfully.

Seed germinates readily and seed-grown plants will bear within 3 years, sometimes sooner.

A programme to identify selections that show superior attributes such as disease resistance, consistent bearing, fruit size and other beneficial properties would greatly benefit the industry.

BIOTIC CONSTRAINTS

Lemon aspen can suffer from several insect and diseases. These include

- Pink wax and other scales and associated sooty moulds
- Pimple psyllid and associated sooty moulds
- Macadamia nut borer

Organic methods of control can be successfully used as there are currently no registered preparations for pest control in *Acronychia* spp. Good orchard practices such as maintaining high organic matter, fertile, healthy soils, appropriate canopy management, regular harvest and orchard hygiene are the best measures to minimise the impacts of pest problems.

Integrated pest and disease management practices such as designing orchards to include windbreaks, refuges and corridors for beneficial insects and insectivorous birds will assist in protecting orchards against severe pest problems.

HARVESTING, PROCESSING AND MARKETING

Fruit is generally produced from January to mid-April, although the season is later in the south. For most producers, harvesting of fruit is done by hand. The use of nets under trees has also been trialled successfully. In the future, it is expected that the increase in cultivation will lend itself to greater mechanisation of the harvesting process.

Once picked, the fruit needs to be sorted, graded, washed with sanitiser and refrigerated straight away. It will be kept in the refrigerator for up to 3 weeks. For processing, fruit needs to be frozen as soon as possible; and it will maintain its colour and taste and can be stored this way for up to 2 years. Processors require the fruit in two forms generally – frozen whole fruit and frozen puree (which has been prepared by putting fruit through a brush finisher).

It is used in an increasing range of value-added products, including sauces, dressings, jellies, chutneys and relishes. Lemon aspen is a common flavouring in juice, mineral water, cordial and fruit wine. In its dried form, lemon aspen can also be ground into a spice mix.

Lemon aspen also has superior antioxidant capacity compared to blueberry, renowned worldwide as the 'health-promoting fruit'. Antioxidants are believed to hold a number of benefits for human health, potentially preventing and delaying diseases such as Alzheimer's disease, autoimmune and cardiovascular disease, cancer and diabetes. Formulating products and marketing strategies that promote these attributes are needed. Food service remains its primary market, although the number of value-added products which contain lemon aspen is increasing.

YIELD

Yields vary with variety and climate but generally increase each year after planting. Mature trees have been known to produce up to 80 kg. Generally, it can be expected that a 5-year-old tree will produce up to 40 kg of fruit a year.

CONCLUSIONS AND RECOMMENDATION

Lemon aspen has appealing flavour and texture. It provides health benefits including anti-oxidants, folate and iron and is growing in popularity in both the food service and retail product industries.

There is a need for improving planting material through both selection and breeding and improving harvesting methodologies. There is still lack of volume, but plantation areas are increasing. Further development of both products and marketing will develop the industry.

CHAPTER **10**

Cultivation of Lemon Myrtle
(*Backhousia citriodora*)

Gary Mazzorana and Melissa Mazzorana

CONTENTS

BOTANY

Lemon myrtle, botanically named '*Backhousia citriodora*', is an Australian Native rainforest tree. *Backhousia citriodora* was named by a German botanist, Baron Ferdinand von Müller (curator of the Melbourne Botanical Gardens), in 1853 as a tribute after James Backhouse (1794–1869), an English Botanist. James Backhouse was an early Quaker missionary who visited Australia during 1832–1838, making observations of Australian flora and fauna as well as writing about Australian society at the time.

'Lemon scented myrtle' was the primary common name until the shortened trade name 'lemon myrtle' became common which was created by the Native Foods Industry to market the leaf for culinary use. This common name reflects the strong lemon smell of the crushed leaves.

Backhousia citriodora ('lemon myrtle') is a flowering plant in the family Myrtaceae, genus *Backhousia*.

It is a medium-sized tree that can reach heights ranging up to 30 m with a dense canopy. The lemon-scented leaves, which are 5–12.5 cm long, are dark green, glossy and lance-shaped. The leaves are highly fragrant when crushed and are used commercially to extract the lemon flavour.

The small, fluffy, cream-coloured abundant flowers of the lemon myrtle grow in clusters at branch tips from late spring to early summer, depending on its geographical location, and emit a delightful lemon perfume in the evenings.

Lemon myrtle has the world's highest and purest source of natural citral (90%–98%) – the oil that gives lemon its characteristic flavour. Other 'lemony'

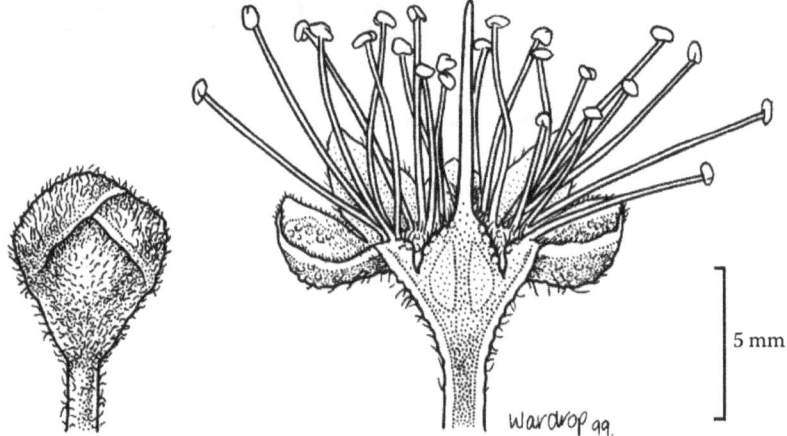

Figure 10.1   Botanical illustration of *Backhousia citriodora* – bud and flower. (Illustrated by Catherine Wardrop. Copyright Royal Botanic Gardens and Domain Trust.)

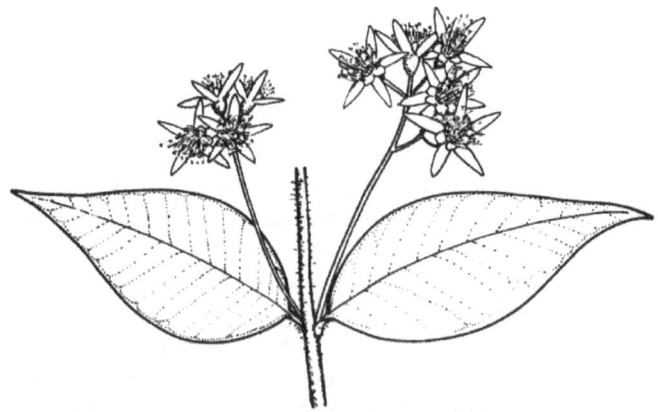

Figure 10.2   Botanical illustration of *Backhousia citriodora* – flowering branch. (Illustrated by Catherine Wardrop. Flora of NSW, Copyright Royal Botanic Gardens and Domain Trust.)

Figure 10.3   (See colour insert.) Photo of lemon myrtle (*Backhousia citriodora*) leaf. (Photo by Gary Mazzorana. Copyright Australian Rainforest Products.)

essential oils come from citrus (3%–10% citral), tropical verbena (74%), lemongrass (75%) and the lemon scented tea tree (80%). Refer to Figures 10.1 and 10.2 for botanical illustrations of the leaf, bud and flowers and Figure 10.3 for lemon myrtle (*Backhousia citriodora*) leaf.

TRADITIONAL USE

Indigenous Australians have long used lemon myrtle, dating back to pre-1788, where anecdotally it has been claimed to be used as a healing plant in the form of medicine and also as cuisine flavouring. It has been documented that the Barabam, Djabugandji, Inawongga and Nawagi people used lemon myrtle.

Lemon myrtle essential oil was first distilled in 1888. A German company, Schimmel & Co., made the first known distillation of lemon myrtle from a tree in south-east Queensland. Schimmel & Co. analysed the species for chemical properties, identifying the compound citral for the very first time. Lemon myrtle oil contains well over 50 different chemical compounds.

The 10 major chemical compounds that are analysed commercially are shown in a typical gas chromatograph analysis (Table 10.1).

Today, typically most manufacturers and end-users are looking for the citral component, which is made up of the major compounds of neral and geranial, but also minor compounds such exo-isocitral, cis-isocitral and trans-isocitral.

The preferred citral chemotype, which is commercially cultivated in Australia for essential oil extraction, is typically 93%–98% with an oil yield of 2%–3% from fresh leaf.

Table 10.1   Essential Oil Gas Chromatograph Analysis of Lemon Myrtle

Certificate of Analysis Method: Area Normalisation Based on ISO 7609:1985

Sample: Oil of *Backhousia citriodora* F. Muell. Lemon Scented Myrtle
Client: Australian Rainforest Products

No.	Component	Batch No. 710; Area %	As 4941–2001; Range (%)
1	6-methyl-5-hepten-2-one	0.3	Trace–2.9
2	2,3-dehydro-1, 8-cineole	0.2	Trace–0.9
3	Myrcene	0.6	Trace–0.7
4	Linalool	0.6	Trace–1.0
5	Exo-isocitral	0.7	Trace–2.0
6	Citronellal	0.2	Trace–1.0
7	*cis*-Isocitral	2.6	Trace–2.7
8	*trans*-Isocitral	4.1	Trace–4.3
9	Neral	38.8	32.0+
10	gEranial	48.0	44.0+
	Citral total	94.2	

Source: Analysis report provided by Gary Mazzorana. Copyright Australian Rainforest
         Products.

In 1889, Joseph H. Maiden reported the potential use of lemon myrtle for commercial production.

Around the time of World War I, the wild bush harvesting of lemon myrtle from its indigenous locations began winding-down.

During World War II, a national drinks company, Tarax, used the essence of lemon myrtle to flavour drinks when lemon essence was in short supply. With the cessation of the war, supplies of lemon essences were restored and lemon myrtle was again neglected, and production subsequently declined – only to be revived again in the late 1980s.

In the late 1980s and through the 1990s, lemon myrtle was rediscovered as a potential crop. A small number of committed Australian pioneers located in Queensland and the north-east region of New South Wales set about reintroducing the Native Australian lemon myrtle for potential food, essential oil, herb and tea production, by establishing commercial plantations.

Sales of lemon myrtle tea have increased significantly over the years and current market sales place lemon myrtle tea as being the predominant export commodity within the industry. It is currently used in its pure form, but also as a blend with green, black and herbal teas.

Lemon myrtle is used as a food flavouring agent due to its refreshingly clean, citrus aroma and taste, which makes it one of the most versatile and well-known Australian native food flavours available. Lemon myrtle is sometimes referred to by the culinary industry as 'queen of the lemon herbs' or 'more lemony than lemon'.

Lemon myrtle has a distinctive flavour that can best be described as a fragrant, sweet, yet spicy, lemon, lime and lemongrass flavour. According to the 'Australian

Native Flavour Wheel', a sensory description of the aroma and flavour of lemon myrtle is 'a lemon lolly aroma, perfumed with some menthol notes'. Lemon myrtle can be used in recipes as a substitution for lemon or citrus flavour. As you only need to use very small amounts, it makes the product very price competitive.

In contrast to many other spices, lemon myrtle can be used fresh, dried, as an essential oil or in a water-soluble form.

It is versatile enough that it is suited for both savoury and sweet dishes.

Because lemon myrtle resembles the flavour of citrus so closely but lacks the associated acidity citrus contains, it is especially useful in recipes that are milk or cream based and makes for a wonderful replacement. It imparts the desired strong lemony flavour, yet would not cause dairy products to curdle during processing.

Lemon myrtle is typically used as flavouring in various foods such as stir-fries, pastas, cheesecakes, breads, scones, biscuits, confectionary, cheeses, yoghurt, ice creams and sorbets. In addition to its use in prepared dishes, lemon myrtle is a good choice to add flavour to dukkah, spice rubs, salt-blends, marinades and sauces for meats, poultry and fish, salad dressings, mayonnaises and dips. It can even be used as a flavouring agent or herbal infusion in hot or iced tea and as a syrup in soft drinks or mixed in cocktails and liquors.

In order to maintain optimal quality, the dried lemon myrtle leaves should be stored in a cool, dry, dark environment that is free of humidity and fluctuating temperatures.

Lemon myrtle 100% essential oil is listed as a therapeutic product under the Australia Therapeutic Goods Administration Act (TGA): the concentration of that oil does not exceed 10 mg per g of the preparation.

Lemon myrtle is used as an ingredient in cosmetics and personal care items such as soaps, therapeutic lotions, lip balms, body butters, massage oils, shampoos and conditioners, shower gels, deodorants and toothpaste. The added benefit for using lemon myrtle in these industries, in addition to its desirably fresh fragrance, is that lemon myrtle contains high antibacterial and antifungal properties, which in turn acts as a natural preservative for these products and enhances shelf-life.

The aromatherapy and perfume industries recognise the value of fragrances for health maintenance and lifestyle. The use of lemon myrtle in these products is said to be uplifting, calming and can aid in better sleep.

It is well documented that the essential oil obtained from lemon myrtle contains antimicrobial compounds. For this reason, it is an ingredient in sanitising products such as hand wash, washing powders and household-cleaning agents. The oil is believed to possess the ability to repel fleas and also breaks down the embryonic layer of flea eggs, killing them, hence a feature in some chemical-free pet shampoos. It is also a pleasant insect repellent against mosquitoes, beetles, cockroaches, houseflies and, to some extent, dust mites.

Lemon myrtle oil is 16 times more effective than phenol, a chemical antibacterial agent. This is considerably more than tea tree and thus is considerably more effective than tea tree oil in controlling a broad range of microorganisms. In addition, lemon myrtle is known for its ability to purify and cleanse the air – it is known to kill bacteria, fungus and odours. Lemon myrtle oil infusions can be used as air fresheners in places such as wardrobes, shoe cabinets, cars and caravans, to name a few. It is good for preventing mould in these locations. In tests against fungus, it has proved to be very active, more so than camphor.

More recently, lemon myrtle has been discovered for its extensive range of medicinal uses and for its health benefits. The School of Biomedical Sciences, Charles Sturt University, Wagga Wagga, NSW, found that the essential oil distilled from the leaf of lemon myrtle has strong antibacterial, antifungal and antiviral properties of all the Australian essential oils. The Rideal–Walker test assesses the antimicrobial activity of a plant by examining the phenol coefficient of a number of essential oil components such as citral, using the bacteria *Salmonella* as a test organism. The test (in which the higher scores are best) awards lemon myrtle oil a coefficient rating of 16, tea tree a score of 11, while eucalyptus citriodora scores 8.

Lemon myrtle has been identified as the richest source of calcium with a content of 1583 mg/100 g DW out of 14 Native Bush foods tested as given in the RIRDC publication (09/133). Lemon myrtle also has a great antioxidant capacity. Lemon myrtle exhibits high oxygen radical absorbance capacity, contains vitamin E, magnesium, zinc, iron and lutein (a carotenoid compound important for eye health).

Lemon myrtle 100% essential oil is a food-grade product, as opposed to other oils such as tea tree oil, which are non-ingestible. It has been said that lemon myrtle can be helpful in reducing the incidence of common cold, influenza, chest congestions and bronchitis; strengthening immunity; and calming indigestion and other irritable digestive disorders, although further research and trials are required within this area. Lemon myrtle has been used as a treatment for herpes – relieving cold sores and warts, and can be applied as a topical antiseptic for stings or skin conditions, such as acne, psoriasis, rashes, neuro-dermatitis, itching, tinea and candidiasis, and for headaches. Further, one of its modern applications is in the treatment of throat disorders caused by infection, overuse or irritation, either via a throat gargle or lozenges.

Lemon myrtle is currently being used as a nutraceutical ingredient for an anti-arthritic joint formulation and in herbal immune lozenges.

Prior to the advent of myrtle rust, the flower industry used lemon myrtle as foliage filler to enhance bouquets with the additional benefit of filling the room with the unmistakable fragrance. Many within the florist industry selected to use lemon myrtle in floral arrangements due to its quality of appearance and durability, outlasting most other green fillers.

The beauty of lemon myrtle is its versatility and endless uses. Applications of lemon myrtle are continually evolving with the advancements in research and development. Lemon myrtle is now being used in new areas for its health benefits and also by using lemon myrtle as a natural preservative in substitution of artificial preservatives opens itself for use in shelf-life extension. New methods of extraction are being explored, such as CO_2 and innovative cellular extraction technology; these new methods are expected to gift the industry with an exciting new range of applications.

Due to the hard work of many people involved in the industry, lemon myrtle is now one of the most well-renowned and most versatile of all the Australian native plants.

ECONOMIC POTENTIAL

Within the last 20 years, the lemon myrtle industry has grown from a small domestic cottage industry, where growers were selling their handmade products locally at markets and stalls. It became commercialised on the Australian domestic market, sold in supermarkets in a range of foods, and in a variety of different health and hygiene products. To date, the industry is producing 575–1100 tonnes per annum in lemon myrtle leaf product and up to approximately 8 tonne in lemon myrtle essential oil for the international and domestic markets.

Lemon myrtle is now estimated to have an industry value in excess of $15 million per annum.

A large proportion of these sales are through the export of this commodity to numerous countries, including Europe, United States and Asia, who are the current leading buyers of lemon myrtle.

The market is registering a sharp growth with each passing year, and currently demand for lemon myrtle has outstripped Australia's capacity to supply. Due to the ongoing increase in sales, other countries, such as Malaysia, have established plantations in an attempt to meet the ever-growing demand.

The native food industry considers lemon myrtle as an important crop of the future, as continuous research indicates that there is potential to use lemon myrtle as a natural food preservative, replacing artificial, chemical preservatives, amongst its already vast applications.

Through continuous research and with the discovery of new uses, the economic potential of lemon myrtle looks extremely promising.

AGROCLIMATIC REQUIREMENTS

Lemon myrtle is a somewhat rare tree species, indigenous to moist, isolated locations along the subtropical and tropical east coast of Australia, in a range of altitudes from 50 m to over 800 m above the sea level.

Naturally found nowhere else in the world, the plant is now grown commercially in plantations in many regions of Australia including Northern New South Wales, parts of South-East, Central and North-East Queensland, Victoria, South Australia and Western Australia.

Typically, lemon myrtle is a hardy plant that handles all but the poorest drained soils. The trees grow best during the warmer (summer) months of the year, and are best suited for areas that are frost-free. However, it will also tolerate cooler climates, provided the plants are protected from frost when young.

Although sensitive to dry conditions as a small plant, once established can tolerate and recover quickly from extremely dry conditions.

In recent years, lemon myrtle plants have been grown commercially in the full tropical climate of Malaysia, and are currently performing well. However, at this stage, given that the trees have only been planted for a short period of time, the future forecast for their performance is unknown, as is their potential for resistance against pests and diseases outside the plant's indigenous environment.

AGRONOMY

Lemon myrtle was grown in commercial plantations originally on very fertile, nutrient-rich soil in Northern N.S.W and South-East QLD. They do require minimal inputs to achieve high yields. The plant is capable of being very productive and can thrive in a range of less fertile soils, provided that the soil preparation and nutrient management parameters are met.

Traditionally, the belief was that the trees need to be grown in fertile soils. This has since been disproved as they are grown commercially and have adapted to a range of soil types, including acid sulphate soils.

The nutritional requirements for lemon myrtle are not well documented. However, the trees respond best when well mulched and a regular fertilising regime is in place, preferably while using organic-based, slow-release, fish and seaweed fertilisers. The use of commercial blended fertiliser has also given good results, but further research needs to be conducted on the optimum levels of nitrogen phosphorus potassium (N.P.K) to reap the maximum benefit. Current information suggests optimum pH levels are in the range of 5.5–6.5.

Soil analysis and leaf testing should also be preformed prior to planting and application of any fertilisers.

Regular pruning of the trees from a young age is recommended to establish both lateral and vertical biomass. Pruning will also lessen the threat of wind damage, as they are prone to breaking off at the base of the trunk, until they reach maturity.

Lemon myrtle, although reasonably tolerant to dry conditions, grows best in areas with high rainfall (approximately 800 mm per annum and above) but will respond well in dry areas with the implementation of an irrigation system.

Irrigation is vital for optimal growth particularly from first planting to an age of 2 years during dry conditions. Overwatering should be avoided, as they are also sensitive to waterlogging for extended periods.

Plantation layout needs to be carefully considered, taking into account maximisation of sunlight to both sides of the trees. A north-facing aspect with a north/south planting is ideal.

There are numerous ways in which lemon myrtle is planted in relation to layout. Typically, in Northern New South Wales, the most common is a 3–3.5 m row spacing, with trees being planted 1–1.5 m along the row. Row spacing at 3 m × 1.2 m will give you approximately 2500 trees per ha. Mechanical harvesting has been specifically designed to suit this spacing configuration and cuts both sides of the tree,

Figure 10.4   (See colour insert.) Photo of a typical lemon myrtle commercial plantation layout and row plantings. (Photo by Gary Mazzorana. Copyright Australian Rainforest Products.)

leaving the tree in an 'A' frame shape. Alternate planting designs can be used to suit different harvesting techniques. One of these is to plant the trees much closer together in multiple rows, of 2 or more, similar to that of tea production, where a harvester only cuts the top of the trees.

Maintenance of trees is very important, in particular weed control in newly established plantations; as with most trees, lemon myrtle performs best without the competition from weeds. There are several methods used in weed control. The most common is heavy mulching, which has a number of other benefits including moisture retention, and also the breaking down of this mulch will boost soil fertility by adding organic matter to the soil. The use of weed matting is another alternative. Once trees reach maturity, due to their dense canopy, they have little competition from weeds, and rarely hand weeding is required, particularly prior to harvesting to eliminate any contamination. Slashing/mowing of the inter-row is recommended. The use of a side throw mower is preferred as the cut material is transferred back under the trees and breaks down to form organic matter (Figures 10.3 and 10.4).

GENETIC IMPROVEMENT AND PROPAGATION

Seeds of lemon myrtle have a low germination rate (0%–4%). Seedlings of lemon myrtle go through a juvenile growth stage before developing a dominant trunk. Young trees grown from seed also have an initial sprawling nature with a low branching habit. This causes problems in commercial plantations, with regard to mechanical harvesting, as low branches are generally not cut to avoid soil contamination of the product.

Vegetative propagation by rooted cuttings is the most effective way of producing planting stock for commercial plantations. These cuttings are taken from mature trees, thus bypassing the juvenile stage.

The use of tip cutting propagation is the favourable method as it provides a true to type, consistent product for commercial production. This form of propagation is expensive and also has a variable strike rate of 60%–90% depending on the clone stock material selected and weather conditions. Tissue culture has also been used and, as advancements are made, could in the future prove to be the preferred method.

During the commercial establishment of the lemon myrtle industry, much research took place, including CSIRO collecting seed and clones from all known naturally occurring trees; these were deposited in a genebank site near Beerburrum QLD in 1995–1996. Many varieties underwent gas chromatograph testing, and from these results, the industry has selected what is commonly called 'Limpenwood' variety, as the preferred commercial stock due to its ability to produce high percentage oil yields, high citral content and maximum biomass production.

Due to the myrtle rust caused by the fungus (*Uredo rangelii* now identified as *Puccinia psidii*) in 2010, the industry has set about conducting trials to ascertain the varieties that are less susceptible and showing higher resistance to rust. Currently, the Lemon Myrtle Industry is conducting trial plantings, using 21 selected plant types of lemon myrtle, to further monitor and gauge the severity on the different clones being evaluated. The findings from these trials will enable the industry to take a look at the more resilient plants to further develop a rust-resistant commercial variety.

BIOTIC CONSTRAINTS (PESTS AND DISEASES)

The trees developed a natural defence to various plant pathogens and leaf-eating bugs, by the evolution of tiny round oil sacs throughout the entire leaf, which contain high citral content, which is a natural insect repellent. However, at certain times of the year, particularly when the plants are flowering, the Monolepta beetles can be found. Predominantly these beetles attack the flower; however, rarely, they do minor damage to young leaves, but this has no significant commercial fallouts.

Lemon myrtle was extremely unique and even in commercial plantations had, in the past, not been affected by any major disease; under specific conditions, rarely, it was affected by the sooty mould, which can be easily controlled.

In 2010, a new disease called myrtle rust, which is believed to be introduced to Australia from South America, affects the Myrtaceae family including lemon myrtle. The rust has caused a substantial damage in terms of production and costs in association with the use of fungicides for its control. This in turn has made growing lemon myrtle chemically free or organically more difficult, as there can be up to a 50% loss in production as a result of the rust.

HARVESTING, PROCESSING AND MARKETING

In plantation cultivation, the tree is typically maintained as a shrub by regular harvesting from the top and sides. It is important to retain some lower branches when pruning for plant health, and to eliminate soil contamination on the product.

Harvesting lemon myrtle for leaf can be undertaken all-year round, although if it is harvested during warmer months it leads to quicker plant regeneration and higher annual biomass recovery.

Harvesting techniques vary, depending on plantation layout and size. Larger commercial plantations are mechanically harvested with specifically designed equipment to suit the layout, whereas smaller plantations are generally harvested manually.

There are numerous ways to process dried lemon myrtle leaf product. Large commercial operations generally have mechanised systems in place, specifically designed to separate the leaf from the stem, and then dry the leaf efficiently and hygienically without damaging the leaf's natural essential oil composition. The most critical factor in the drying process is not using excessive heat, as this dissipates the volatiles in the leaf, reducing its flavour profile and quality.

Drying can be achieved using specifically designed, fully automated, computerised drying systems, right down to very basic rack drying methods. These basic drying methods are not as effective in terms of quality, as the product is exposed to ambient air temperature and humidity and without the consistency of a controlled environment, the moisture levels in the dried product will be inconsistent.

Microwave drying technique with advancements in technology is also a favourable option, although energy efficiency is much lower than using computerised fluid airbed drying systems. One of the benefits of microwave drying is its ability to reduce the microorganisms within the product, thereby lowering the plate count.

Evaluation of combining both the fluid air-bed drying system with microwave drying is being explored, as this gives the potential of making a drying system that is energy efficient along with giving lower microbial plate counts.

Once the product has reached its desired moisture content, the product is then cut or milled to meet the end-user's specific requirement. Final moisture levels are dependent on the application. For example, dried tea product with a maximum moisture content of less than 10% is preferred.

In large commercial processing plants, the equipment used are custom-designed hammermills, herb/tea cutting equipment, vibratory separating equipment, conveyor systems and dust-extracting units which allows the manufacturer to process in excess of 200 kg per hour. In smaller operations, the process is still performed manually and without the use of specialised equipment.

The packaging options for lemon myrtle products vary. It is generally dependant on consumer requirements, but also must meet the desired quality of the product, which in turn will extend the shelf-life of the products. The packaging must also be able to withstand shipping and handling conditions, taking into account export, where the product will be subject to huge fluctuations in temperatures. Currently, the most common packaging for dried product is polywoven plastic–lined bags or plastic bag–lined cardboard cartons. Recent research conducted into alternate packaging for storage of dried product has indicated that the use of a multi-layered foil bag is a far superior material for packaging with regard to extending shelf-life and quality retention. Shelf-life of dried lemon myrtle is very sensitive to correct temperature control and can be stored best in a dark, dry environment.

Good manufacturing processors ensure there are batch control procedures in place for traceability and that samples are taken for microbiological analysis to ensure it always meets the required standards for each customer.

The production of lemon myrtle 100% essential oil is different to that of processing dried product: once the fresh leaf material is harvested, it can be directly placed into a distillation unit and no separation or drying is required.

Lemon myrtle essential oil is extracted from the leaves and young branches of lemon myrtle trees by steam distillation. Leaves are freshly harvested and distilled immediately to ensure the extraction of the highest quality oil.

In large commercial operations, the harvested leaves are placed into large stainless steel vessels, and placed into the distillation units. Steam is introduced and the lemon myrtle's volatile elements (essential oil) are extracted from the leaf with the steam. The condensation process turns this vapour mixture into liquid (water) and essential oil. The essential oil floats on top of the water and is separated off. For more efficient distillation, the product can be cut or mulched into smaller particles, enabling more kilograms of product to be placed into the distilling vessel and in turn yielding higher volumes.

After distillation, the remaining leaf residue can be used as composted mulch, which can dramatically reduce the input costs of fertiliser and soil conditioners.

The ideal storage and packaging material for lemon myrtle oil is nothing but dark glass containers; however, stainless steel is the next best alternative and is commonly used to store large volumes. The oil is generally sold and shipped in stainless steel, aluminium and fluorinated plastic containers. For long-term storage, it is recommended to use stainless steel or glass; if stored in aluminium or plastic, an undesirable reaction can occur which will reduce the quality of the oil.

Marketing of lemon myrtle during commercial establishment in the early 1990s predominantly involved selling lemon myrtle products at a micro-level. Individual growers each marketed their own products within their local community, at markets and local retail outlets.

Following mass plantings, in the mid- to late 1990s, lemon myrtle was in oversupply. Between 2004 and 2011, some growers moved away from the industry due to a lack of market for their products. Other growers improvised and sought new, unexplored avenues such as providing foliage to the cut flower industry, which proved to be a significantly large market. This continued until the advent of myrtle rust in 2010, which, within a 3-week period, crumbled this market. This prompted innovative growers to look for new outlets and markets for their product once again.

The focus then shifted to capitalise on new and existing researches on lemon myrtle uses and benefits, to see if this would lead to any potential new application.

The research findings of lemon myrtle had unearthed some exciting developments, particularly with regard to the antibacterial and antimicrobial properties of the oil, which opened up markets in the cosmetic industry and personal hygiene products.

Originally, the marketing involved celebrity chefs advertising the products on television and individuals attending trade shows, which initially helped establish lemon myrtles recognition both domestically and internationally.

As the popularity and knowledge of lemon myrtle's uses grew, so too did the market. This led to a surge in the international market, in particular, the use of the product as a tea, tea blend, food flavouring and with applications in the nutraceutical market.

There were problems initially relating to the overseas marketing for the lemon myrtle industry in that the product had no HS number – International harmonised commodity description, coding system or Generally Recognised as Safe (GRAS). Thus, a number of issues had to be addressed, such as toxicology reports, creating nutritional panel data and meeting the guidelines outlined by Food Standards Australia New Zealand (FSANZ).

There were further hurdles for the industry to export into Europe. The novel food regulations required a proof that lemon myrtle was used for human consumption to a significant degree within the European community prior to the 15th of May 1997. With Rural Industries Research and Development Corporation's (RIRDC's) assistance and funding, along with industry contribution, eventually it was accepted by the European Union countries. As a result, Europe has become the prominent market today.

YIELD

Yields will vary with variety, plantation fertility, management and environmental conditions. Accurate production figures are not currently available. However, there are some figures, which report yields of leaf production per tree, green weight, are in excess of 8 kg per annum. In the backdrop of these figures and an estimated 2500 trees per ha (depending on plant density), expected yields should be in excess of 20,000 kg per ha per annum.

Yields for lemon myrtle oil production on the same basis, at a 2.5% recovery, should yield 500 kg of oil per hectare per annum.

CONCLUSIONS AND RECOMMENDATION

The lemon myrtle industry has come a long way in a short period of time, considering it is a relatively new commercial crop and also given the added hurdles associated with the development of myrtle rust faced by the industry.

Lemon myrtle currently is in an enviable position, with a continuous growth for the past 5 years. In both dried product and oil sales, it has reached the point where demand is almost met by the available supply.

Major growth in the Industry is required, with the establishment of new plantations in order to further expand and open new potential, along with an ongoing commitment to research and development, will lead the way to new and more lucrative markets.

BIBLIOGRAPHY

Ahmed, A.K., Johnson, K.A., 2000. Turner Review No. 3. Horticultural development of Australian native edible plants. *Australian Journal of Botany* 48, 417–426.

Buchaillot, A., Caffin, N., Bhandari, B., 2009. Drying of lemon myrtle (*Backhousia citriodora*) leaves: Retention of volatiles and color. *Drying Technology* 27, 445–450.

Burke, B.E., Baillie, J-E, Olson, R.D., 2004. Essential oil of Australian lemon myrtle (*Backhousia citriodora*) in the treatment of molluscum contagiosum in children. *Biomedicine and Pharmacotherapy* 58, 245–247.

Doimo, L., 2001. Iso-citrals and iso-geraniols in lemon-myrtle (*Backhousia citriodora* F. Muell.) essential oils. *Journal of Essential Oil Research* 13, 236–237.

Hayes, A.J., Markovic, B., 2002. Toxicity of Australian essential oil *Backhousia citriodora* (Lemon myrtle). Part 1. Antimicrobial activity and in vitro cytotoxicity. *Food and Chemical Toxicology* 40, 535–543.

Konczak, I., Zabaras, D., Dunstan, M., Aguas, P., Roulfe, P., Pavan, A., 2009. *Health Benefits of Australian Native Foods: An Evaluation of Health-Enhancing Compounds*. Rural Industries Research and Development Corporation, RIRDC Publication No. 09/133, Canberra, Australian Capital Territory, Australia.

Smyth, H., 2010. *Defining the Unique Flavours of Australian Native Foods*. Rural Industries Research and Development Corporation, RIRDC Publication No. 10/062, Canberra, Australian Capital Territory, Australia.

Southwell, I.A., Russell, M., Smith, R.L., Archer, D.W., 2000. *Backhousia citriodora* F. Muell. (Myrtaceae), a superior source of citral. *Journal of Essential Oil Research* 12, 735–741.

Sultanbawa, Y., Williams, D., Chaliha, M., Konczak, I., Smyth, H., 2015. *Changes in Quality and Bioactivity of Native Foods during Storage*. Rural Industries Research and Development Corporation, RIRDC Publication No. 15/010, Canberra, Australian Capital Territory, Australia.

Wilkinson, J.M., Hipwell, M., Ryan, T., Cavanagh, H.M.A., 2003. Bioactivity of *Backhousia citriodora*: Antibacterial and antifungal activity. *Journal of Agricultural and Food Chemistry* 51, 76–81.

Zhao, J., Agboola, S., 2007. *Functional Properties of Australian Bushfoods*. Rural Industries Research and Development Corporation, RIRDC Publication No. 07/030, Canberra, Australian Capital Territory, Australia.

CHAPTER 11

Cultivation of Muntries (*Kunzea pomifera* F. Muell.)

Fazal Sultanbawa

CONTENTS

INTRODUCTION

A fruit that has been compared to apples and sultanas, muntries is a versatile fruit that tastes like a spicy apple and appears like a sultana when dried. It was an economically and socially important food among aboriginal communities and was even used as an item of barter (Clarke, 2007). Early settlers used it as an ingredient in pies, jams and sauces. It was introduced and grown in England in 1889 (Elliot et al., 1993) but suffered a period of obscurity until recently. Interest in growing muntries has increased in the last decade, with new growers in South Australia and Victoria and expansion into New South Wales and Queensland as well. Many workers consider it to have a similar potential as that of macadamia, which has grown from a little known tree nut to a $100 million industry today (Holt, 2005).

BOTANY

Genus *Kunzea* includes about 30 species that are endemic to Australia (Baines, 1981). *Kunzea pomifera* is a prostrate shrub spreading up to 2–3 m across, with stiff, glossy leaves and cream, feathery flowers which give rise to small pink to purple tinged, downy fruit around 1 cm in diameter. Their flowers are self-incompatible and therefore rely on insects for pollination and fertilisation. The fruits are predominantly formed on 1-year-old wood, at the ends of the branches (Hele, 2001; Page, 2004) (Figures 11.1 and 11.2). Flowering has been reported from September to January and fruiting from October to April, with peak availability from February to April (Ryder et al., 2008). Wild collected fruits represent a small proportion of marketed fruits.

The native habitat is the sandy, coastal soils in the cool and dry mallee region of South Australia and Victoria. Soils in these regions tend to be alkaline, with a pH of ~8. However, plants can also be found growing 150 km interior from the coast, indicating its adaptability. Dispersal by birds or even people from local communities in the past has been postulated to explain the occurrence of such populations.

TRADITIONAL USE AND ECONOMIC POTENTIAL

Muntries obtain their name from the aboriginal word for it, one of 400 such words that have been included in the English language. Other names include emu apples, native cranberries, munthari, muntaberry or monterry (Ryder et al., 2008). Among aboriginal communities, fruit was eaten fresh as well as dried or made into a paste after drying and grinding. It was a part of their preserved food reserve for winter.

Figure 11.1 Photo of a muntries (*Kunzea pomifera*) shrub with fruits. (Photo by Yasmina Sultanbawa. Copyright University of Queensland.)

Figure 11.2   (See colour insert.) Photo of a muntries (*Kunzea pomifera*) shrub with drip irrigation. (Photo by Yasmina Sultanbawa. Copyright University of Queensland.)

It was also an item of trade among communities and was used as exchange for tools like axe heads (Clarke, 2007).

The prostrate, low-growing habit makes muntries a good option for windy areas, as a ground cover, land reclamation or even commercial plantings; however, where trellises are used, wind tends to break stems. It is also tolerant to soil salinity, which makes it suitable for areas which are unsuitable for most horticultural crops. The variety, *K. pomifera* Rivoli Bay muntries, has been granted plant breeders' rights (Plant Breeders' Rights, 1999).

Information on production and returns is scarce and quite variable from year to year, reflecting the lack of an organised marketing structure. Current annual national production is low, being estimated at 3–5 t/year, based on anecdotal evidence and statements made by growers to the media (ABC Landline, March 2007, www.abc.net.au/landline/). Commercial farms are family run, and therefore, it is not easy to obtain a full estimate of costs and returns on investment. A yield of 2 kg/plant at the peak tree age of 5 years under trellised management is possible. Prices fluctuate within the range $8–12 per kg in an 'average' year, with prices rising as high as $35 per kg for fresh fruit to $60 per kg for frozen fruit (Hele, 2001; Ryder et al., 2008). The expansion of markets and value additions will have to drive increases in production.

Muntries (and other native foods) have received attention in recent years because of their health-promoting properties. Levels of antioxidants, phenolics, vitamins, pigments and other beneficial chemicals have been recorded at high concentrations compared to currently available foods that are considered as good sources. The level of total phenolics in muntries was 2.5 times that of blueberry, which is currently considered a good source. Phenolics have a high correlation with antioxidant capacity (Netzel et al., 2006), which is associated with protection against cancers, ageing and degenerative diseases. Therefore, promotion of the health-enhancing properties of muntries is a key strategy in expanding the production and utilisation of this valuable resource.

CROP ESTABLISHMENT AND MANAGEMENT

Planting from seed leads to too much variation in the stand; hence, the usual planting materials are either cuttings or grafts. Page (2004) reported that 8 cm long semi-hardwood cuttings taken from terminal stems and treated with a 50% ethanol solution of IBA at 3000 ppm gave good rooting, usually over 70%. Grafting onto rootstocks of *K. ambigua* gave success rates over 50%. No graft incompatibility was observed and plants remained productive even after 3–4 years (McKenzie, 1984; Page, 2004).

One of the key constraints of production is the difficulty of harvesting, because of the plant's prostrate habit. Planting cuttings, while being easier and cheaper, has the disadvantage of difficulty in harvesting. Plants are therefore trained onto 1 m high trellises (Figure 11.2). Grafts are costlier but give more erect growth and also enable the training of stems on to a low trellis. Planting a variety of selections in the same row is strongly recommended, to overcome self-incompatibility within 'clones' and to ensure good pollination and fruit set (Page, 2004).

Most current plantings are done as row plantings on polythene mulch, with drip irrigation underneath as a source of water and for fertigation. Soil pH of 6–8 is considered suitable. Inter-row spacing is governed by the type of machinery used and can range from 1.5 to 3 m. Within-row spacing can also vary from 1 to 3 m, according to the availability of planting material and the desired yield. Intensive plantings of 2 × 1 m are also possible. As growth proceeds, plants are trained onto wire trellises fixed to 1 m high posts. Wires can be fixed at 20 cm vertical increments to hold the stems as they grow (Hele, 2001).

BIOTIC CONSTRAINTS

Birds were considered the main pests in some areas; however, their potential as pests remains to be seen if large-scale production materialises. Netting may have to be used if damage levels escalate. Some dieback has been reported, attributed to

Pythium and *Phytophthora*, with solutions ranging from good drainage to the use of chemicals like phosphorous acid (Hele, 2001).

GENETIC IMPROVEMENT

Self-incompatibility precludes the generation of pure lines, which are the basis of producing hybrids. *K. pomifera* is classified as a 'facultative outcrosser' and has mechanisms to prevent growth of the pollen tube in the style (Page, 2004). However, considerable variation has been observed in a number of economically important characters such as time for onset of flowering, weight, diameter and amount of soluble solids in fruit, indicating the possibility that selection for desirable characters is possible. However, the problem of self-incompatibility discussed earlier has to be borne in mind, as plantations will need to have a genetic diversity to ensure fruit set.

Interspecific compatibility with *K. ericoides* and *K. ambigua* (which has an erect habit) has been reported, offering the possibility of obtaining hybrids with erect habit (Page, 2004).

CONCLUSIONS AND RECOMMENDATIONS

Muntries offers potential as a future commercial crop, to diversify Australian agriculture and to create global interest in a unique native fruit with benefits to consumers worldwide. The occurrence of health-promoting factors like the phenolics in the fruit must be exploited to increase awareness and attract investment in order to elevate the status of muntries from native food to an internationally traded product.

Considerable research effort has been extended by state and national organisations. Basic scientific work, such as plant propagation, pollination, crop management systems and chemical composition, have been done. Preliminary molecular screening of different populations using RAPD analysis (Page, 2004) has provided insights into population genetics and would be of value in planning a crop improvement programme. More work needs to be done in areas such as planting material selections, plant nutrition, harvesting and value addition. The importance of a concerted marketing effort cannot be overemphasised, as with any new crop, demand will be the key driver of production.

BIBLIOGRAPHY

Baines J.A., 1981. *Australian Plant Genera*. The Society for Growing Australian Plants, Picnic Point, New South Wales, Australia.

Clarke, P.A., 2007. *Aboriginal People and Their Plants*. Rosenberg Publishing, Dural, New South Wales, Australia.

Elliot, R.W., Jones, D.L., Blake, T., 1993. *Encyclopaedia of Australian Plants Suitable for Cultivation*, Vol. 6 (K-M). Lothian Press, Port Melbourne, Victoria, Australia, pp. 15.

Graham, C., Hart, D., 1997. *Prospects for the Australian Native Bushfood Industry.* Rural Industries Research and Development Corporation, RIRDC Publication No. 97/22, Canberra, Australian Capital Territory, Australia.

Hele, A., 2001. *Muntries Production.* Australian Native Produce Industries Pty Ltd, Primary Industries and Resources SA, Adelaide is South Australia, Australia.

Page, T., 2004. *Muntries: The Domestication and Improvement of Kunzea pomifera* (F. Muell.). Rural Industries Research and Development Corporation, RIRDC Publication No. 03/127, Canberra, Australian Capital Territory, Australia.

Plant Breeders Rights, 1999. http://www.ipaustralia.gov.au/get-the-right-ip/plant-breeders-rights/.

Ryder, M., Latham, Y., 2005. *Cultivation of Native Food Plants in South Eastern Australia.* Rural Industries Research and Development Corporation, RIRDC Publication No. 04/178, Canberra, Australian Capital Territory, Australia.

Ryder, M., Latham, Y., Hawke, B., 2008. *Cultivation and Harvest Quality of Native Food Crops.* Rural Industries Research and Development Corporation, RIRDC Publication No. 08/019, Canberra, Australian Capital Territory, Australia.

CHAPTER 12

Cultivation of Native Pepper (*Tasmannia lanceolata*)

Christopher D. Read

CONTENTS

INTRODUCTION

Native Pepper (*Tasmannia lanceolata*), also referred to as Tasmanian or Mountain Pepper, is one of the more popular native Australian food ingredients to have appeared on the culinary horizon in the last 20 years. At present, most of the world supply is derived from unmanaged stands of self-regenerated plants, which are scattered through the landscape, many on private land or forestry concessions and are subject to all the natural variables of the wild environment. The move to managed orchard production is a pressing imperative for a variety of reasons, including a reduction of the natural variability in fruit production and increased uniformity of quality and flavour from selected varieties to create opportunities for mechanical production techniques, including harvesting, and, from the perspective of an Australian horticulturist, to establish 'Native Pepper' as an Australian or Tasmanian speciality, at least in the medium term.

Although it is mostly the pepper berry, known for its fruity, berry flavours and sharp spicy heat which is marketed for domestic or restaurant use, there is a small but growing market for pepper leaf (dried and ground) for food manufacture. This product, cheaper and more readily available than berries, offers exciting prospects for use in more technical applications such as food preservation, cosmetic manufacture, flavour extracts, nutritional supplements and veterinary products.

The approach taken here is to examine what is known about the species' ecological preferences and derive from these some basic production and agronomic guidelines, which might help to establish and manage a commercial Native Pepper orchard.

BOTANY AND DISTRIBUTION

Tasmanian Native Pepper (*T. lanceolata*) is one of a small group of plants from the family Winteraceae, which had its origins on the ancient Gondwana supercontinent. As the land mass broke up and was distributed around the Pacific, the family was carried along with it, so that today Native Pepper and its relatives occupy the Pacific seaboards of Australia, South America, New Zealand and some archipelagos to Australia's north. There are seven genera and, in the Australian genus of *Tasmannia*, about seven species.

T. lanceolata is a medium-sized shrub or a small tree, typically found as an understorey species in rainforest and wet sclerophyll forests in south-eastern mainland Australia, Tasmania and several of Tasmania's coastal islands, where reliable rainfall and cloudy, protected conditions suit its tender growth habit. Within its natural range, the plant colonises readily after disturbance, such as the formation of openings in the rainforest canopy or on cleared land. It is an 'early succession' coloniser, though seed is quite slow to germinate and may require some form of pretreatment, possibly partial digestion by birds, which utilise the berries as a food resource in late autumn. Currawongs, Australian ravens and Rosellas are often seen feeding on the trees in late autumn and frequently drop pellets of regurgitated fruit and seed beneath perching points, so that it is common to see clusters of Native Peppers adjacent to fallen logs and tree stumps, which also provide protection from browsing, desiccation and physical damage (Figures 12.1 and 12.2).

In terms of microclimate, it is usual to find the plant on relatively well-drained soils (in the natural, higher rainfall range), and not in permanently wet or waterlogged soils as they are susceptible to root rots and other soil-borne diseases.

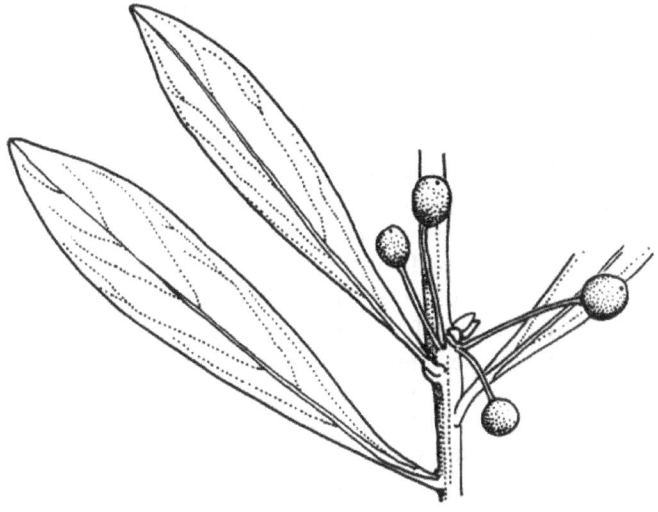

Figure 12.1 *Tasmannia lanceolata* – fruits and leaves. (Illustrated by David Mackay. Flora of NSW, Copyright Royal Botanic Gardens and Domain Trust.)

Figure 12.2 (See colour insert.) Photo of a Native Pepper (*Tasmannia lanceolata*) tree and its fruits. (Copyright Christopher Read/Diemen Pepper.)

The physiology of the plant is best suited to what can be called 'mesic' environments where the extremes of temperature, moisture availability, radiation levels and access to nutrients are moderated by physical location, microclimatic factors and accompanying vegetation. These are the conditions found where the Native Pepper commonly grows, beneath a rainforest canopy, at moderately high altitudes and in regions of higher average rainfall and frequent cloudy days.

These site characteristics give us a clue to the conditions in which *T. lanceolata* is most likely to do well and can be used as the basis for an 'agronomic best bet' for bringing the species under cultivation.

Native Pepper is dioecious: the male and female flowers are borne on separate plants, so we can loosely refer to plants as being either 'male' or 'female', though in every other respect they are similar. The only easy way to distinguish them is when they are in flower or fruit, though occasionally female plants do not bear fruit.

As flower buds begin to separate from the shoot tips in spring, it is possible to distinguish between the male and female flowers. Male flowers tend to be plumper, while female flower buds are often rather flattened and asymmetrical. As they open, female flowers reveal a whorl of four to six pale yellow petals, around a small green bump representing the gynoecium, which, if fertilised, will develop into a berry. Male flowers are more showy, releasing a spray of short stamens, each ending in a bright yellow anther (Figures 12.3 and 12.4).

After fertilisation, the fruit begins to develop during the summer and ripens in early autumn, a relatively long period for berries generally, though typical of many alpine berry–bearing species at high altitudes. During this period, there seems to be a tendency for berries to drop freely under stressful conditions such as days of high temperature, extreme drought and occasional sudden summer frosts.

Shoot and leaf developments occur either concurrently with flowering or immediately afterwards; the timing of this varies greatly between trees. A number of shoots typically develop from the apical bud cluster and may extend from two to more than a dozen nodes, depending on the clone and the conditions during spring–early summer. These leaves mature slowly and by midsummer, growth typically ceases and the leaves mature, thicken and darken. Leaves usually remain on the plant for up to 2 years, though this characteristic is also quite variable.

Figure 12.3 Photo of a Native Pepper (*Tasmannia lanceolata*) female flower. (Copyright Christopher Read/Diemen Pepper.)

Figure 12.4 Photo of a Native Pepper (*Tasmannia lanceolata*) male flower. (Copyright Christopher Read/Diemen Pepper.)

TRADITIONAL USES

Most historical references to humans using Winteraceae family relate to herbal and medicinal applications. New Guineans (probably also Solomon Islanders) used the pounded leaves of a local species for treatment of 'diseased spots' on the skin of pigs, and *Drimys winteri* (locally known as 'canelo') was used in central Chile and Brazil as a treatment for cholic and cattle itch and as a 'stomach tonic'. *Pseudowintera axillaris*, a New Zealand relative, was used by Maori people as a stimulant for skin diseases, venereal diseases and stomach ache and the leaves were chewed to relieve toothache.

Europeans took their cue from these practices during the early exploration of the New World. With *D. winteri*, from South America, it was Captain Winter, commander of one of the ships under Drake's command, who first used the bark to relieve scurvy among his crew, and soon the bark was exported to Europe as 'winter's bark', where it was used for a time as an aromatic tonic and antiscorbutic. Soon supplies became scarce and winter's bark was substituted with an unrelated species and then gradually fell from favour. In the same way, the bark of *P. axillaris* (Horopito) was used by pioneering New Zealanders as a quinine substitute and the sap was used for treating skin diseases. In recent years, a complementary therapy using an extract of horopito leaf for management of *Candida* in humans has been successfully marketed in New Zealand and the United States.

In Australia, unfortunately, there is no convincing evidence of indigenous use of the Australian species either as a food resource or medicinal herb, though according to eminent Australian ethnobotanist, Dr. Beth Gott, there were two aboriginal names ascribed to the local species in an early Tasmanian journal, which does imply that it may have had some economic importance.

In the late 1800s, European interest in turning native flora to 'useful purpose' was high and was promoted by the work of the curator of the Technological Museum in Sydney, Joseph Maiden, who published *'Useful Plants of Australia'* in 1889. There he mentioned *T. aromatica* (currently *T. lanceolata*) as a potential pepper or allspice substitute, noted its resemblance to winter's bark and referred to it as a possible source of a 'succulent, though insipid fruit'.

In the second half of the twentieth century, Australian 'bush tucker' writers began referring to Australian herbs, roots, fruits and berries as an 'unexplored' resource, and Native Pepper was mentioned as a flavouring spice for preparation of meat and savoury dishes by several writers (Cherikoff, 1989; Cribb and Cribb, 1975; Low, 1988).

By the late 1980s, there was enough interest and 'bicentennial sentiment' to induce innovative chefs, food technologists and researchers to experiment with the more accessible species, and 'Tasmanian Mountain Pepper' was an obvious candidate – small black berries which dry to a size and shape roughly similar to conventional black pepper and a sharp spicy taste that made them easy to adopt and adapt.

At the same time, a research project in Tasmania identified the extract of the dried leaf as a potential flavour product, so from that point, the basis for the herb *and* spice market for Native Pepper was established.

On the technical side, plant biochemists became interested after it was discovered in the early 1960s that *T. lanceolata* contained high concentrations of a relatively rare (in nature) compound – the hot-tasting sesquiterpene polygodial. Polygodial has been found not only in a handful of species, including most of the Winteraceae family, but also in an African tree species and in 'water pepper', a widespread annual herb used to make a spicy relish in Japan. The compound has a number of interesting biological properties, the most important of which are its hot taste to humans and antibiotic activity, which have been widely reported and offer an interesting avenue for research and potential markets for the products.

ECONOMIC POTENTIAL

At the present time, most of the commercially traded pepper berry fruits and leaves are sourced from stands of naturally established trees on private land in Victoria and Tasmania and on leased public land in Tasmania. Only one other species of *Tasmannia* has been used to any degree as a 'native food' ingredient – *T. stipitata* – 'Dorrigo pepper', naturally found on the northern tablelands of NSW, where conditions, though warmer than Tasmania, are also relatively benign.

The economic potential for *T. lanceolata* is hard to gauge, as the current market interest has been developed in tandem with a gradual increase in production capacity and prices are highly sensitive to changes in supply and demand. Current production and consumption is around 7 t per annum of berries and about 3 t dried leaf, of which probably half (by value) is exported to other western markets, principally Europe and North America. At this time, there is almost no trade in Asian markets, though these offer an obvious opportunity to absorb increased production in the future.

In any case, new production projects *must* include investment in market development to avoid the sort of industry volatility seen in, for example, the tea tree oil industry in the 1990s and early 2000s, where poorly planned commercial production resulted in price collapse, to below the cost of production in many instances.

FRUITING BEHAVIOUR

As mentioned earlier, current production is largely from natural stands. Gathering of the fruit by hand or with the use of simple harvest aids and the scattered nature of the plant stock limits the rate of harvest, particularly in years when yields are low and many female trees carry few berries, sometimes none at all.

Across the regions where the tree is common, yields of fruit vary sharply, year to year, varying from almost none to very highly productive seasons in which almost every female tree bears a heavy crop. This characteristic is not unusual in the natural situation, very heavy seed or fruit crops often occur every few years, called 'mast years' as well as years with lower yields or the complete absence of fruiting in the years between. There are a number of theoretical explanations for this, the most plausible and relevant explanation being that the plants respond to resource availability – in favourable seasons, plants invest heavily in reproduction or, at least, undergo some form of 'switching' from vegetative to reproductive effort.

If resource availability is indeed a significant factor in flowering and fruit production, then there might be scope for manipulating the yield of fruit, most likely by adjusting nutrition and water relations to 'smooth out' environmental effects.

This certainly points to a need to better understand source/sink relations within the plants, their response to nutrient and water availabilities and to investigate other mechanisms influencing flower initiation, pollination, fruit development and retention. Very little research has been undertaken in these areas to date, and each of these issues has the potential to dramatically alter the fruiting behaviours of the plants.

Besides the response of the plant to conditions, season to season, there are also the external factors which affect pollination success. Although there is only limited published reference to the pollination mechanisms of *Tasmannia* species, it seems probable that pollen transfer is both animal (insects, spiders, birds) and wind mediated. More study is needed to establish just how mobile pollen is, just how *necessary* it is (there is some suggestion of plants setting fruit in the complete absence of male plants) and if there are issues of compatibility between individuals.

AGRONOMY

At the present time, a simple calculation of typical single plant yields and current market size suggests that the entire global consumption of pepper berries in 2014 could be met by production from about three or four hectares of 6-year-old female plants. Present yields from natural populations, while variable, will continue to contribute to supply for the medium term. Therefore, new orchard production units will need to be matched to market growth during the 5–8 years required before they commence fruit production, in order to avoid a potential oversupply situation in the future.

The following discussion does not consider the opportunities offered by other production paradigms, such as enrichment planting, polyculture gardens or simple scattered plantings under forest canopies. These systems will certainly allow more complex and 'natural' ecological arrangements but will not address the pressing issue of harvest labour. Nevertheless, the aforementioned ecological discussion points to some issues that still apply to these production systems.

If the pepper berry is to be economically viable, then commercial production must include mechanical harvesting strategies, as the current cost of hand harvest, even in the orchard situation, is not sustainable in an Australian context. The approach taken will determine basic parameters such as plantation layout, tree density and clonal selection for ease of harvest.

Varieties, Clones and Selections

Since the species is wild, dioecious and presumably open pollinated, every individual wild plant represents its own 'varietal' type – a fantastic and daunting opportunity for clonal selection! Even a casual observation of the natural population reveals a spectrum of gross characters – leaf size and shape, canopy structure, stem colour and architecture and fruiting habit which will be valuable when selecting vigorous, reliable, heavy bearing, easily managed stock for propagation. Then there are less obvious, but also important characteristics such as pest and disease resistance, response to pruning, pollination compatibility, fruit size and flavour and ease of harvest, which might need to be assessed over an extended period of cultivation. For these reasons, a selection and propagation programme should include as wide a variety of selections as possible, many or most of which may ultimately be rejected for some reason or other.

There is an opportunity to develop a broad database in this area, but this will require some coordinated research in a long running programme to determine optimal parameters for as many characteristics as possible – the development of an 'ideotype'. It will also require early-uptake orchardists to accept the likely attrition of unsuitable selections and to ensure the continued addition of fresh varietal material selected from the wild population.

A study conducted in Tasmania in 2000–2001 attempted to gauge the variability in relation to one parameter – polygodial concentration in the leaf, using more than 400 randomly chosen individuals from just one site in north-east Tasmania (Menary et al., 2003). This work revealed a more than 20-fold variation in polygodial concentration in leaf extract, around a median level of about 2% w/dw leaf. Given that this compound confers the characteristic hot flavour in the leaf and berry, it is clearly important to include this parameter in any future selection programme. It is also likely that other important characters will show the same capacity for variation in the wild population.

At the present time, there are only a few 'selections' available through native plant nurseries. Obviously, the need to know the gender and some of the characteristics of the selection is of fundamental importance, but there is little guidance available from the nursery industry or existing pepper producers that would aid a start-up enterprise. This author has some thirty 'good bet' selections under cultivation at the time of writing, but only a handful of those have reached fruiting maturity, and it is likely that many or most will prove to be unsuitable for commercial production within the next 8–10 years.

Site Selection

Following the logic of the ecological discussion, it is reasonable to assume that orchard production will succeed best where the climate is cool, temperate with reliable rainfall, soils are well drained and exposure to hot conditions and bright sunshine is moderated by aspect, shelter, overstorey and geographical location. Beyond that it should be feasible, using inputs such as irrigation, fertiliser, shade, shelter and attentive management, to stray outside the natural ecological zones of the species, as has been achieved with many other horticultural crops.

Orchard Establishment

Issues of plantation layout, time of planting and site preparation can follow existing practices for perennial horticulture and tree crop production, but it is worth mentioning that layout should accommodate any intended mechanical management such as mowing, fertilising, harvesting and mechanical pruning. If tractor access is required, row spacing should allow for tree spread – typical canopies are 1–1.5 m wide in unpruned 8-year-old trees, and headland provision should be adequate.

Plants are usually propagated by cuttings in winter and spring. Callusing and root development in fresh cuttings takes between 6 and 8 weeks under mist and with bottom heat, before they can be potted into 50–65 mm pots and held for planting in autumn, when they should be robust enough to withstand transplant shock to ensure the maximum opportunity to establish sturdy roots before the following summer. At this time, there is no proven ratio or spacing for inclusion of male plants as pollinators. Pollen is mobile, as mentioned earlier, and there are no known issues of compatibility between selections, though all of this needs further research. A best bet (again) would be to ensure plenty of variety among male plants that are spatially well distributed among the female selections.

Selection of suitable stock is, as explained earlier, an area in which each grower will need to undertake their own 'research and development'. Commercial nursery operators will have views about suitable selections and should be able to identify the provenance, gender and some characteristics of their plant stock.

Orchard Maintenance

Depending on the prior vegetation on the site, competition from 'weeds' – any other plant besides the pepper – will be the most challenging issue for newly established plants. Irrigation and fertilisation around the plants will encourage this, so it is important to have a workable strategy to manage competitive species. Weed mats, hand mowing and weeding or mulching offer the best solutions. While there are a couple of herbicide products which can be used in pepper production, part of the market appeal of Native Pepper is its wild and natural environment, so as a general approach, it would be preferable to avoid all herbicide applications if possible.

Irrigation of Native Pepper is likely to be necessary in almost every situation, even within its natural distribution, in order to maximise early growth and canopy development. As the plant normally grows in regions where 1500–2500 mm rainfall is distributed in a Mediterranean-type climatic pattern (cool wet winters; warmer, drier summers), the critical period for irrigation will be during the summer months. Mean monthly precipitation in natural locations in Tasmania and Victoria during the spring and summer growing period can be as high as 150–200 mm (e.g. Waratah, Tasmania; data for 1883–2014). This can be used as a rough guide in calculating the supplementary irrigation likely to be required for minimum natural growth, but with a species such as this, close attention to soil moisture levels and plant's response will be essential throughout the growing season.

The method of water application will depend on local circumstances, soil type, wind patterns, water availability and economics. While trickle or micro-spray systems are more economical to operate, they do have some disadvantages: installation and maintenance costs are the key ones. If the site and location are commonly exposed to hot drying winds in summer, it may be worthwhile to consider some form of overhead system, either as the principal means of irrigation or as an atmosphere-modifying strategy, in tandem with other forms of water application.

Recent agronomic work conducted at the University of Tasmania (Matthew Wilson, pers. comm.) has shown some interesting trends with differing application rates for the macronutrients, nitrogen, phosphorous and potassium. This work (to be published shortly) should provide some useful guidance for fertilisation of cultivated plants. Nevertheless, there is much we do not understand about the nutritional demands of Native Pepper, and this is an area in which continued glasshouse and field research is a high priority.

PRUNING AND HARVEST

At present, berry harvest is conducted largely by hand, with the aid of various crude implements and catching systems. Since the cost of harvest labour, as noted earlier, contributes more than half of the total market cost of the product, there is an urgent need to introduce some form of mechanical harvest, based on a row crop system, or at least in conjunction with a canopy management strategy. Many cultivars respond well to pruning and quickly develop new shoots from wood which is 3 or 4 years old, sometimes older. This enables the development of a canopy surface that is quite amenable to raking or plucking systems or simply pruning for the purpose of gathering foliage for leaf production.

Drying Procedures

For both leaf and berry, the drying of the fresh material is a critical part of the process, costly, logistically and mechanically challenging and essential for the production of a superior quality dried product.

At the present time, producers of Native Pepper berry and leaf supply a product which far exceeds Australian food safety standard parameters, in contrast to the paradigm for the international market for spices.

The most important aspect of production (after careful harvest, transport and handling of course) is the complete and efficient drying of the product. Rapid reduction of water activity to a level below which bacteria, fungi, yeasts and moulds cannot proliferate will minimise the microbial load in the dried material. Further reduction will help destroy many microorganisms completely; however, many spore-forming organisms are not destroyed under normal drying conditions: so the initial period of drying provides the best opportunity to minimise the final microbial load. Slow drying of fresh product, especially berries, which contain high sugar levels, allows propagation of moulds and yeasts on the surface of the fruit, leading ultimately to high levels in the finished product.

As a general rule, the initial part of the drying cycle should be characterised by lower temperatures (<30°C) with rapid air movement and maintenance of low relative humidity in the airstream. As the water activity falls, temperature can be raised somewhat to further hasten the process. Finally, the product should contain 6%–10% moisture (as determined by sample), which should ensure good microbiological control and long shelf life.

With leaf, the tendency for 'stewing' of young or immature leaf can result in a discoloured product, so it is important to harvest only tougher, mature leaf and to avoid over-temperature drying conditions early in the cycle. The finished result should be a bright, green leaf with a pungent, herbaceous fragrance.

Cleaning, Packaging and Storage of Dried Product

Each producer will develop their own methods for removing extraneous plant material (e.g. leaves, sticks, buds and seeds from pepper berries and twigs from pepper leaf), and the usual precautions against accidental inclusion of metal, glass and plastic material must be attended to.

Clean, dry product should be stored carefully in closed, impervious packaging and held in dark, dry conditions. This is necessary to prevent resorption of moisture during storage, introduction of airborne microbial contaminants and infestation by insects and other invertebrates. Dark conditions are especially critical for leaf products, as it is the photodegradation of green pigments and flavour compounds which leads to fading and browning of the leaf and loss of flavour and shelf life.

Storing the product in a closed container will also minimise the loss of flavour and fragrance volatiles over time, again maximising shelf life. The use of impervious materials to prevent contamination by other flavour volatiles in the storage facility, especially where many different flavouring products and herbs and spices may be stored together, is another option.

Marketing

For effective marketing of an expensive spice or herb, such as the Native Pepper, a carefully planned approach is essential. Intending producers or traders should develop a strong and targeted marketing plan, identifying their target market and developing appealing packaging and promotional material which meets statutory requirements and which helps consumers understand the product and how and where to use it.

Native Pepper has been recognised as a traditional food by the Australasian Food Standards Authority (Food Standards Australia and New Zealand), and there are published data on the nutritional values of the products (Read, 2012). Similarly, there are many recipes and suggestions for use that can be incorporated into marketing and packaging material to inform the consumer.

There have been several exciting developments in the potential application of leaf and berry products in other areas besides simply as herb and spice flavourings, especially focusing on nutritional properties (Konczak et al., 2010), and the widely reported antimicrobial activity of polygodial, a bicyclic, unsaturated sesquiterpene aldehyde (Kubo et al., 2005). Much more work is required in these areas, basic science and product development, before they offer significant commercial opportunities.

Yields

At this time, there is very little firm indication of potential yield per hectare or even per plant, as most of the current production is obtained from unmanaged older plant stock. Translating this into expectations for managed stock in an orchard system is not helpful, though as a general indication, it appears reasonable to expect yields of 3–4 kg fresh berries from vigorous bushes around 6–7 years old and increasing as the canopy size increases. Pruning and management for mechanical harvest will limit canopy surface area and therefore impose a ceiling on mature plant yield, dependent on the characteristics of the clone or variety.

Based on this and the approximate plant densities of 3000–4000 plants per hectare, we can expect that fresh yields of 9–16 t per hectare would be achievable.

Typical recovery of clean, dried berries from fresh fruit is 20%–23%, so production budgets based on yields of approximately 2–3 t dried berries should be achievable around 6 years after establishment.

CONCLUSIONS AND RECOMMENDATION

In conclusion, although there is no significant production of either berries or leaf presently generated from managed orchards of Native Pepper, there is a reasonable basis for proceeding with investment in commercial ventures, subject to careful consideration of the risks involved in an untested model.

The lack of established varietal choices, agronomic uncertainties and research gaps, no proven mechanical harvesting machinery and the lack of detailed yield data are risk factors which should be incorporated into investment proposals.

Furthermore, the relationship between market supply and demand is untested, and returns based on present-day prices and demand may prove difficult to reproduce as the scale of production increases.

Clearly, research is urgently required in production, (specifically agronomy, reproductive biology and harvest technology) and in market development, commercial applications of berries and leaf products. In the latter area, there are identified opportunities for high value applications in food preservation, flavour extracts, cosmetics, nutritional and therapeutic applications. Funding for this work will require commitment from existing and new participants, from research institutions and government to take it to a commercial level and to convert what is presently a tentative, speculative enterprise conducted by a few enthusiasts to a new, commercial opportunity for Australian horticulturists. The potential benefits for regional development, production diversity and export income are substantial and can be achieved in the medium term, provided care is taken to develop markets as well as production capacity.

REFERENCES

Cherikoff, V., 1989. *The Bush Food Handbook*. Ti Tree Press, Boronia Park, New South Wales, Australia.

Cribb, A.B., Cribb, J.W., 1975. *Wild Food in Australia*. Angus and Robertson, Sydney, New South Wales, Australia.

Konczak, I., Zabaras, D., Dunstan, M., Aguas, P., 2010. Antioxidant capacity and phenolic compounds in commercially grown native Australian herbs and spices. *Food Chemistry* 122, 260–266.

Kubo, I., Fujita, K., Lee, S.H., Ha, T.J., 2005. Antibacterial activity of polygodial. *Phytotherapy Research* 19, 1013–1017.

Low, T., 1988. *Wild Food Plants of Australia*. William Collins Pty. Ltd., Sydney, New South Wales, Australia.

Menary, R.C., Dragar, V.A., Thomas, S., Read, C.D., 2003. *Mountain Pepper Extract, Tasmannia lanceolata: Quality Stabilisation and Registration*. Rural Industries Research and Development Corporation, RIRDC Publication No. 02/148, Canberra, Australian Capital Territory, Australia.

Read, C. 2012. *Nutritional Data for Australian Native Foods*. Rural Industries Research and Development Corporation, RIRDC Publication No. 12/09, Canberra, Australian Capital Territory, Australia.

CHAPTER 13

Cultivation of Quandong (*Santalum acuminatum*)

Ben Lethbridge

CONTENTS

INTRODUCTION

Among native Australian 'bush foods', quandong or the 'Native Peach' is considered one of the more promising plants that could make it to the world stage as a dessert fruit. Currently, it is used as a dried or frozen fruit as an ingredient in 'an enormous number of products' (Herde and Herde).

BOTANY

Quandong is an Australian native shrub or tree belonging to the family Santalaceae. The tree's natural habitat is in southern Australia, ranging from New South Wales, south Australia and western Australia with pockets in Victoria, Southern Queensland and Northern Territory. Like other members of this family, it is a semi-parasite attached to the roots of a host plant for additional water and nutrients.

Quandong can be semi-parasitic on more than one host at a time: often members of nitrogen-fixing trees like *Acacia* and *Casuarina* are the honourable hosts. Other species of the *Santalum* genus occur in Asia and the Pacific Islands. There are six Australian species found on the Australian mainland. Three of these have been/are harvested for sandalwood (species *album*, *spicatum* and *lanceolatum*).

The habit is quite variable, ranging from a 1 m high shrub to a 6 m tall tree. The leaves are olive green coloured, 3–9 cm long, drooping and sometimes sickle shaped. The small, greenish flowers exhibit a raceme inflorescence, appearing during the

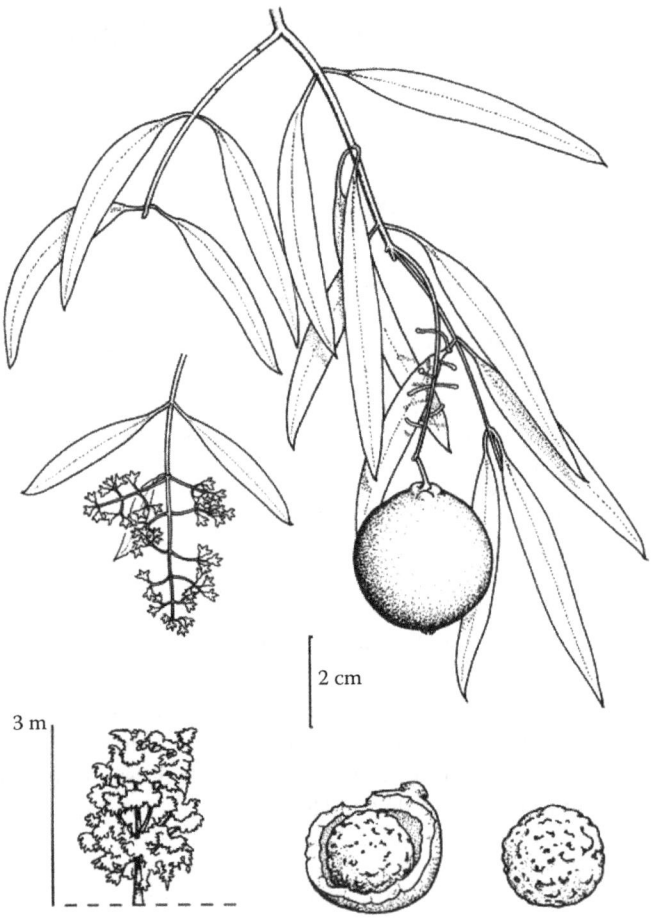

Figure 13.1   Botanical illustration of quandong (*Santalum acuminatum*). (Taken from Stewart, K. and Percival, B., *Bush Foods of New South Wales – A Botanical Record and an Aboriginal Oral History*, Royal Botanic Gardens Sydney, Sydney, New South Wales, Australia, 1997; Illustrated by David Mackay. Copyright Royal Botanic Gardens and Domain Trust.)

months of October–March. Insects, including bees, native bees, wasps and flies, appear to be the main insects for pollen distribution. Ripe fruits are 15–25 mm with a shiny, yellow to red skin. Its flesh is white to cream, about 3–5 mm thick and was taken as food by Aboriginals and Europeans alike. The red fleshy fruit layer (epicarp + mesocarp) surrounds the hard, woody shell (endocarp). The kernel (endosperm + embryo) is covered by a thin woody parchment layer (Cromer and Woodall, 2007). Even though fruits are small, it is expected that with breeding and selection improvements in size may lead to desired flesh thickness and consistency, as seen in other domesticated fruits. Some progress has been made; two named cultivars have been released for commercial cultivation (Figures 13.1 through 13.3).

Figure 13.2 (See colour insert.) Photo of a quandong (*Santalum acuminatum*) tree with fruits. (Copyright Ben Lethbridge.)

Figure 13.3   (See colour insert.) Photo of quandong (*Santalum acuminatum*) fruits. (Copyright Ben Lethbridge.)

TRADITIONAL USE AND ECONOMIC POTENTIAL

The fruit is referred to by several names such as Sweet Quandong, Wild Peach, Desert Peach and Native Peach as well as Aboriginal names such as Guwandhuna, Gutchu, Wanjanu, Mangata, Goorti and Wadjal, among others. Aboriginal communities ate the fresh fruit or dried it for later use and then reconstituted it by soaking in water. The kernel was also extracted and used for various medicinal purposes (Williams 2010). The fruit was also a favourite among early European explorers and settlers for jams and pies. The fruit is usually dry in texture and tart tasting and has a variable sweetness. The vitamin C content is higher than in oranges.

CSIRO Australia has been conducting research on propagation, orchard management and commercial cultivar development since 1973. The Rural Industries Research and Development Corporation (RIRDC) views this crop as one of the most important native food crops currently being developed and had committed core funds for further research up to 2013.

Production is predicted for grafted varieties to begin at year 4 with increasing yields of 0.5 kg per annum until around year 15. Quandong production has declined significantly from its peak in 2001 (25 t). Roughly, a third comes from commercial plantings and the remainder from wild harvest. Prolonged drought has been a significant factor affecting limiting production. Current production (2014) has been around 7 t with 90% from cultivation (RIRDC, 2013). Quandong has the advantage that the fruit matures in spring (August–December), when few other local fresh fruits are available in the market. Its greatest potential will probably be best realised in inland areas where the water quality (salinity) and harsh conditions could be very limiting to other fruit crops (Sedgley 1982, Hewson and George 1984, Herde and Herde 2007).

AGROCLIMATIC REQUIREMENTS

Quandongs require a climate with high light intensity and low relative humidity and will grow in a range of soil types with pH variations and high salinity, root diseases will be more prevalent in poorly drained soils. Quandong exhibits high tolerance to saline water and drought. Growing quandong in a medium with as much as 200 mol m^{-3} Cl$^-$ had little effect on growth in the short term (up to 7 weeks), but extended treatment for 14 weeks with 200 mol m^{-3} Cl$^-$ reduced growth (Walker, 1989).

AGRONOMY

A significant challenge to orchard production of the quandong is the horticulturally unique semi-parasitic nature of the tree. The parasitism is not specific, but species with extensive surface rooting, allowing significant root connections (haustoria) to the quandong and roots at depth for expanded water access will be advantageous. Species with adaptive features for low evapotranspiration such as needle-like, blue-grey-coloured leaves would be preferred. The nutritional status of the host, if in the normal range, is a less significant factor. Nitrogen-fixing species such as *Acacias* and *Allocasuarina* may offer benefits in poor fertility soils and therefore may be better hosts than other species. The drought tolerance of Quandong has been attributed to its ability to maintain more negative water potentials than the host (Loveys et al., 2001).

GENETIC IMPROVEMENT AND PROPAGATION

Varieties with useful attributes have been selected from seedling orchard trees grown from seeds of trees from a wide geographic distribution and/or backyard trees with proven performance. No formal breeding programmes have been conducted to combine all the useful properties into a single variety.

Propagation by seed is often slow with large varietal differences. Soaking or vacuum infiltration with gibberellic acid, wetting and drying cycles or cracking the endocarp have shown to be advantageous with some varieties (Cromer and Woodall, 2007; Loveys and Jusaitis, 1994). Media reports claim that simulation of the natural mulch layering of quandong seed can also be effective (www.abc.net.au/gardening/stories/s2577682.htm).

Quandong rootstocks are propagated from seed. Clonal trees can be grafted in the nursery, with scion collected from semi-ripe material from new seasons growth (October to December) or pre-bud burst (July–August) or cinctured material in September–October. Quandongs are best field grafted during July–September. The great susceptibility of quandong rootstock to root rot diseases, leading to losses during transplantation, overwatering and collar rot, suggests that rootstocks from other species in the *Santalum* genus may be useful. Some progress has been made on this front, but it is too early to make any definitive recommendations.

BIOTIC CONSTRAINTS

Phytophthora or *Pythium* species causes root rotting and can be a problem for poorly managed plantations. Girdling of the trunk by a commonly used primary host, creeping boobialla (*Myoporum parvifolium*) can lead to collar rot and slow death of the quandong.

Another major problem is the quandong moth (*Paraepermenia santaliella*), as the larvae damage fruit. Previously, a registered systemic dimethoate-based insecticide was used. The registration has lapsed and reasonable control is now obtained with a pyrethrum-based insecticide delivered after fruit set and just before fruit ripening. It is good orchard practice to remove any fallen fruit. Parasitic wasps have been observed infecting the larval stage of quandong moth and research into these biological control agents could be a useful. Interestingly, quandong associated with white cedar (*Melia azedarach*) as a host is known to accumulate sufficient azadirachtin-like chemicals to deter attack by the quandong grub (Beth Byrne, Waite Institute, University of Adelaide); this also offers another avenue for control.

Other problems include mites, black spot of leaves (due to high humidity) and leaf-eating caterpillars, and these can be controlled by standard fungicide/pesticide treatments.

HARVESTING, PROCESSING AND MARKETING

Harvesting is done by hand and often over a long ripening period. Selection of varieties for a narrow ripening period would be advantageous for commercial production. Most quandongs are sun-dried (the fruit is naturally low in moisture) or fresh frozen.

There are no organised marketing strategies for this fruit. The fruit is popular in most tourist-oriented establishments.

CONCLUSIONS AND RECOMMENDATIONS

The quandong industry has had to cope with many challenges during its short development (Lethbridge 2004, Hele 2006, Cooper 2010, Bonney 2013). Other fruit crops often boast centuries of orchard trials. The quandong's apparent hardiness has been an attractant to many new horticultural industry pioneers. Managing the unusual semiparasitic habit and fragile root system has presented many challenges not anticipated initially. Research into alternative rootstocks may hold the answers. Many growers have made successful businesses out of seedling-based enterprises, but there are many inherent problems in relying on this approach as this industry starts to compete with mainstream fruit production. To move from a niche industry into broad horticultural production, much finer control of fruit production will be necessary. Some significant steps have been taken, but there are still challenges, so credit should go to those willing to tread the unbeaten path for this iconic species of southern Australia.

REFERENCES

Bonney, N., 2013. *Jewel of the Australian Desert: Native Peach (Quandong)*. Neville Bonney, Tantanoola, South Australia, Australia.

Cooper, N., 2010. Lynray quandong (*Santalum acuminatum*) orchard establishment. In: Eldridge, D.J., Waters, C. (eds.), *Proceedings of the 16th Biennial Conference of the Australian Rangeland Society, Bourke*. Australian Rangeland Society, Perth, Western Australia, Australia.

Cromer, E.L., Woodall, G., 2007. Breaking mechanical dormancy in Quandong using silica gel and enhancing germination response using Gibberellic acid. No 28, Alcoa World Alumina Australia, Perth, Western Australia, Australia.

Hele, A., Latham, Y., Ryder, M., O'Hanlon, M., Lethbridge, B., 2006. Quandong Production. Government of South Australia Primary Industries and Resources SA Factsheet No: 17/03.

Herde, G.J., Herde, I.L., 2007. Practical plantation planning and establishment for Santalum growers, http://www.nectarbrook.com/quandong/tree.html.

Hewson, H.J., George, A.S., 1984. *Santalaceae, Flora of Australia*, Vol. 22. Australian Government Publishing Service, Canberra, Australian Capital Territory, Australia.

Lethbridge, B., 2004. DOOR for Quandong Production – A feasibility study. RIRDC Web Publication No W04/111, Canberra, Australian Capital Territory, Australia.

Loveys, B.R., Jusaitis, M., 1994. Stimulation of germination of quandong (*Santalum acuminatum*) and other Australian native plant seeds. *Australian Journal of Botany* 42, 565–574.

Loveys, B.R., Loveys, B.R., Tyerman, S.D., 2001. Water relations and gas exchange of the root hemiparasite *Santalum acuminatum* (quandong). *Australian Journal of Botany* 49, 479–486.

RIRDC, 2013. Rural Industries Research and Development Corporation, Publication No. 14/120, Canberra, Australian Capital Territory, Australia. http://www.exploroz.com/Members/58567.500/6/2011/The_Quandong The_Blog.aspx.

Sedgley, M., 1982. Preliminary assessment of an orchard of Quandong seedling trees. *Journal Australian Institute of Agricultural Science* 48, 52–56.

Stewart, K., Percival, B., 1997. *Bush Foods of New South Wales – A Botanical Record and an Aboriginal Oral History*. Royal Botanic Gardens Sydney, Sydney, New South Wales, Australia.

Walker, R.R., 1989. Growth, photosynthesis and distribution of chloride, sodium and potassium ions in salt-affected quandong (*Santalum acuminatum*). *Australian Journal of Plant Physiology* 16, 365–377.

Williams, C., 2010. *Medicinal Plants in Australia*, Vol. 1: Bush Pharmacy. Rosenberg Publishing Pty Ltd., Dural, New South Wales, Australia.

CHAPTER 14

Cultivation of Riberry
(*Syzygium luehmannii*)

Rus Glover

CONTENTS

INTRODUCTION

Riberry is sometimes known as the small-leaved lilly pilly and the clove lilly pilly. Several species of lilly pilly grow in a variety of landscapes in Australia. They are attractive shrubs or trees and have ornamental value due to the shape of leaves, pink or light green colour of the leaf flush and general appearance. Fruits are edible and also serve as food for birds and other animals.

BOTANY

Riberry is a medium-sized to large tree which in its native habitat grows mainly in littoral and subtropical rainforest on the east coast of New South Wales (NSW) and southern Queensland (from Kempsey, NSW, to the Sunshine Coast, Queensland) in Australia. Some isolated populations have been found further north.

The small, glossy, lance-shaped leaves are pink/red when young. They are opposite, simple, entire, lanceolate to ovate, 4–5 cm long and drawn out to a long prominent drip tip. Their leaf stalks are 2–3 mm long.

The flowers form in November through to early December. They are in small panicles at the ends of branchlets, half the length of the leaves or less. The white or cream petals form in fours or fives, 1.5 mm long. The stamens are about 2–5 mm long (Figures 14.1 and 14.2).

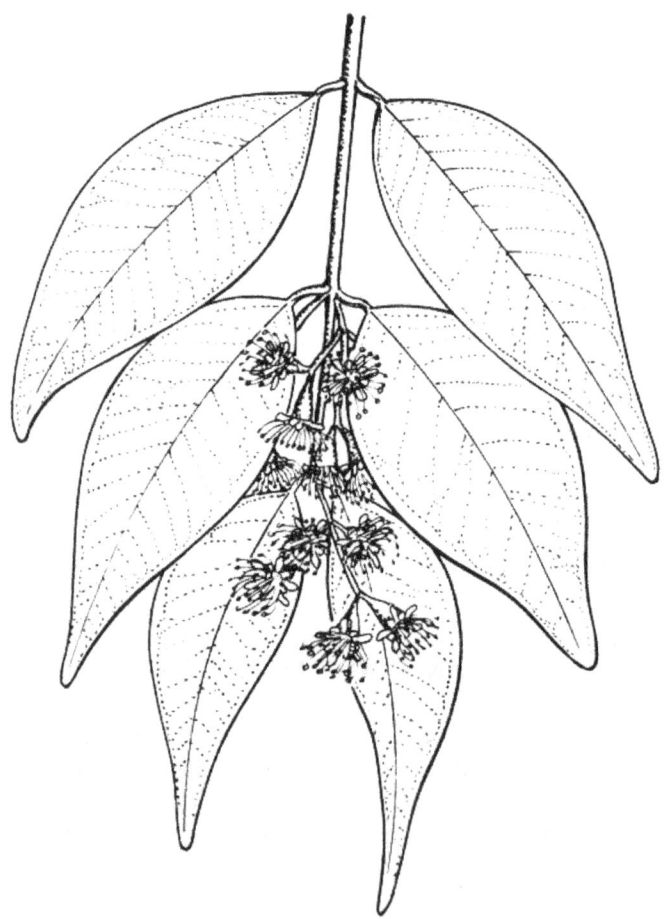

Figure 14.1 Botanical illustration of *Syzygium* (genus drawing) – flowering branch. (Illustrated by David Mackay. Flora of NSW, Copyright Royal Botanic Gardens and Domain Trust.)

Figure 14.2 **(See colour insert.)** Photo of a flowering riberry (*Syzygium luehmannii*) tree. (Photo by Rus Glover. Copyright Woolgoolga Rainforest Products.)

Figure 14.3 Botanical illustration of *Syzygium luehmannii* – fruits and leaves. (Illustrated by David Mackay. Flora of NSW, Copyright Royal Botanic Gardens and Domain Trust.)

The fruit is a pink to red berry, 6–15 mm long and 5–10 mm diameter with white flesh and matures from December to February. The fruit contains a single seed and some selections propagated vegetatively produce a large number of seedless fruits (Figures 14.3 and 14.4).

In the wild, trees occasionally reach 30 m in height and 90 cm in trunk diameter. The tree's crown is dense with small leaves, above a tall straight trunk. Large trees are buttressed at the base. The bark is red brown, light grey or pinkish grey with soft papery scales. In cultivation and grown by cuttings, the tree usually grows to 5–7 m and can be multi-trunked.

Figure 14.4 (See colour insert.) Photo of a riberry (*Syzygium luehmannii*) tree with fruits. (Photo by Rus Glover. Copyright Woolgoolga Rainforest Products.)

TRADITIONAL USE

Historic references from Victoria, NSW and Queensland state that aboriginal people in Australia regularly ate the fruits. The fruit has been reported to be one of the first fruits consumed as jam and cordials by early colonists of Australia. The Australian Botanical Garden in Sydney has reported that the lilly pilly was one of the first edible plants to be noted during Captain Cook's visit to Australia in 1770.

ECONOMIC POTENTIAL

The estimated annual production of riberry, as reported in 2012, was 4–5 t, with an estimated farm gate value of AU$100,000. As of 2013, about 70% of production was owned by a company that grows trees in subtropical properties in NSW and Queensland, which collectively total 6000 trees on 60 ha. The company was established to create a critical mass in terms of supply of fruit and to create economies of scale for the purchase of inputs. As of 2013, there were about five growers in the industry considered commercial producers: some of them participate in the riberry company previously mentioned. The balance of riberry production comes from smaller-scale producers (10–20), who value-add their fruit and distribute primarily through farmers' markets. Wild harvested fruit is a minor supply source and mainly a supplementary source of product when yields of cultivated fruit are low (Clarke, 2012).

High production costs in the past have dictated that high-value niche products be produced, but the potential now exists for riberry to enter mainstream food and beverage production with economies of scale and reduction in production costs to attract processors in using the fruit.

In addition to its potential as a fruit tree, riberry also has the potential as an ornamental, for use in domestic, bush revegetation, park, roadside and other situations. It is fast growing, has attractive foliage especially during leaf flushes and produces fruit that are eaten by birds.

AGROCLIMATIC REQUIREMENTS

Grown in a plantation situation, riberry will perform best when agroclimatic conditions are similar to its endemic range, that is, a subtropical climate. However, they can tolerate quite low temperatures in winter and mild frost, particularly after establishment. No damage has been shown to occur with temperatures as low as −2°C as long as they are not continuous. High summer temperatures, whilst flowering and particularly during fruiting, can be a problem.

Riberries grow well in both sandy and clay-based soils. The pH in natural conditions varies from 4.5 to 5.5. Plants grow well in soils with a pH of 5.5–6.5, which can allow for effective nutrient uptake when a fertilisation programme is used. Sandy soils benefit from a reasonably high level of organic matter, minimising the need for irrigation.

Clay soils need to be moderately drained (planted on a sloping field or a small mounding of rows, 250–500 mm high).

AGRONOMY

Plantation layout needs to take several variables into account: slope and field orientation. Northerly aspects and slopes maximise sunlight, especially during winter, particularly on the east coast of NSW/Queensland. Maximising sunlight should be a priority in any locality:

- Ideally, orientation of rows should be north/south to provide equal amount of sunlight to all plants all-year round. This is not always possible due to slope and/or land shape. A balance needs to be found between being able to effectively mow/slash and access between rows with machinery, minimise erosion and maximise sunlight.
- Inter-row spacing needs to allow for tractor access for slashing, tree maintenance and harvesting.

Monocropping riberries can increase problems particularly with scales and sooty moulds and consideration should be given to designing plantings that incorporate other species, whilst retaining ease of harvesting and maintenance.

As riberries flower and fruit during spring and summer, protection from prevailing winds during this period must be taken into account. Native windbreaks can add biodiversity value to the plantation area, reduce moisture requirements and provide habitat for insect predators and flower pollinators.

Currently, there is only limited information available on the water requirements of riberries. Some observations and factors that should be considered are as follows:

- Excessive irrigations, leading to extended periods of waterlogging, should be avoided. Riberries will, however, perform well in sandy soils even with extended periods of natural waterlogging.
- Riberries grown within their natural range will grow, flower and fruit without irrigation – plants still need watering during establishment. Some plantations are producing well with no permanent irrigation system.
- Extremely dry soil conditions should also be avoided.
- Within their range, plants normally experience a wet summer/autumn and dry winter/spring rainfall pattern.

During the fruiting period, adequate moisture is necessary, particularly during periods of high temperatures. Fruit can suffer sunburn that expresses itself as a white to clear shrunken portion on the fruit that extends all over the fruit and makes it unsaleable. Adequate moisture during these periods can limit damage.

As with water requirements, at the moment, little is known about the nutrient requirements of riberries. As experience with the crop accumulates, firmer guidelines and objective assessment techniques, such as leaf analyses, will come into common use. Until then, some general principles should be kept in mind when designing a fertiliser programme:

- Many Australian native plants are intolerant to high phosphorus levels, so a relatively low phosphorus fertiliser, suitable for natives, should be used.
- In the first season following establishment, the aim should be to maximise growth to develop a good plant structure and produce several flushes of next season's fruiting wood. Therefore, frequent but relatively small fertiliser applications, particularly of nitrogen, potassium, phosphorus and trace elements, are desirable.
- Once established, restricting nutrient availability may be necessary to avoid excessive vegetative growth in spring and stimulate flowering. Therefore, it may be advisable to avoid or restrict fertiliser applications in spring.
- After harvest, vegetative growth, which will form next year's fruiting wood, should be encouraged. A complete micronutrient foliar spray may be advisable in late winter/spring as flowers are being initiated.
- Experience has shown that trees benefit from applications of good-quality compost after fruiting has finished in February and in winter/early spring.

GENETIC IMPROVEMENT AND PROPAGATION

Several selections have been made from naturally occurring or seed-grown planted amenity trees. These have been made on various qualities such as 'seedlessness', size of fruit and flavour and propagated vegetatively to provide a known selection for commercial production.

There exists a need to carry out a scientific breeding programme to continue the selection process as different properties and associated products become important.

BIOTIC CONSTRAINTS

Riberries have several pests that can affect growth, fruiting and the fruit itself:

- Leaves and fruits can be attacked by *Monolepta australis*, the red-shouldered leaf beetle.
- Scale insects and associated sooty moulds can affect photosynthesis, bud formation, flowering and fruiting.
- Macadamia nut borer can attack fruits, particularly varieties with a lot of seed.
- Both grey-headed and black flying foxes (fruit bats) can feed on fruits and damage foliage.
- Some native birds, notably king parrots, can also feed on the ripe fruits.
- There are several so far unidentified beetles that can interrupt pollination through feeding on flower stamens.
- Myrtle rust has been shown to infect new growth on a very limited number of selections, but so far the effect of this has been very minimal.

Most pests and diseases can be controlled through organic methods, plantation physical protection and hygienic practices including maintaining soil health. Apart from the treatment for myrtle rust, there are currently no registered preparations for pest control in *Syzygium* spp. Integrated pest and disease management practices including designing plantations with windbreaks, refuges and corridors for beneficial insects and insectivorous birds will assist in buffering them against a majority of pest and disease problems.

HARVESTING, PROCESSING AND MARKETING

Harvesting in the past was done by hand, but this is only possible when trees are quite young. Several growers now harvest into specially made nets placed under the trees during the fruiting period. When ripe the fruit drops off the trees and is 'swept' from the nets daily (Figure 14.5).

Fruit needs to be cleaned, sized, sorted, washed in a sanitiser, packed and frozen immediately. Fruit will store for short periods of up to 2 weeks if placed in refrigerated cool rooms but needs to be frozen (−18°C to −24°C) to be kept for longer periods. Frozen fruit will remain in good condition for over 2 years.

Processors require the fruit to be either frozen whole, pureed through (usually) a brush finisher or in some cases dehydrated or freeze-dried. The state of the fruit reflects the type of product being produced.

Riberry fruit is basically not a fresh fruit (although many people enjoy the fruit fresh off the tree), and it is processed as compotes, salsas, chutneys, glazes, comfits, as an ingredient in fruit bars, yoghurt, cereals, fruit drinks, cordials, jams and conserves.

Marketing relies mainly on the taste of the fruit as it has a refreshingly tart, spicy flavour that has a hint of cloves and cinnamon. Fruit can vary in taste depending on plant selection. Good flavoured variants tend to have higher amounts of certain

Figure 14.5 Photo of riberry (*Syzygium luehmannii*) trees with specially made nets placed under them to harvest the fruits. (Photo by Rus Glover. Copyright Woolgoolga Rainforest Products.)

essential oils like myrcene (occurring in bay leaves), pinene (occurring in pine trees), ocimene (occurring in Brazilian cherries), limone (occurring in citrus, especially lemons) and phellandrene (occurring in ginger), among others. The range of essential oils present reflects the complexity of riberry's flavour. Other points that can be considered are the health benefits as it is high in antioxidants and folates and is very low in sugar. It is also identified as a native food/bush food and is produced in a clean and green production system.

YIELD

Yields increase as trees mature but can vary with variety and climate. Mature trees have been known to produce up to 70 kg per season. Cutting grown plants can produce fruit from the second year of growth and have been shown to produce

3–5 kg of marketable fruit per tree in the third year. Some selections take several years to fruit consistently. Yields in areas outside their natural range have not yet been quantified.

CONCLUSIONS AND RECOMMENDATIONS

Demand for riberry fruit has increased over the last couple of years as processors have developed products that are considered mainstream. Growers are streamlining production systems to reduce growing costs and can supply a product that is not regarded as expensive and only suitable for a high-value niche market.

Volume and price are still limiting factors for very large-scale processing, but plantations are increasing in size and new plantings are taking place. The industry is in a position where year-round supply can be guaranteed and several shipments have been exported.

Several areas of the production chain, from tree selection through to innovative product development, need to be more intensely researched.

REFERENCE

Clarke, M., 2012. *Australian Native Food Industry Stocktake*. Rural Industries Research and Development Corporation, RIRDC Publication No. 12/066, Canberra, Australian Capital Territory, Australia.

CHAPTER 15

Production of Wattle Seed
(*Acacia victoriae*)

Lyle Dudley

CONTENTS

INTRODUCTION

Acacias are key tree species in the Australian landscape and are found in all states. They are an important pioneer species and because of their nitrogen-fixing ability, they have been used in regeneration of degraded land. Seeds collected from over 120 species of *Acacia* have been used as food by indigenous communities over the centuries; some of the more common ones are *A. coriacea*, *A. colei*, *A. pycnantha* and *A. murrayana*.

BOTANY

Acacias belong to the subfamily Mimosoideae of the family Fabaceae. The habit of *Acacia victoriae* Benth. ranges from a medium-sized bush to a small tree, reaching 5 m in height. Several common names have been recorded, such as

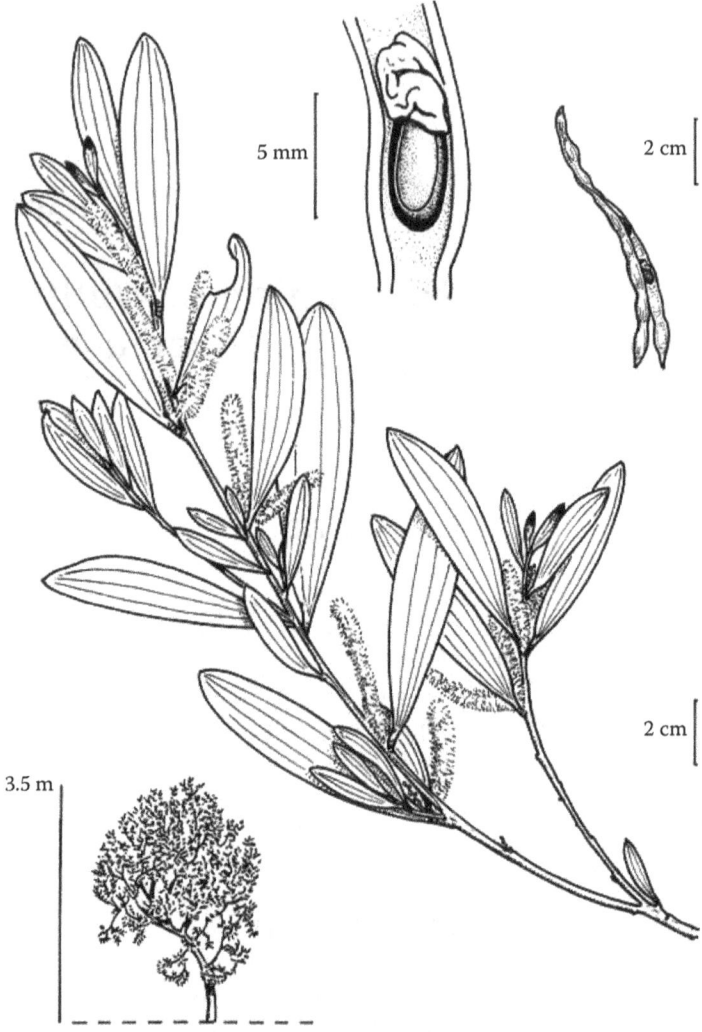

Figure 15.1 Botanical illustration of *Acacia longifolia* subsp. *Sophorae*. (Illustrated by David Mackay. Flora of NSW, Copyright Royal Botanic Gardens and Domain Trust.)

elegant wattle, gundabluie, prickly wattle and bramble wattle (Maslin et al. 1998; PlantNet, 2002). Flowering occurs in spring and continues through early summer. Flowers are yellow and fragrant and occur in clusters on the spiny branches. Pods are pale coloured and the small seeds (~0.5 cm in diameter) are brown in colour. The tree is deep rooted, enabling it to survive droughts. Studies have also been conducted to identify the variations within the species (Ariati et al., 2007) (Figures 15.1 through 15.3).

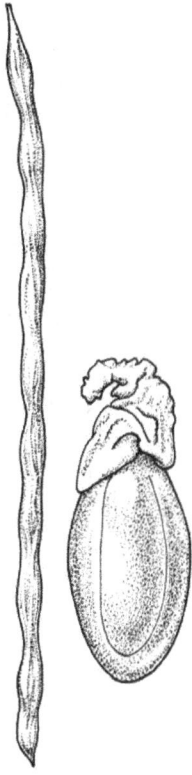

Figure 15.2 Botanical illustration of *Acacia longifolia* var. *longifolia* seed and pod. (Illustrated by Catherine Wardrop. Copyright Royal Botanic Gardens and Domain Trust.)

TRADITIONAL USE AND ECONOMIC POTENTIAL

Australian aboriginal people use wattle seed by first milling it into a flour, making a dough and then placing it on open coals to make a kind of bread. As a member of the pea family, the flour has a pea flavour. Immature green seed are also used by cooking lightly over coals.

In the early 1980s, work was done to add value to seed by roasting, which resulted in a hazelnut/coffee flavour (Forbes-Smith and Paton, 2002). Since then, the product has been used across the food industry in bakery products and beverages. The seed is also used in the cosmetic and pharmaceutical industries. Seedpods have been reported to have antitumour properties (McDonald et al., 2006) and have been used in research to treat cancers at the Anderson Cancer Centre in Phoenix, Arizona, United States.

The tree has also been identified as being suitable for rehabilitating saline environments (Marcar and Crawford, 2004). The leaves are also used as fodder,

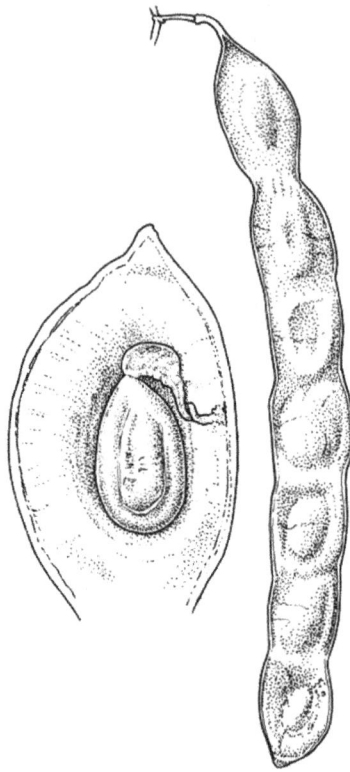

Figure 15.3 Botanical illustration of *Acacia fimbriata* seed and pod. (Illustrated by Lesley Elkan. Copyright Royal Botanic Gardens and Domain Trust.)

especially in times of drought. Attempts to grow the species in other regions such as the arid areas of the Middle East have also been successful.

NUTRITIONAL STATUS

Konczak et al. (2009) reported the nutritional status of wattle seed. Selenium which is required for the antioxidant enzymes to function was found in wattle seed (31.7 μg/100 g, dry weight), and this may be promoted as source of this essential mineral. According to this report, it is also a good source of vitamin E and other minerals including zinc, magnesium, calcium, iron, phosphorus and potassium. Selection for nutritional value can further improve this trait and enhance its value to bush communities.

TREE ESTABLISHMENT AND MANAGEMENT

Plants are established with seed. Seeds are collected in early summer and germinated, to be planted out as tube stock in the following autumn. Seed have to be treated with boiling water and left overnight. Seed are then planted into a native soil mix and then covered by a thin layer of fine gravel (Bonney, 1994). After germination, leave only one healthy plant per tube and harden seedlings in direct sun. Planting out can be done in autumn. Deep ripping of planting sites in the previous autumn helps autumn/winter rains to soak into the soil. Direct seeding is also an option, and fire, chemicals or mechanical disturbance of the soil can be used to improve survival.

Thrall et al. (2005) and his team at the Centre for Plant Biodiversity Research (CSIRO) have demonstrated that inoculating wattle seed with the symbiont soil bacterium *Bradyrhizobium* increases survival and growth rates. *Bradyrhizobium* is present in natural wattle-growing soils but is usually absent in farmland or in lands under reseeding or revegetation programmes. Work by the Victoria Department of Primary Industries and other collaborators including the CSIRO has shown that wattle seed inoculated with *Bradyrhizobium* had establishment rates 2–5 times better than untreated seed. Glasshouse trials have also demonstrated similar advantages to treated seed.

A. victoriae will start flowering after 2–4 years but has a relatively short lifespan of 10–15 years; therefore, for commercial plantations, replanting must be done in a planned manner to avoid 'off' years. Plants subject to water stress prior to flowering but followed by irrigation (usually drip) tend to flower and seed better, a practice which can be used to induce flowering. Pruning to shape trees is beneficial, as it results in a tree architecture that facilitates harvesting (see section 'Genetic Improvement' for ideal tree type). Fertilisers can also be used through the irrigation system to improve yields in plantations.

Seed in pod is harvested in late spring through early autumn, depending on region, season and altitude. Pods must be fully mature and dry with a hard seed before collection. Trees are usually shaken or hit with sticks to release pods, which are collected on tarpaulin or shade net spread under each tree. Hand harvesting can involve 'finger stripping' or hand removal of pods or removal of all the above-ground biomass followed by separation of seeds and pods. There has been investment in various harvesting and seed-cleaning machines, such as mechanical shakers, vacuum harvesters, hand sieves, gravity tables and handheld olive harvesters. Pods do not shatter uniformly and therefore, mechanical threshing is necessary to extract seed.

Commercial plantations were established in Riverland, south Australia, western districts of Victoria and other areas like Junee in New South Wales. Marketing has generally been done through produce companies. In the early days, Emu Bottom Biscuits with ANZAC Wattle Seed Biscuits were served on Qantas flights. Later, Dick Smith's 'Bush Foods Breakfast Cereal' included wattle seed. The number of products incorporating wattle seed as an ingredient has increased in recent years, following interest in native foods because of health benefits. Simpson and Chudleigh (2001) reported an economic analysis of a plantation based on 625 trees/ha and a yield of 1.25 t/ha/year and concluded that the crop would be profitable only if production costs could be kept around $200 (AUD) (Figures 15.4 and 15.5).

Figure 15.4   (See colour insert.) Photo of a wattle seed tree. (Photo by Yasmina Sultanbawa. Copyright University of Queensland.)

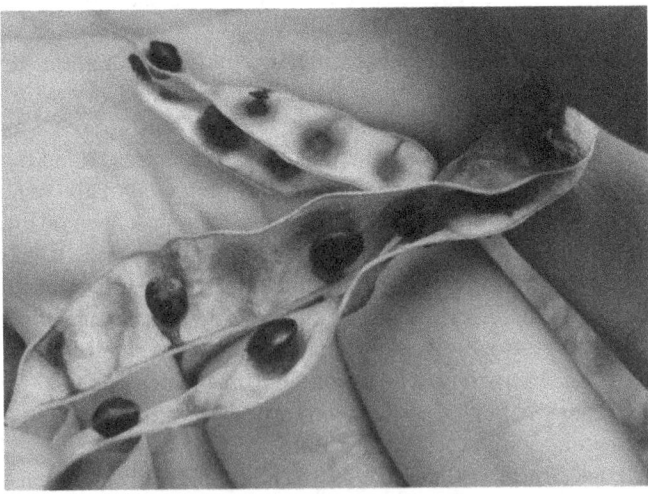

Figure 15.5   (See colour insert.) Photo of wattle seedpods and seeds. (Photo by Yasmina Sultanbawa. Copyright University of Queensland.)

GENETIC IMPROVEMENT

Plant selection based on some of the characteristics like yield, salinity tolerance, resistance to serious insect pests, high protein or other beneficial properties in the seed requires research and development. Currently, because most seeds are obtained from wild sources, the emphasis on the requirements for sustainable plantations has not received the required investment. Olsen (2002) suggested that plant selection should focus on short, compact and erect trees which bear their seeds near the top of the plant, have a high ratio of seed production to vegetative growth and are suited to growing at high density, to facilitate mechanical harvest, because the cost of harvest is the key limiting factor for profitability.

BIOTIC CONSTRAINTS

Acacia spp. have insects that attack leaves (e.g. fireblight beetle, *Acacicola orphana*) and stems (Wattle Goat moth, *Xyleutes eucalypti*). Psyllids have emerged as serious pests in recent years. Aphids and sap suckers are minor pests. A more serious practical problem in wattle plantations is the presence of caterpillars (e.g. bag-shelter moths, *Ochrogaster lunifer*). Care must be taken to avoid contact with these caterpillars as severe itching can result.

CONCLUSIONS AND RECOMMENDATIONS

As a collector from the wild, I believe that there is sufficient opportunity for investment in large-scale plantations. This would especially be true if the potential as an anticancer treatment materialises. There is also the current growing interest in native foods as an ingredient with health and nutrition benefits. Small-scale planta-tions, as can be found today, often find it difficult to compete with wild harvest in 'tough' years such as a drought. Mechanisation and economies of scale must be developed and can only be achieved sustainably with large-scale plantations. The importance of genetic improvement of planting material was mentioned earlier and remains a critical area needing investment.

The industry has had its share of fly-in investors who have set up plant nurseries to encourage small business investment and promised to buy back product; however, the realities of growing trees and producing seed in a tough growing environment such as the Australian bush along with recent global economic uncertainties have taken its toll.

REFERENCES

Ariati, S.R., Murphy, D.J., Gardner, S., Ladiges, P.Y., 2007. Morphological and genetic variation within the widespread species *Acacia victoriae* (Mimosaceae). *Australian Systematic Botany* 20, 61.

Bonney, N., 1994. *What Seed Is That? A Field Guide to Identification, Collection and Germination of Native Seed in South Australia.* Neville Bonney, Adelaide, South Australia, Australia, pp. 324.

Forbes-Smith, M., Paton, J.E., 2002. Innovative products from Australian native foods: A report for the Rural Industries and Development Corporation. Rural Industries Research and Development Corporation, RIRDC Publication No. 02/109, Canberra, Australian Capital Territory, Australia.

Konczak, I., Zabaras, D., Dunstan, M., Aguas, P., Roulfe, P., Pavan, A., 2009. *Health Benefits of Australian Native Foods: An Evaluation of Health-Enhancing Compounds.* Rural Industries Research and Development Corporation, RIRDC Publication No. 09/133, Canberra, Australian Capital Territory, Australia.

Marcar, N.E., Crawford, D.F., 2004. *Trees for Saline Landscapes.* Rural Industries Research and Development Corporation, RIRDC Publication No. 03/108, Canberra, Australian Capital Territory, Australia.

Maslin, B.R., Thomson, L.A.J., McDonald, M.W., Hamilton-Brown, S., 1998. *Edible Wattle Seeds of Southern Australia.* CSIRO Publishing, Collingwood, Victoria, Australia, pp. 108.

McDonald, J.K., Caffin, N.A., Sommano, S., Cocksedge, R., 2006. *The Effect of Post Harvest Handling on Selected Native Food Plants.* Rural Industries Research and Development Corporation, RIRDC Publication No. 06/21, Canberra, Australian Capital Territory, Australia.

Olsen, G., 2002. Broadscale production of wattle seed to address salinity. *Conservation Science Western Australia Journal* 4, 185–191.

PlantNet, 2002. *Acacia victoriae.* http://plantnet.rbgsyd.nsw.gov.au.

Simpson, S., Chudleigh, P., 2001. *Wattle Seed Production in Low Rainfall Areas.* Rural Industries Research and Development Corporation, RIRDC Publication No. 01/08, Canberra, Australian Capital Territory, Australia.

Thrall, P.H., Millsom, D., Jeavons, A.C., Waayers, M., Harvey, G.R., Bagnall, D.J., Brockwell, J., 2005. Studies on land restoration: Seed inoculation with effective root-nodule bacteria enhances the establishment, survival and growth of Acacia species. *Journal of Applied Ecology* 42, 740–751.

PART II

Food and Health Applications

CHAPTER 16

Alternative Medicines Based on Aboriginal Traditional Knowledge and Culture

Donna Savigni

CONTENTS

INTRODUCTION

Traditional, complementary and alternative medicine (T/CAM) is a major industry in Australia and other Western nations, and there is an escalating push from the community and healthcare professionals for more education and research into traditional medicines (Fennell et al., 2009; Patwardhan and Patwardhan, 2006). Melnick (2006) pointed out the irony of labelling these therapies 'alternative' given their rich historical traditions that have formed the basis of many modern therapies. It should be remembered that botany and medicine were closely allied until the eighteenth century when they were separated by advances in science (Dufault et al., 2001).

There is a growing demand for herbals with therapeutic value that are available without prescription (McClatchey and Steven, 2001). Nearly half the population in many industrialised nations regularly use T/CAM already, and the proportion is as high as 80% in developing countries (Bodeker and Kronenberg, 2002; Melnick, 2006). Although some people are undoubtedly becoming dissatisfied with the limitations of orthodox medicine, this increasing demand for T/CAMs largely reflects the changing beliefs, needs and values of today's society (Parris and Smith, 2003). Indeed, Astin (1998) found that the majority of T/CAM users feel these alternatives more closely align with their own values, beliefs and philosophical orientations towards health and life. Moreover, many people have deep concerns about the detrimental side effects and toxicity of potent conventional medicines and have developed a respect for knowledge based on centuries of herbal use. However, the medical profession has been slow to embrace these therapies, with critics saying there is not enough good science to justify the use of herbs and that regulations guaranteeing that herbs can fulfil their claims are not strict enough (Carson et al., 2006; Dufault et al., 2001). Orthodox medicine requires evidence of efficacy before pharmaceuticals can be developed, endorsed and supplied to the public. This means rigorous basic and clinical research is absolutely necessary. The effects of potential treatments need to be thoroughly investigated through placebo-controlled, randomised, scientifically valid experiments, and the subsequent data published in peer-reviewed journals (Parris and Smith, 2003; Trachtenberg, 2002).

Recently, there has been an increased appreciation of the value of ethnomedical knowledge as a guide to identifying new therapeutic agents, and a heightened interest in their modes of action (Heinrich and Bremner, 2006; Kong et al., 2003; Melnick, 2006). Over half of all drugs have been derived wholly or partly from plants (Cragg et al., 1997; Melnick, 2006; Newman and Cragg, 2007). Furthermore, 80% of these compounds have had an ethnomedical use identical or related to the current use of the active elements of the plant (Fabricant and Farnsworth, 2001). Despite this impressive statistic and the fact that Australian Aboriginal people maintain the oldest culture on Earth, very little research has been undertaken to evaluate the therapeutic potential of any plants in the Aboriginal pharmacopoeia (Locher and Currie, 2010).

TRADITIONAL ABORIGINAL MEDICINE

Traditional medical practice within Aboriginal Australia encompasses a holistic worldview which perceives good health as a complex system involving interconnectedness with the land, awareness of spirit and ancestry, and social, mental, physical and emotional well-being of both the individual and the community. Indigenous Australians traditionally believe that ill health is caused by either a natural physical cause, a spirit causing harm, or sorcery (Oliver, 2013). Notwithstanding the cultural diversity between different Aboriginal groups, communities and individuals, the socio-medical system of health beliefs held by Aboriginal people in general emphasises social and spiritual dysfunction causing illness and many of the Aboriginal medical belief systems have similarities despite being from different parts of the country (Maher, 1999). European colonisation and the subsequent displacement and disconnection of people from their traditional lands and families have significantly impacted traditional practices, including the use of traditional medicine (Oliver, 2013). However, despite the lifestyles of Australian Aboriginal people undergoing immense changes and stresses in that time, they have been somewhat successful in maintaining much of their culture in the face of periods of cultural suppression. With special reference to healthcare, Aboriginal people have passively resisted many Western beliefs and practices and the 'traditional' beliefs of Aboriginal people have adapted to the changing circumstances in which they live. So while Aboriginal viewpoints of treatment may have changed since colonisation, there has been less change in beliefs regarding the causes of illness (Maher, 1999).

In traditional Aboriginal societies, any condition thought to have been caused by supernatural intervention was dealt with by traditional healers. Traditional healers were usually males who have special capabilities and who can employ various techniques such as physical contact, counselling, healing songs and chants to treat patients. Specific symptoms were generally treated by bush medicines, which included herbal preparations, diet, rest, massage and external remedies such as ochre, smoke, steam and heat. A 1994 survey revealed that, in most regions of the Northern Territory, nearly a quarter of Aboriginal people had used bush medicine during the previous 6-month period, and that its decreased use was likely because Western medicine was easier to access, not because of a lack of faith in the efficacy of bush medicine (Maher, 1999).

Aboriginal people living in traditional communities had an encyclopaedic knowledge of Australian plants and animals and of the seasonal changes of their environment. The Aboriginal awareness of life cycles of animals and plants was gathered over thousands of years and was understood by everyone in the community. Despite not being written down, it was shared from generation to generation through example, song and dance so that every child learnt the significance of the natural signs. This traditional knowledge was specific for each local area and covered not only food resources, but extended to the medical uses of various plants. At one stage, all members of the family knew the names, location, structure, value and application of medicinal plants (ACNTA, 1988; Isaacs, 1987; Pearn, 2005). Isaacs (1987) suggested that the greatest repository of centuries' worth of botanical knowledge and experience lies with Aboriginal women rather than men. Theoretically, everyone in a community knew the names and locations of all useful plants, but gender roles were well defined in traditional communities and it was the women who were the gatherers of plant foods and herbal medicines while the men were preoccupied with preparing for the hunt. While men knew all the plants, it was the women who knew the finer details of their respective uses (Lassak and McCarthy, 2001; Reid and Betts, 1979).

Even today, because of their close relationship with the land, many Aboriginal people have a broad and intimate ecological knowledge and still practise medicinal ethnobotany (Pearn, 2005). Interestingly, nomadic tribes confine their use of bush medicines to plants within their respective areas and do not necessarily recognise plants outside tribal boundaries as being of medicinal value, such that the same plant growing in different areas is frequently used therapeutically in completely different ways by different tribal groups (ACNTA, 1988). A characteristic feature of Australian ethnobotany is that a given plant may be used non-specifically for various purposes, including a range of medicinal preparations from different plant parts, as a food source or to make tools or adornments (Pearn, 2005). In general, Australian medical ethnobotany is based on preventative medicine and is characterised by the 'broad-spectrum' use of multiple plant species to treat symptoms rather than the causes of diseases (Pearn, 2005).

TRADITIONAL PLANT-BASED REMEDIES

At the time of colonisation, Aboriginal people were relatively healthier than most Europeans (Cribb and Cribb, 1981; Lassak and McCarthy, 2001; Low, 1990; Ngaanyatjarra et al., 2013). Apart from living in a land largely free of disease and benefiting from a better diet, more exercise, less stress, a more supportive society and more harmonious worldview (Low, 1990), this was likely due to their sophisticated system of healthcare based on healing methods honed over tens of thousands of years (Ngaanyatjarra et al., 2013). It has been suggested that, apart from colds and digestive upsets, and malaria in Northern Australia, infectious diseases were relatively unknown until the arrival of the first white settlers (Lassak and McCarthy, 2001). Nevertheless, Aboriginal people often needed bush medicines to treat commonplace ailments (Low, 1990). As in most parts of the world, plants played a major role in healthcare. The majority of plants were used as a 'first-aid' response to wounds, burns, bites and stings where ease of access was a priority (Packer et al., 2012). Thus, plants that required minimal preparation and were readily accessible endure as the most prevalent bush medicines (Packer et al., 2012). Most plant preparations were obtained from leaves, bark, flowers, seeds, pods and/or roots (Locher and Currie, 2010). Traditional herbal remedies were mostly based on topical treatments such as washes, rubs or ointments. In addition, the use of 'smoke treatment' as a healing agent was widespread. A fire was lit within a small pit, branches from the medicinal plant were placed on the hot coals and the sick person was placed on them and bathed by, and inhaling, the fumes produced (Ghisalberti, 1994). Before European contact and the introduction of the billycan, Aboriginal people had no way of boiling water. Therefore, most traditional medicines were not 'decoctions' in the true sense (Ghisalberti, 1994). Instead, infusions of fresh plant material that had been chewed, mashed or ground up first and steeped in water in wooden bowls were very commonly used as body washes ('bogeys'), or else, in restricted cases, drunk as a tea (Ghisalberti, 1994; Reid and Betts, 1979). Alternatively, the pounded plant material was made into a paste and mixed with animal fat to make an ointment which was smeared over the body (Ghisalberti, 1994). Relatively few medicinal plants were taken internally as Aboriginal people have always retained an extensive and detailed knowledge of botanical toxicology (Pearn, 2005).

Traditional Aboriginal remedies varied between clans. Just as there was no single Aboriginal language, there was also no single Aboriginal pharmacopoeia (Low, 1990). However, in general, most herbal remedies used reflect the types of ailments that Aboriginal people experienced prior to contact with Western society. Chiefly, these included headache, coughs and colds, wounds, sores, skin disorders, diarrhoea and digestive complaints, aches and pains, rheumatism, bites and stings, fevers, sore eyes, toothache and earache (Low, 1990). In addition to the remedies discussed herein, native plants have also been used by Aboriginal people to treat many other conditions, including diabetes, piles, female menstrual complaints and as lactation stimulants, contraceptives and abortifacients (Clarke, 2007; Lassak and McCarthy, 2001).

It must be noted that there was a lot of crossover. Many plants had different uses for different Aboriginal groups in different regions of Australia. Aboriginal people lived as separate populations in widely varied geographical areas of Australia with different botanical profiles. Thus, these different groupings developed different ethnopharmacologies, dependent on the plants available and the environmental stresses of the area (Cock, 2011). Further, many plants possess a range of medicinal properties and were used to treat numerous ailments, even within a single community. The same plant was often used to treat many conditions, including symptoms that may be part of very different disease states. For example, a plant may be traditionally used to treat headache, but headaches may be due to stress, tiredness, migraine attack or a brain tumour, so it was the symptom, not the possible underlying biological cause, that was treated. Few plants were used for very specific purposes and this is thought to be related to the traditional nomadic lifestyle of Aboriginal people. That is to say, when on the move, it was better to concentrate on a range of plants with a broad spectrum of uses as the specific plant required for a particular ailment may be a long way away just when it was needed (Latz, 1995). Moreover, the season of the year could restrict the availability of useful plants as the effectiveness of traditional medicines was known to depend on using the appropriate plant parts or plants at a particular stage of maturation (Lassak and McCarthy, 2001; Latz, 1995).

EXAMPLES OF PLANTS USED AS ABORIGINAL
TRADITIONAL MEDICINES

This summary chapter is not, nor can it be, an exhaustive catalogue of native plants used by Aboriginal people as traditional medicines. Rather, the following highlighted examples are intended to showcase the vast range of plant-based medicines traditionally used by Aboriginal people to treat common ailments. The selected examples have had some level of scientific evaluation and their traditional use been validated in controlled laboratory tests. For a more comprehensive list of traditional Aboriginal plant-based remedies, the reader is directed to books such as those by Isaacs (1987), Low (1990), Lassak and McCarthy (2001) and Clarke (2008) and more local works like those of Smith (1991), Young (2007), Packer et al. (2012) and the Aboriginal communities of the Northern Territory of Australia (ACNTA, 1988). Most examples given are from tropical or desert regions as comparatively less is known about traditional medicinal plants from southern regions due to knowledge loss within indigenous communities through the impact of European colonisation (Clarke, 2007).

Headache

As its vernacular name suggests, 'Headache vine' (*Clematis glycinoides*) was used to relieve headaches by both Aboriginal people and bushmen (Lassak and McCarthy, 2001; Williams, 2013). A reportedly dramatic remedy, the aromatic leaves of this rainforest species were crushed and the fumes inhaled, resulting in an incredibly painful but ultimately effective reaction. The uncomfortable sensation of breathing in the ammonia-like fumes has been described as 'the head exploding, the eyes watering and intense irritation of the nasal passages' such that the initial headache was quickly forgotten (Williams, 2013). Given the scale of discomfort due to violent side effects, this bush remedy was possibly used only as a last resort. Moreover, sometimes headache vine was totally ineffective, seemingly due to variability in the active constituents. Indeed, Everist, who experienced both the irritant and analgesic effects of *C. glycinoides* firsthand, found that only young leaves or mature leaves with vigorous sappy growth were effective (Everist, 1974). The 'Small clematis' (*Clematis microphylla*) has a wider distribution than *C. glycinoides* and shares its irritant properties. Its practical use as an anti-inflammatory was recorded in the *Australian Medical Journal* in 1931 (Cleland, 1931). The irritant effects of headache vine are shared by other members of the *Clematis* genus, with many species having similar reputations worldwide. Various other indigenous cultures have employed *Clematis* species as a diuretic and to treat many different ailments, including decongestion, sexually transmitted infections and skin diseases and ulcers (Grieve, 1931; Hao et al., 2013). In recent years, the anti-inflammatory properties of *C. glycinoides* and *C. microphylla* have been partly explained by the discovery that ethanol extracts of leaves of both species inhibit the activities of cyclooxygenase-1 (COX-1) and 5-lipoxygenase, enzymes involved in the inflammatory response (Li et al., 2003, 2006). The toxin protoanemonin, which causes itching, rashes and blistering on contact with the skin or mucosa, has been isolated from almost all *Clematis* species, including *C. glycinoides*, accounting for its irritant properties (Chawla et al., 2012; Southwell and Tucker, 1993).

Another Australian tropical rainforest plant traditionally used by Aboriginal people to treat headaches is 'Snakevine' (*Tinospora smilacina*) (Clarke, 2007; Dobson, 2007). Endemic to Northern Australia, this climber also grows in beach forest, monsoon forest and gallery forest as well as open woodland. Snakevine has a milky latex sap with a cooling property and headaches were reportedly alleviated by wrapping the mashed stem around the head (Clarke, 2007, 2008; Smith, 1991). *T. smilacina* was also wound around affected limbs to ease the pain of leprosy, stonefish wounds, rheumatism and other inflammatory-related ailments (Lassak and McCarthy, 2001). The anti-inflammatory properties of *T. smilacina* have been confirmed in laboratory tests: methanol extracts of the stem and leaves displayed antiplatelet activity (Rogers et al., 2001). In related studies, ethanol extracts of the stem were more effective than aspirin at inhibiting the activity of the pro-inflammatory enzyme COX-1 (Li et al., 2003). Interestingly, Snakevine was also reportedly used to counter snake venom, and it has been suggested that its effectiveness in doing so is attributable to the presence of columbin (Hungerford et al., 1998).

Plants of the genus *Eremophila* have played a very important role in Australian Aboriginal traditional medicine, especially in desert regions (Ghisalberti, 1994). Commonly known as 'Fuschia bush', 'Emu bush', 'Poverty bush' or 'Kerosene bush' (among others), different species of *Eremophila* were widely employed to treat a range of medical conditions, including headache (Sadgrove et al., 2013; Singab et al., 2013). For example, the Barkindji people of the Darling River district in Western New South Wales as well as Aboriginal people in the Northern Territory treated headaches by tying a small bunch of heated *E. bignoniiflora* leaves and/or branches to their forehead (Clarke, 2007; Sadgrove et al., 2013; Smith, 1991). In other areas, headaches were alleviated by drinking aqueous decoctions of the leaves of *Eremophila* species, including *E. gilesii*, *E. freelingii* and *E. longiflora* (Ghisalberti, 1994; Singab et al., 2013; Smith, 1991). The headache-relieving action of these four species is supported by the detection of bioactive compounds that inhibit platelet activity and modulate serotonin release in extracts of leaves of these plants (Grice et al., 2003; Rogers et al., 2000, 2001).

Coughs and Colds

The large number and variety of bush medicines Aboriginal people traditionally used to relieve the unpleasant symptoms of the common cold, including a sore throat, cough, congested nose, mild fever and ill-defined body pains, suggest that this ailment was always as widespread in Australia as everywhere else in the world (Cock, 2011; Lassak and McCarthy, 2001; Packer et al., 2012). Aromatic plants feature heavily among those used to treat colds and clear the head and nasal passages. With some plants, including several species of *Melaleuca* (*M. symphyocarpa*, *M. cajuputi*, *M. hypericifolia*) and *Centipeda* (*C. cunninghamii*, *C. minima*), *Mentha australis* and *Prostanthera cineolifera*, leaves were simply crushed in the hands and sniffed or rubbed on the nose or wound around the head (Lassak and McCarthy, 2001). At other times, leaves of these and other *Melaleuca* species (*M. uncinata, M. quinquenervia, M. alternifolia, M. cajuputi*) or of *Eremophila* species (*E. cuneifolia, E. fraseri, E. longifolia, E. freelingii*) were boiled and the steam inhaled or the decoction drunk to alleviate symptoms (ACNTA, 1988; Lassak and McCarthy, 2001; Carson et al., 2006). *E. freelingii* could also be used as a pillow for a sick head or placed into the perforation of the nasal septum (Lassak and McCarthy, 2001). The medicinal properties of some of these plant species are probably directly related to their cineole-rich leaf oils (Cock, 2011). For instance, 1,8-cineole ('eucalyptol'), a common volatile component of numerous essential oils of *Melaleuca, Eucalyptus* and *Prostanthera*, is known to reduce the swelling of mucous membranes and loosen phlegm so as to make breathing easier. In fact, it is so effective that it has become a standard additive to cough mixtures and elixirs, inhalants and cough lozenges the world over (Curir et al., 1995; Lassak and McCarthy, 2001). Mainly due to its high 1,8-cineole content, eucalyptus oil is arguably Australia's most significant and best known contribution to medicine. The antibacterial activity of oils from many eucalypt species has been confirmed in numerous recent studies (Cock, 2011; Marzoug et al., 2011; Mulyaningsih et al., 2011; Wilkinson and Cavanagh, 2005). Similarly, studies have confirmed the antimicrobial, antiviral and anti-inflammatory effects of 'tea tree' oil from *M. alternifolia*, supporting its traditional and commercial uses (Carson et al., 2006). The effectiveness of bark and leaf decoctions of some eucalypt and *Acacia* species used to treat sore throats and coughs is also likely due to their astringent properties. Tannins present cause proteins to be precipitated, offering relief by way of stemming the secretion of excess fluid from mucous membranes in the throat (Lassak and McCarthy, 2001). Additionally, the antimicrobial properties of tannins are well documented, with some studies suggesting a possible link between relative astringency and antimicrobial activity (Locher and Currie, 2010; Scalbert, 1991).

Aboriginal people living in the Northern Territory still eat a whole soft, ripe fruit of *Morinda citrifolia* when they are affected by a sore throat, bad cough or cold (ACNTA, 1988; Low, 1990; Smith, 1991). Other records describe different plant parts used for a variety of internal and external remedies for a range of complaints, including the use of a rootbark infusion as an antiseptic (ACNTA, 1988; Lassak and McCarthy, 2001). Apart from containing vitamin C, the traditional use of this plant to resist respiratory tract infections has been validated by the finding that the fruits contain proven antibacterial agents, including acubin, L-asperuloside and alizarin (Wang et al., 2002). Moreover, the fruits possess analgesic and tranquilising activities and aqueous extracts of dried roots exhibited significant, dose-related, central analgesic activity in mice (Wang et al., 2002; Younos et al., 1990). In fact, the analgesic efficacy of the root extracts was 75% as potent as morphine, but not addictive and displayed no side effects (Wang et al., 2002; Younos et al., 1990). Anti-inflammatory activity has also been established in fruit powder (Li et al., 2003).

Aboriginal people from Ltyentye Purte (near Alice Springs) traditionally boiled fresh leaves of 'Pentye pentye' (*Pterocaulon sphacelatum*) and inhaled the vapours to relieve colds and influenza (Smith, 1991). Further north, older community members at Elliot know this plant as 'Manyanyi' (Smith, 1991). Here, fresh leaves are crushed in the hand and the vapours inhaled to clear a blocked sinus and to provide relief from head colds and influenza. To prolong the effect, a small plug of crushed leaves may be inserted into the nasal cavity and left there (Smith, 1991). *P. sphacelatum* is commonly called 'Fruit salad plant' or 'Applebush', because the crushed leaves smell sweetly of fruit salad, or Granny Smith apples (Low, 1990). Aromatic oils are known to stimulate cells lining the throat to secrete more lubricating fluid, easing the irritation that causes coughing (Low, 1990). Aboriginal people actually found the stronger smelling, related ragwort, *P. serrulatum*, to be a superior medicinal plant to *P. sphacelatum* and used it in very similar ways to treat colds and flu (ACNTA, 1988; Dobson, 2007; Low, 1990). However, there have been more scientific studies examining the bioactivity of *P. sphacelatum*. For example, forty Australian medicinal plants (not including *P. serrulatum*) were screened for antiviral activity and the crude ethanolic extracts of the green aerial parts of *P. sphacelatum* were the most active against poliovirus type 1 (Semple et al., 1998). This non-enveloped, single-stranded RNA virus belongs to the viral family responsible for respiratory and central nervous system infections in humans and *P. sphacelatum* inhibited the production of the cytopathic effect caused by this virus invading host cells by more than 75% (Semple et al., 1998). Antiviral activity–guided fractionation of the extract of *P. sphacelatum* has since yielded the antiviral flavonoid chrysosplenol C, which belongs to a group of compounds known to potently and specifically inhibit the replication of rhinoviruses, the most frequent cause of the common cold (Semple et al., 1999). Besides cold and flu, decoctions of leaves and twigs of *P. sphacelatum* were also used to treat other respiratory, skin and eye infections (ACNTA, 1988; Lassak and McCarthy, 2001; Smith, 1991). However, these traditional uses have not yet been completely scientifically validated as no antibacterial activity was detected against eight common bacteria (Palombo and Semple, 2001), but there was some effect against a fast-growing *Mycobacterium* in a related study searching for new agents to combat tuberculosis and other diseases caused by mycobacteria (Meilak and Palombo, 2008). Several bioactive compounds, including 4,5-epoxy-13-hydroxy-β-caryophyllene, 14-hydroxy-β-caryophyllene, coumarins and a flavanone, have been isolated from *P. serrulatum* (Macleod and Rasmussen, 1999).

Sores, Wounds and Skin Disorders

Due to their wide distribution, *Eremophila* species were well known to, and highly valued by, Aboriginal people. It has been estimated that of the 70 odd plant species used by the central Australian Aboriginal people for medicinal purposes, about one-third were either *Acacia* or *Eremophila* species (Ghisalberti, 1994; Singab et al., 2013). For example, leaves of *Eremophila bignoniiflora* were boiled and used as a medicinal wash to treat skin disorders, as well as colds, fever and headaches (Sadgrove et al., 2013; Smith, 1991). Essential oils extracted from leaves demonstrated moderate to high activity against Gram-positive microorganisms, but mostly only low activity against tested Gram-negative bacteria (Sadgrove et al., 2013).

Since Gram-positive bacteria are normally associated with skin conditions, this is consistent with known traditional applications (Sadgrove et al., 2013). Results indicated that antimicrobial activity was related to the less abundant, more polar essential oil constituents and not the major components fenchyl, bornyl or nerol acetate (Sadgrove et al., 2013). However, given that essential oils or volatile monoterpenoid compounds present in the leaf material would be removed or substantially reduced if *E. bignoniiflora* topical applications were prepared as in contemporary descriptions, it is unlikely that essential oils would account for any antimicrobial effects (Sadgrove et al., 2013); that is, unless decoctions were, in fact, unboiled infusions traditionally used as a poultice mixed with animal fat, as suggested by Ghisalberti (1994). In that case, leaf essential oils are likely to be dissolved and would be transported onto the skin following topical applications (Sadgrove et al., 2013). Decoctions of other *Eremophila* species, including 'Kerosene bush' (*E. sturtii*) and 'Kangaroo bush' (*E. duttonii*), were likewise traditionally used as antiseptic washes for infected cuts, open sores and painful ears, or *E. duttonii* could be gargled for initial treatment of sore throats (ACNTA, 1988; Smith, 1991). In a screening study of 39 traditional Australian medicinal plants, *E. duttonii* displayed the strongest antibacterial activity, with Gram-positive bacterial growth inhibited within 1–2 hours in some cases (Palombo and Semple, 2001). Two serrulatane diterpenes were identified as the compounds principally responsible for the observed antibacterial activity of *E. duttonii* in tested extracts (Ghisalberti, 1994; Smith et al., 2007). Similarly, two other serrulatane diterpenes isolated from *E. sturtii* inhibit the inflammatory enzymes COX-1 and COX-2 and exhibited bactericidal activity against *Staphylococcus aureus* (Liu et al., 2006). Additionally, the phenylethanol glycoside verbascoside, isolated from *Eremophila* species used to treat sores and wounds, has been shown to possess many biological activities, including antimicrobial and analgesic properties (Ghisalberti, 1995; Singab et al., 2013).

The 'Northern black wattle' (*Acacia auriculiformis*) was also used in various ways, and still is in some traditional societies. In the Northern Territory, leaves are steeped in water and the liquid, along with a few softened leaves, rubbed over the skin as a cleansing wash for cuts and bad sores (Smith, 1991). In addition, the legumes are rubbed vigorously onto the skin, with some water, to produce a rich lather used to relieve itchy skin and are apparently particularly good for skin that has been affected by the irritant hairs of stinging caterpillars (Smith, 1991). A methanol extract of *A. auriculiformis* phyllodes inhibited growth of Gram-positive bacteria *Staphylococcus aureus* and *Streptococcus pyogenes*, as well as that of the Gram-negative bacterium, *Escherichia coli* (Pennacchio et al., 2005). This supports the traditional use as a routine antiseptic cleanser and as a treatment for allergy rash. Several saponins have been isolated from this species, but it is not clear whether they are responsible for the observed antibiotic activity (Pennacchio et al., 2005). Additionally, a decoction of the roots of this plant was applied to relieve pains (Low, 1990). *A. auriculiformis* is a known source of alkaloids, which are much more important in Western medicine than aromatic compounds or tannins (Low, 1990). Among its many traditional applications, *Acacia tetragonophylla* was also used for wounds, including circumcision, and extracts of the leaves displayed antimicrobial activity against Gram-positive bacteria (Cribb and Cribb, 1981; Lassak and McCarthy, 2001; Palombo and Semple, 2001; Reid and Betts, 1979).

'Lolly bush' (*Clerodendrum floribundum*) has several local Aboriginal names, including 'Dutji', 'Molorrk' and 'Marbordalla', indicative of its wide use and versatility as an important traditional medicine in Northern Australia (ACNTA, 1988; Smith, 1991). Preparations of crushed young leaves and branchlets are boiled in water and, after cooling, the strained liquid is used as a lotion on sores and itchy or scaly skin (ACNTA, 1988; Smith, 1991). Small amounts of the same liquid can be splashed in the eyes to relieve tired or sore eyes or taken orally for very bad cases of diarrhoea or internal pain such as bad headaches and backache (ACNTA, 1988; Smith, 1991; Webb, 1969). All of these traditional uses can be considered anti-inflammatory and are supported by the finding that extracts of this plant potently inhibit xanthine oxidase, an enzyme involved in inflammation (Sweeney et al., 2001). The genus *Clerodendrum* also contains many other plant species that are used medicinally in other countries and extracts or constituents of various species have been found to possess multiple biological activities (ACNTA, 1988; Sweeney et al., 2001).

Another common complaint was scabies, a highly contagious skin disease caused by *Sarcoptes scabiei*, a mite which burrows into the skin, causing intense irritation (Ghisalberti, 1994). The inner bark and the kinos of several eucalypt species (e.g. *E. miniata*, *E. opaca*, *E. tetrodonta*) were widely utilised to control itching and heal sores caused from scratching (ACNTA, 1988; Smith, 1991). Kinos are red, gummy wood exudates, notably astringent due to their high tannin content (von Martius et al., 2012). Their use as traditional medicinal agents is widespread, particularly in relation to healing sores (Locher and Currie, 2010). As mentioned previously, tannins have excellent antimicrobial activity to combat infections (Locher and Currie, 2010; Scalbert, 1991). However, tannins also complex with proteins and polysaccharides, aiding the healing of wounds and inflammations by producing an impervious layer under which the natural healing processes can occur (Haslam, 1996; Locher and Currie, 2010). In addition, many of the monomeric flavanoid and phenolic constituents isolated from kinos have also been shown in various studies to have a range of pharmacological activities, including antioxidant, anti-inflammatory and antibacterial properties (Akiyama et al., 2001; Cowan, 1999; Haslam, 1996; Locher and Currie, 2010).

Digestive Complaints

Prior to European colonisation, most Aboriginal cases of diarrhoea and indigestion were caused by overeating or a sudden change of diet (Low, 1990). Plants were used to treat digestive issues to great effect. In Northern Australia, an Aboriginal remedy for diarrhoea became so popular with bushmen that it is still known to this day as 'Dysentery bush' (Low, 1990). The fruits of (*Grewia retusifolia*), known as 'jelly-boys' in Queensland and 'dog's balls' elsewhere, are useful for treating diarrhoea and dysentery (Lassak and McCarthy, 2001; Webb, 1959). Alternatively, these conditions and other stomach upsets can be treated by simply chewing the leaves, or by drinking infusions prepared from roots, leaves and/or stems (ACNTA, 1988;

Lassak and McCarthy, 2001; Smith, 1991). While plant parts used differ between regions, all methods of preparation describe how they first must be cooked between hot stones, mashed and then soaked or boiled in a small amount of water (ACNTA, 1988; Lassak and McCarthy, 2001; Smith, 1991). Additionally, this plant has been used to treat boils, sores, cuts, toothache, sore eyes and headaches (ACNTA, 1988; Smith, 1991). *G. retusifolia* is very rich in mucilage (ACNTA, 1988; Lassak and McCarthy, 2001), which has known demulcent properties: it forms a soothing film over mucous membranes to prevent irritation to the nerve endings to relieve minor pain and inflammation (Morton, 1990). Mucilage from many plants used as traditional medicines has been found to possess biologically active principles, accounting for their beneficial effects on burns, wounds, ulcers, external and internal inflammations and irritations, diarrhoea and dysentery (Goyal, 2012; Morton, 1990).

Scrapings of the inner bark of the 'Cluster fig' (*Ficus racemosa*) were soaked in warm water to make a bath and drink for diarrhoea patients in Arnhem Land (Low, 1990). Other records name *Ficus opposita* used in the same way in the same region (ACNTA, 1988; Smith, 1991). In Arrernte Country, the dried seeds of the 'Bush fig' (*Ficus platypoda*), which contains bioactive compounds, are ground up and given to sick children (Afifi et al., 2014; Dobson, 2007). The reduction in the severity of diarrhoea by the bark of *F. racemosa* has been established in rats (Mukherjee et al., 1998). Ethanol extracts significantly inhibited the frequency of defecation and the wetness of the faecal droppings, as well as intestinal propulsive movement and pooling of fluids and electrolytes (Mukherjee et al., 1998). This efficacy of *F. racemosa* bark extract as an antidiarrhoeal agent is thought to be due to tannins present, which causes the intestinal mucosa to be more resistant and reduce secretions (Low, 1990; Mukherjee et al., 1998). As tannins bind to proteins and polysaccharides, gut secretions are hindered, thus protecting the underlying mucosa from toxins and other irritants in the bowels (Haslam, 1996; Locher and Currie, 2010). Additionally, tannins have antimicrobial, antiviral and antiparasitic activities, including against gut pathogens (Singh et al., 2003). Bark extracts are also anti-inflammatory, potently inhibiting COX-1 activity (Li et al., 2003).

Ocimum sanctum (or *O. tenuiflorum*) has many vernacular names, highlighting its range of renown. Variously known as 'Sacred balm', 'Sacred basil' and 'Holy basil', Aboriginal people of Northern Queensland called it 'Bulla bulla' or 'Mooda' and drank an infusion of the leaves for fever, dysentery and general sickness (Lassak and McCarthy, 2001; Low, 1990). While *O. sanctum* does grow naturally in Northern Australia, it is also native to the Indian subcontinent where it has been a sacred and important medicinal herb, called 'Tulsi' in Hindi, since ancient times (Godhwani et al., 1987; Khanna and Bhatia, 2003; Low, 1990; Singh et al., 2007). Because of its reputation as a virtual cure-all, it has been extensively studied for its biological activities. Both a methanol extract and an aqueous suspension of *O. sanctum* partially delayed castor oil–induced diarrhoea in rats (Godhwani et al., 1987). The same early study showed that extracts of this herb possess anti-inflammatory, analgesic and antipyretic activities and that, given its antidiarrhoeal activity, it is possible that these effects may be attributed, at least in part, to inhibition of the biosynthesis of prostaglandins (Godhwani et al., 1987). Since then, numerous studies have verified its antibacterial activity (Mahomoodally et al., 2010; Mishra and Mishra, 2011), and analgesic and anti-inflammatory effects in rodents (Khanna and Bhatia, 2003; Singh and Majumdar, 1999). Bioassays suggest that the antioxidant and anti-inflammatory properties are primarily due to eugenol and flavonoids (Kelm et al., 2000). One study reported antimicrobial activity against enteric pathogens, including *Candida albicans*, one of the causative agents of diarrhoea in immunocompromised hosts (Geeta et al., 2001). Additionally, the fixed oil isolated from seeds of this plant also displays anti-inflammatory, analgesic, antibacterial and antipyretic activities in rats (Singh et al., 2007).

At the other end of the scale, constipation was treated by many plants with laxative effects. For example, Barcoo Aboriginals used a decoction of the fruits of *Eremophila bignoniiflora* (Lassak and McCarthy, 2001). Especially in cases of extreme sickness, the fruits were eaten or a concoction prepared from the berries was used as a strong bowel purgative, while just the leaves were used where only a mild laxative was required (Ghisalberti, 1994; Sadgrove et al., 2013). The presence of mannitol might partly explain the laxative properties of the infusions (Ghisalberti, 1994).

Similarly, eating raw fruits of the 'Bush tomato' (*Solanum ellipticum*) is a traditional cure for constipation in Arrernte Country (Dobson, 2007). If too many of these fruits are eaten without cooking, diarrhoea ensues, so eating them raw has a laxative effect (Dobson, 2007). This purgative effect is likely due to the presence of solanine, potentially toxic glycoalkaloids found in species of the *Solanum* genus (Hornfeldt and Collins, 1990).

Aches and Pains

Muscle cramps are one of the many ailments treated with the 'Lemon scented grass', *Cymbopogon ambiguus* (also 'Unanam', 'Kalpalp', Marrkan' or 'Aherre-aherre') (ACNTA, 1988; Dobson, 2007; Low, 1990; Young, 2007). Leaves (or any part of the plant, so long as the leaves are green) are crushed between stones or cut into small pieces and placed in boiling water (ACNTA, 1988). Solid matter is removed and the liquid rubbed onto the body, or else a small amount is drunk (ACNTA, 1988). This plant is very highly regarded as a bush medicine and is one of the few which are dried for storage so that it can be used whenever necessary (ACNTA, 1988). Its traditional use as a headache treatment is supported by studies showing whole plant extracts displaying strong inhibition of platelet aggregation and serotonin release and the compound principally responsible for this anti-inflammatory activity has been identified as eugenol (Grice et al., 2011; Rogers et al., 2001). The presence of essential oils such as camphene and borneol help explain its use as a decongestant (ACNTA, 1988). No antibacterial activity was detected against the microorganisms that have been tested (Palombo and Semple, 2001).

'Narrow-leaf fuchsia bush' (*Eremophila alternifolia*) is regarded as very potent and has even been described as 'number one medicine' by Aboriginal communities of Kaltukatjara and surrounding outstations in the Northern Territory (ACNTA, 1988). It is considered one of the few medicinal plants to be worthwhile storing and has many uses, including as a decongestant, expectorant and analgesic and as a remedy for general malaise (ACNTA, 1988). For internal pain, decoctions or infusions of dried or fresh leaves are taken orally several times a day, as well as rubbed on the skin in the same form, or as a paste (ACNTA, 1988). The presence of verbascoside in *E. alternifolia* leaves could partly account for some of its medicinal qualities as this compound has been shown to have a range of pharmacological activities including antibacterial, antioxidant and analgesic effects (Alipieva et al., 2014). Relevantly, it can reduce hypersensitivity to nerve pain when administered orally (Isacchi et al., 2011). Verbascoside has also been shown to dilate blood vessels and reduce blood pressure, as well as increasing heart rate, contractile force and coronary perfusion rate, which may explain how leaf infusions helped to induce sleep and promote general well-being in patients (Pennacchio et al., 1995, 1996a, 1996b, 2005; Pennacchio and Ghisalberti, 2000; Singab et al., 2013).

Eremophila bignoniiflora was also used as an analgesic to treat other types of pain, using the same procedure described for headaches but applying the plant material to the bodily area affected (Sadgrove et al., 2013). As previously stated, anti-inflammatory bioactive compounds have been detected in extracts of leaves of these plants (Grice et al., 2003; Rogers et al., 2000, 2001). Additionally, extracts can block neuronal voltage-gated Ca^{2+} channels which may have a role in neurovascular associated disorders, including nerve pain (Rogers et al., 2002).

Another Aboriginal treatment for rheumatism and other aches and pains, including broken bones, was an infusion of the mashed roots of 'Beach bean' or 'Wild Jack bean' (*Canavalia rosea*), known as 'Windi' or 'Yugam' in two Aboriginal languages (Lassak and McCarthy, 2001; Low, 1990; Webb, 1969). The guanidine alkaloid canarosine was isolated from extracts of the aerial parts of *C. rosea* and shown in vitro to cause inhibition of the dopamine D_1 receptor (Pattamadilok et al., 2008). Dopamine is a neurotransmitter implicated in modulating pain perception (Wood et al., 2007). Abnormal dopaminergic systems have been clearly demonstrated in painful clinical conditions, like burning mouth syndrome, fibromyalgia and restless legs syndrome, and further evidence suggests a role for dopamine in chronic regional pain syndrome and painful diabetic neuropathy (Wood, 2008). Nerve pain in a rat model has been shown to be significantly diminished by a dopamine D_1 receptor–selective antagonist (Coffeen et al., 2008).

Rheumatism

Aboriginal people of the Pilbara and Gascoyne regions of Western Australia used the crushed leaves of 'Mindharri' or 'Minjaarra' or 'Minjarri' (*Stemodia grossa*) soaked or boiled in water as a body rub for rheumatism and headaches (Lassak and McCarthy, 2001; Low, 1990; Reid and Betts, 1979; Webb, 1969; Young, 2007). This traditional use has been validated by the discovery that extracts of aerial parts of this plant exhibited high xanthine oxidase inhibitory activity, thereby limiting inflammation (Sweeney et al., 2001). It is sometimes called 'Vicks bush', due to its strong smell and was also used as an inhaler for colds as well as an ant repellent (Young, 2007).

In the Kimberley region of Western Australia, 'Konkerberry' (*Carissa lanceolata*) latex was rubbed in as a liniment for rheumatism (Lassak and McCarthy, 2001; Low, 1990; Reid and Betts, 1979). The oily sap was obtained by chipping the whole plant, including roots, into small pieces (Lassak and McCarthy, 2001; Webb, 1969). Other Northern Australia Aboriginal groups used the mashed and boiled roots as a mouth wash to relieve pain or a decoction of leaves and branches to treat the symptoms of colds and flu (Smith, 1991). Extracts from other *Carissa* species have been shown to have antioxidant, anti-inflammatory and analgesic properties in bioassays and in rodent and bird models (Alam et al., 2014; Bhaskar and Balakrishnan, 2009; Galipalli et al., 2014; Ngulde et al., 2013; Patel, 2013; Rao et al., 2005; Woode et al., 2007), but so far, no published studies have validated the traditional use of *C. lancleolota* against pain and inflammation. However, triterpenes and steroids are probably present in this plant (Simes et al., 1959) and since many of these compounds have anti-inflammatory activities (Bhaskar and Balakrishnan, 2009; Saleem, 2009; Salminen et al., 2008), this could be an avenue worth exploring to account for the traditional value of *C. lanceolata* in targeting inflammatory disorders. Additionally, eudesmanes carissone, dehydrocarissone and carindone have been isolated from the dichloromethane extract of the wood and been shown to have antibacterial activity against *Staphylococcus aureus*, *Escherichia coli* and *Pseudomonas aeruginosa* (Lindsay et al., 2000) and 2′-hydroxyacetophenone and carinol isolated from roots were found to possess significant antibacterial activity against these three pathogens as well as *Bacillus subtilis* (Hettiarachchi et al., 2011).

Victorian Aboriginals used a hot infusion of the roasted bark of 'Blackwood' (*Acacia melanoxylon*) to bathe rheumatic joints (Lassak and McCarthy, 2001). The bark is rich in tannins, which have antioxidant and anti-inflammatory properties (Haslam, 1996; Lassak and McCarthy, 2001; Li et al., 2003). Some inhibition of xanthine oxidase activity has been reported, even though only extracts of the leaves and stems were tested (Sweeney et al., 2001). Tellingly, patents detailing methods of using *Acacia* species to alleviate inflammatory-associated diseases and conditions and patents have been filed (Chang and Tung, 2013; Jia et al., 2006, 2012, 2013). Bioactive compounds such as flavonoids, alkaloids and phenolics have been detected in extracts of aerial parts of this plant and strong antioxidant activity has been demonstrated (Luis et al., 2012).

'Tickweed' or 'Spiderflower' (*Cleome viscosa*) was traditionally used to treat an array of ailments, including rheumatism (Lassak and McCarthy, 2001; Low, 1990). The whole plant was mashed and used externally to relieve rheumatism, swellings as well as headaches, colds, ulcers and open sores (Cribb and Cribb, 1981; Lassak and McCarthy, 2001; Low, 1990; Packer et al., 2012). The seeds were eaten to cure fever and diarrhoea (Lassak and McCarthy, 2001). This herb is also found in many other parts of the world, where it is commonly known as or 'Wild mustard', and has been similarly used as a medicinal agent in traditional communities (Mali, 2010). Given the extent of folklore claims as to its beneficial effects, this herb has been explored scientifically to validate its potential as a curative agent (Mali, 2010). Many of the ethnomedical uses have been justified in studies which verify analgesic (Ahmed et al., 2011; Parimaladevi et al., 2003a), antimicrobial

(Sudhakar et al., 2006; Williams et al., 2003), anti-inflammatory (Bawankule et al., 2008), antipyretic (Parimaladevi et al., 2003b), antiemetic (Ahmed et al., 2011) and antidiarrhoeal properties (Mali, 2010; Parimaladevi et al., 2002), among others. Numerous bioactive compounds have been isolated that explain these biological activities (Bawankule et al., 2008; Mali, 2010). Further, as one of its vernacular names hints, this plant contains mustard oils, renowned for their irritant effects, which can draw blood to the afflicted area and stimulate the release of the body's own anti-inflammatory agents (Low, 1990).

Stings and Bites

'Swamp lily' (*Crinium pedunculatum*) is a herbaceous plant found in swampy ground and floodplains, along banks of rivers, tidal creeks and streams and along most coasts from Northern New South Wales to Queensland and an island off the Northern Territory coast (Lassak and McCarthy, 2001). The main stems, roots and bulbs were crushed, soaked for a day, strained and applied to inflamed skin conditions such as marine and insect stings. Otherwise, the sap from under the outer layer of the stem was rubbed onto the body, acting as an emollient to soothe the skin (McCarthy and Ratcliffe, 2009; Webb, 1959). Alternatively, the thin film of skin under the outer layer of the bulb was peeled off and applied directly to the stung area (McCarthy and Ratcliffe, 2009). This remedy was reputedly very effective and Aboriginal people from Bingil Bay in North Queensland even used *C. pedunculatum* to treat box jellyfish stings (McCarthy and Ratcliffe, 2009). The active principle is thought to be the alkaloid lycorine, which has both antinociceptive and anti-inflammatory properties (Çitoğlu et al., 1998, 2012; Lassak and McCarthy, 2001; Low, 1990).

In the Northern Territory, the thin green bark of the 'White (or Grey) mangrove' (*Avicennia marina*) was placed directly onto stingray stings (ACNTA, 1988). Small pieces could also be chewed or perhaps softened and spat out onto the sting (ACNTA, 1988). Both methods helped relieve the pain and helped heal the injury (Smith, 1991). A recent in vivo study in rats revealed extracts of this plant were as good as morphine at reducing the sensation of pain in the long-term inflammatory phase, although they were not effective in relieving early pain. This significant analgesic response to chronic pain is likely due to antinociceptive compounds, such as steroids, phytoalexins, flavonoids, carboxylic acids, triterpenes and tannins present in the extracts tested (Gandomani et al., 2012). Additionally, *A. marina* has been shown to play a protective role against inflammation in acetic acid–induced colitis in mice, validating its traditional use in treating stings. This anti-inflammatory property might be attributed to the presence of higher levels of decanoic acid, diethylhydroxylamine, pentanoic acid, pyrrolidine, 4-chlorophenyl, thiazolidinones and arabinopyranoside (Rise et al., 2012). *A. marina* extracts also display antimicrobial activities, due, at least in part, to flavonoids and phenols present (Afzal et al., 2011; Khafagi et al., 2003; Zandi et al., 2008). Leaf extracts exhibit potent bioactivity against allergenic fungi, meaning any subsequent skin hypersensitivity would be limited (Afzal et al., 2011). Due to these medicinal properties, *A. marina* was also used to treat conditions such as rheumatism, ulcers and smallpox by Australian Aboriginals, as well as other indigenous peoples of the world (Afzal et al., 2011; Bandaranayake, 1998; Revathi et al., 2014). Toxic metabolites like alkaloids and saponins were not detected in extracts of *A. marina*, explaining its low toxicity (Khafagi et al., 2003).

'Hopbush' (*Dodonaea viscosa*) is another plant traditionally employed by Aboriginal people to treat numerous complaints, including stonefish and sting-ray stings (Lassak and McCarthy, 2001; Low, 1990; Webb, 1969). In Northern Queensland, chewed leaves and juice were applied directly to the wound for several days (Lassak and McCarthy, 2001; Webb, 1969). Amongst its many scientifically proven medicinal properties that correlate with its traditional uses (Lawai and Yunusa, 2013; Pengelly, 1999; Smyth et al., 2009; Teffo et al., 2010), *D. viscosa* has anti-inflammatory activity due, at least in part, to hautriwaic acid (Ghisalberti, 1998a; Salinas-Sanchez et al., 2012). It has also been shown to have analgesic effects in mice, both peripherally and centrally, and can reduce fever (Amabeoku et al., 2001). A mixture of saponins is thought responsible for its anti-exudative and immune-stimulating properties (Wagner et al., 1987). Other bioactive compounds present in *D. viscosa* that could account for the anti-inflammatory activity include diterpenes, saponins and flavonoids. Additionally, the presence of a range of flavones, including quercetin and oleanene glycosides, may contribute to the wound-healing activity. Triterpene saponins and the coumarin fraxetin may be responsible for analgesic properties attributed to this plant (Ghisalberti, 1998b).

Similarly, 'Goat's foot' (*Ipomoea pes-caprae*) had many medicinal uses, including the treatment of marine stings (ACNTA, 1988; Lassak and McCarthy, 2001; Packer et al., 2012; Smith, 1991). In its case, leaves were crushed, heated on hot stones and placed directly onto the affected skin and reapplied as required to stop bleeding and ease pain (ACNTA, 1988; Smith, 1991). Its analgesic and anti-inflammatory properties have been confirmed in extracts (Maria de Souza et al., 2000; Rogers et al., 2000; Vieira et al., 2013). Additionally, extracts of leaves of *I. pes-caprae* have been shown to be clinically effective in limiting dermatitis caused by venomous jellyfish and to neutralise toxic activities of jellyfish venoms (Pongprayoon et al., 1991).

Juice from fronds of the Bracken fern (*Pteridium esculentum*) was used by the Yaegl people of Northern New South Wales and Cape Barren Islanders in Tasmania to treat insect bites and stings (Clarke, 2007; Debboun et al., 2015; Lassak and McCarthy, 2001; Packer et al., 2012). Phenolics such as kaempferol glycosides and chlorogenic acid isolated from *P. esculentum* may help explain the anti-inflammatory and antinociceptive properties ascribed to this plant (De Melo et al., 2009; dos Santos et al., 2006; Rajendran et al., 2014; Tanaka et al., 1993). Additionally, other plant parts have been reputedly used by others as an astringent, a worm medicine and to treat rheumatism and relevant bioactive components have been isolated (Daw et al., 1997; Lassak and McCarthy, 2001; Wohlmuth, 1997).

It is widely acknowledged that plants have little value as antidotes to snake venom (Lassak and McCarthy, 2001). However, some appear to have been used to alleviate the side effects of snakebite, such as pain. For example, *T. smilacina*,

Capparis lasiantha ('Split Jack') and *Nauclea orientalis* ('Leichardt tree') were all used for relief, rather than as a cure (Lassak and McCarthy, 2001). A bark infusion of *N. orientalis* was used to treat snakebite, while the leaves, stems and roots of *C. lasiantha* were mashed and applied externally for snakebite and insect bites and stings to reduce swelling (Lassak and McCarthy, 2001).

Fevers

Fevers were probably rare before infectious diseases like influenza, tuberculosis and smallpox were introduced into Australia after contact with European or Macassan traders (Campbell, 1983, 2007; Low, 1990). Nevertheless, bush medicines were employed to treat fevers associated with malaria, digestive upsets and giving birth (Lassak and McCarthy, 2001). *Alstonia constricta* is more commonly known as 'bitter bark', 'Feverbark' or 'Quinine bark' due to the use of its stem bark as a febrifuge (Lassak and McCarthy, 2001; Low, 1990). The Bitterbark was adopted by colonial doctors to treat fevers and as a general tonic and was even exported to England (Low, 1990). While laboratory tests have shown it has only slight antimalarial activity (Wright et al., 1993), it was said to be a better antiperiodic than quinine and cinchodine, and particularly useful during episodes of influenza or the early stages of typhoid fever (Lassak and McCarthy, 2001). Called 'Lacambie' by Aboriginal people of the Clarence River region (Lassak and McCarthy, 2001), its ability to reduce fevers is attributed to several alkaloids, such as alstonine and alstonidine (Lassak and McCarthy, 2001). Alstonine is a known antipsychotic agent (Elisabetsky and Costa-Campos, 2006), so would have helped to treat anxieties and hallucinations that can accompany fevers. The hypotensive alkaloid, reserpine, the first antipsychotic medication used in modern medicine (Elisabetsky and Costa-Campos, 2006), is found in the root bark, although this part of the plant was not traditionally used by Aboriginal people, as far as is known (Lassak and McCarthy, 2001). The stem bark of another *Alstonia* species (*A. boonei*), which was an African ethnomedicine for many ailments including relapsing fevers, has been shown to possess anti-inflammatory, analgesic and antipyretic activities in rodents, supporting its traditional uses (Olajide et al., 2000).

'Turpentine bush' (*Beyeria leschenaultii*) was a universal Aboriginal remedy, originally used against fevers (Low, 1990; Webb, 1969). Many bioactive compounds have been isolated from parts of this plant or resin coating the stems, including beyerol and other diterpenes and triterpenoid alcohols (Baddeley et al., 1964a,b; Ghisalberti and Jefferies, 1968). Kaurenic acid, which has proven antipyretic and anti-inflammatory activities in rats (Sosa-Sequera et al., 2010), is a diterpenoid detected in a related *Beyeria* species (*B. calycina*) (Croft et al., 1978).

'Native bauhinia' (*Lysiphyllum cunninghamii*) is an important antipyretic and antiseptic used by Aboriginal people in parts of the Northern Territory (ACNTA, 1988). Locally known as 'Wanyarri' (or 'Wanyari'), a piece of the inner bark was boiled up in water and, when cool, a small amount of the strained liquid was taken orally to relieve high temperatures and headaches associated with colds and other illnesses (ACNTA, 1988). Saponins and tannins are present in the inner bark, but no alkaloids were detected (ACNTA, 1988). The roots also contain saponins and preparations are regarded as more effective than any Western medicine in treating boils, sores and scabies (ACNTA, 1988; McDonald, 2006). As yet, no published studies have investigated the constituents responsible for the antipyretic activity accredited to this plant.

A decoction of the inner bark of *Timonius timon* was drunk for 2–3 days in another remedy used by Aboriginal people to alleviate fevers (Cribb and Cribb, 1981; Lassak and McCarthy, 2001; Webb, 1969). Bioactive constituents such as triterpenes (Khan et al., 1993) and iridoid glycosides have been isolated from dried leaves (Erdelmeier et al., 1994), and compounds in these classes often display pharmacological activities, including anti-inflammatory effects (Ghisalberti, 1998b; Küpeli et al., 2005; Salminen et al., 2008). Another *Timonius* species (*T. flavescens*) has been investigated for anti-inflammatory properties and been shown to inhibit lipooxygenase activity in vitro (Chung et al., 2009). Similarly, a cooled infusion of the leaves of 'Narrow-leaf hopbush' (*Dodonaea attenuata*) was sponged onto the forehead and body to reduce fever (Lassak and McCarthy, 2001). Extracts of leaves and terminal branches contain diterpenes and the lactone from hautriwaic acid (Payne and Jefferies, 1973), already mentioned as having medicinal qualities.

Sore Eyes

Sore and infected eyes were a major concern for Aboriginal people, so eye washes were popular and a number of plants were employed for this purpose (Cock, 2011; Lassak and McCarthy, 2001). For example, a strained infusion of the inner bark and sapwood of 'Wild plum' (*Buchanania obovata*) was used as an eyewash and, in at least one reported case, even cured temporary blindness (ACNTA, 1988; Lassak and McCarthy, 2001). Wood decoctions of 'Emu apple' (*Owenia acidula*) and root or stem infusions of *Grewia retusifolia* were also used to bathe sore, infected or tired eyes (Lassak and McCarthy, 2001; O'Connell et al., 1983; Smith, 1991; Webb, 1969), but the active principles of these species has not yet been investigated.

An infusion of bark from the 'Peanut tree' (*Sterculia quadrifida*) was applied to relieve sore eyes or the juice from the inner bark was wrung straight into the affected eyes as in eye-drops (Webb, 1969). Northern Queensland Aboriginal communities used crushed leaves of the same plant to treat wounds (ACNTA, 1988; Lassak and McCarthy, 2001). Water extracts of the leaves have been shown to have bacteriostatic and bactericidal activities against selected microorganisms, supporting its customary medicinal use (Smyth et al., 2009).

Decoctions or infusions of a few native 'Sneezeweeds', *Centipeda cunninghamii*, *C. minima* and *C. thespidiodes*, were used to successfully treat eye inflammation and introduced infections like trachoma ('sandy blight') and purulent ophthalmia (Lassak and McCarthy, 2001; Taylor, 2001; Webb, 1969). In vitro studies have shown that *C. minima* displays broad-spectrum antimicrobial activity, likely due to the presence of artemisia ketone, and is not toxic to humans, affirming its traditional use (Soetardjo et al., 2007; Taylor and Towers, 1998).

Toothache

Toothache was a relatively common ailment, particularly amongst old people, due to a lifetime diet of tough, fibrous food (Cock, 2011; Lassak and McCarthy, 2001). It was treated by a wide variety of plant medications including the 'Toothache tree' (*Melicope vitiflora*, formerly *Euodia vitiflora*). The resinous exudate of the bark was used to fill tooth cavities and relieve pain (Cribb and Cribb, 1981; Lassak and McCarthy, 2001; O'Donnell et al., 2009; Smyth et al., 2012; Webb, 1959). The resin contains several coumarins and alkaloids are present in the bark (as well as leaves and branchlets) (Lassak and McCarthy, 2001; McCormick et al., 1996). Methanol and ethyl acetate extracts of both bark and leaves displayed antimicrobial activity against several bacterial strains, including a methicillin-resistant *Staphylococcus aureus* (MRSA) (O'Donnell et al., 2009; Smyth et al., 2009). It is thought that bioactivity may be due to coumarins present, although those originally isolated (Lassak and Southwell, 1972) do not appear to have been tested yet (Smyth et al., 2009).

In Kaanju traditional medicine, 'Uncha' (*Dodonaea polyandra*) is a favoured medicine for the relief of pain associated with toothache and removal of rotten teeth (Locher et al., 2013). Male healers from the Northern Kaanju (KuukuI'yu) homelands in Cape York Peninsula still use this plant by breaking off one of the terminal branchlets with a leaf attached and applying directly to the mouth or inserting into the hole left after the removal of a decayed tooth (Simpson et al., 2010, 2011). Significant anti-inflammatory effects of extracts of leaves of this plant have been confirmed in a mouse model of acute inflammation (Simpson et al., 2010). As with its close relative *D. viscosa*, which has also been used to treat toothache (Lassak and McCarthy, 2001), it has been speculated that the high content of flavonoids and diterpenoids in this plant may be responsible, at least in part, for its potent anti-inflammatory properties (Salminen et al., 2008; Simpson et al., 2010). Indeed, four new benzoyl ester clerodane diterpenoids have recently been isolated and elucidated and the three tested showed significant anti-inflammatory activities (Simpson et al., 2011).

The flesh of the fruit and inner seeds of the 'Lady apple', or 'Red bush apple' (*Syzygium suborbiculare*) were chewed to relieve toothache (Lim, 2012; Low, 1990; Smith, 1991; Vuong et al., 2014). In Arnhem Land, the pulp of *S. suborbiculare* was squeezed into sore ears (Low, 1990). Extracts of other *Syzygium* species have been shown to have analgesic (Ávila-Peña et al., 2007; Mollika et al., 2014; Quintans et al., 2014) and anti-inflammatory effects (Kandati et al., 2012; Muruganandan et al., 2001; Panahi et al., 2014), although no bioactivity reports could be found for *S. suborbiculare*.

Earache

A leaf decoction of the 'River mangrove' (*Aegiceras corniculatum*), or the juice obtained by squeezing the leaves, was used as eardrops to ease earache (Lassak and McCarthy, 2001; Low, 1990). Leaf extracts display antibacterial activity, including against MRSA (Chandrasekaran et al., 2009). Both leaf and stem extracts of *A. corniculatum* showed significant analgesic activity against chemical- and thermal-induced pain models in rodents (Roome et al., 2011; Singh et al., 2010). Stem extracts of this plant also suppressed early- and late-phase inflammation in vivo mediated via multiple mechanisms of action (Roome et al., 2008b). Extracts possessed pronounced antioxidant and free radical–scavenging activity, thought to account for its ability to inhibit oxidative stress–induced inflammation (Roome et al., 2008a). These anti-inflammatory properties and antinociceptive effects could be due to phenolic constituents, flavonoids, tannins, saponins or any of the other bioactive compounds identified that can scavenge free radicals, chelate metal ions, reduce the respiratory burst in cells and also exert a protective effect against oxidative damage (Bandaranayake, 2002; Poompozhil and Kumarasamy, 2014; Roome et al., 2008a; Torel et al., 1986).

The sap from bruised leaves of *Crotalaria cunninghamii* was reportedly poured into the ears to relieve earache (Lassak and McCarthy, 2001; Reid and Betts, 1979). It is thought that alkaloids are the compounds most likely to account for the bioactivity of this plant, but, compared to other *Crotalaria* species at least, the alkaloid content is low (Fletcher et al., 2008). However, the pyrrolizidine alkaloid retusamine is the main component in extracts of new growth and flowering tops (Fletcher et al., 2008), and monocrotaline is a major constituent of the seeds of most *Crotalaria* species (Pilbeam et al., 1983). Pyrrolizidine alkaloids such as these cause liver and lung damage and so are toxic if ingested too frequently (Roeder and Wiedenfeld, 2009). Aboriginal people probably understood this very well as, from all accounts, they only used *Crotalaria* species in external applications (Lassak and McCarthy, 2001; Smith, 1991).

The boiled root juice of *Dodonaea viscosa* and the fruit pulp of *Syzygium suborbiculare* were also reportedly employed by Aboriginals to alleviate earache (Low, 1990). Both these species have already been discussed in relation to their analgesic and anti-inflammatory properties.

Cancer

It is important to understand that within local communities, conditions referred to as 'cancer' may not be what is clinically recognised as cancer. This is because cancer, as a specific disease entity, is likely to be poorly defined in terms of folklore and traditional medicine (Newman and Cragg, 2005). Cancer is not a well-defined disease state in Aboriginal traditional culture, although it is likely it was recognised as a wasting syndrome and may have had a specific name (Carrick et al., 1996). However, McGrath et al. (2006) asserted there is not even an Aboriginal word for cancer. A problem is that there is little or no historical data on cancer in traditional Aboriginal societies prior to white settlement to be certain (Cunningham et al., 2008; Low, 1990). Furthermore, Aboriginal people most commonly use plants to treat the symptoms of disorders such as gastrointestinal problems, skin infections and respiratory illnesses rather than in the treatment of cancer. It is, therefore, frequently difficult to establish a direct link between traditional plant remedies and true neoplasms

(Heinrich and Bremner, 2006). Nevertheless, reports of Aboriginal people using traditional bush medicines to cure cancer do exist. Perhaps the most famous is the case of *Scaevola spinescens*, otherwise known as 'Maroon bush', 'Currant bush' or 'Prickly fanflower'. This plant was used by Aboriginal people to treat many ailments, including cancer (Ghisalberti, 2004). In 1935, an Aboriginal man named Albert Neebrong was diagnosed with cancer of the tongue and reportedly cured himself using 'Murin Murin'. After hearing this anecdotal evidence, and after painstaking enquiries and being convinced of its authenticity, Police Sergeant Athol Monck promoted the use of *S. spinescens* as a beneficial treatment for many illnesses including skin cancer, melanoma, ulcerative colitis and cancer (Crago, 2011). Monck's advocacy of the plant led to the 1957 instigation of a programme whereby the Chemistry Centre of Western Australia (CCWA) prepared and supplied water extracts of *S. spinescens* free of charge to terminally ill cancer patients (Ghisalberti, 2004). No new patients were added after 1991, but eight patients diagnosed with terminal cancer prior to this, having received approval from their doctors, were still being treated up until at least 2000 (CCWA, 2001; Kerr, 1999). In 1989, a sample of a methanol extract of *S. spinescens* was sent to the National Cancer Institute (NCI) in the United States, and shown to be active in all 60 cell lines tested. However, although the results indicated significant anticancer activity in this material, its lack of selective toxicity for different cell types precluded it from being of great interest. Kerr et al. (1996) identified the presence of several taraxerenes in various extracts of *S. spinescens* and isolated the most abundant, myricadiol. These authors suggested that these compounds may be of potential use as lead compounds for synthetic anticancer agents. However, subsequent studies showed that some taraxerenes themselves have antitumour properties (Takasaki et al., 1999). For example, while myricadiol failed to show any antiproliferative effects against the leukemic and lung cancer cell lines tested, the same study revealed its moderate inhibitory activity against COX-1 and COX-2 (Lee et al., 2006). Given that specific COX-1 and COX-2 inhibitors have demonstrated antiproliferative effects against bladder, prostate and breast cancers in vitro (Farivar-Mohseni et al., 2004; McFadden et al., 2006; Mohseni et al., 2004), myricadiol may be useful against these types of cancer. Alternatively, it may have a potential role in antiangiogenesis therapy (Rahman and Toi, 2003; Zhong and Bowen, 2006).

There is also the interesting tale of *Euphobia peplus*, known variously as 'Radium weed', 'Milkweed' and 'Petty spurge' (Ogbourne et al., 2007). The milky sap has been used since ancient times by various cultures around the world, including postcolonial Aboriginal people and bushmen to treat a range of ailments, including carcinomas (Jassbi, 2006; Scarborough, 1978; Scott and Karp, 1996; Toledo-Pereyra, 1973; Weedon and Chick, 1976). Guided by such ethnopharmacological knowledge imparted by his mother, Dr Jim Aylward and a team led by Professor Peter Parsons from the Queensland Institute of Medical Research (QIMR) successfully identified a potent anticancer agent from the sap of *E. peplus* (Hurst, 2009; Ogbourne et al., 2004). The active principle, ingenol 3-angelate (I3A), formerly known as PEP005 but now called ingenol mebutate, is being developed as a safe topical treatment of non-melanoma skin cancers and precancerous actinic keratosis (Siller et al., 2009). Subsequently, ingenol mebutate has shown potential in the treatment of bladder cancer and leukaemia (Hampson et al., 2005; Ogbourne et al., 2007). However, this plant is not an Australian native.

Apart from these, few published studies could be found that have specifically assessed the anticancer potential of plants traditionally used by Australian Aboriginal people as medicines (Mijajlovic et al., 2006). My own work has used the ethnopharmacological approach to screen extracts of desert plants used by Aboriginal people as traditional medicines to try to identify compounds that may be useful cancer therapeutics. Due to confidentiality agreements in place to protect Aboriginal intellectual property rights, this work remains unpublished, although the restraint has recently been lifted. Screening for cytotoxic activity verified the bioactivity of several of these medicinal plants, scientifically validating what the Aboriginal people have known for thousands of years. In these in vitro studies, the most promising plant extract was the ethyl acetate extract of *Eremophila duttonii*, and the flavonoids quercetin, rutin and luteolin were identified as the constituents most likely to account for some of its observed cytotoxic activity. However, the compound likely responsible for the main cytotoxic actions of this species could not be definitively identified as part of this study, leaving open the possibility that the inhibitory effects were due to a new chemical entity (NCE) (Savigni, 2010).

DRUG DISCOVERY

Bioprospecting for Natural Products

The Australian Concise Oxford Dictionary (p. 135, *The Australian Concise Oxford Dictionary*, 2009, ed. B. Moore. Melbourne, Oxford University Press) defines bioprospecting as 'the search for animal and plant species from which medicinal drugs and other commercially valuable compounds can be obtained'. In the past, natural products (NPs) were the prime source of NCEs. NPs are typically secondary metabolites produced by living organisms in response to the specific requirements of their local environments (Strohl, 2000; Verdine, 1996). Secondary metabolites are an exceptionally diverse range of organic compounds not directly involved in the normal functioning of an organism (Cock, 2011). Although their biological role is not always obvious, they are not formed without reason. Organisms have evolved to synthesise these important survival and propagation factors in response to the specific conditions of their local environment (Strohl, 2000; Verdine, 1996). For example, plant secondary metabolites may be involved in chemical defence by deterring herbivores from browsing or insect pathogens from attacking (Lord, 1987; Wink, 1999). In other cases, the compound may be important for the everyday existence of the organism but have serendipitous activity in an unrelated biological system (e.g. egg-derived proteins) (Colegate and Molyneux, 2008). Secondary metabolites are frequently unique to a particular plant species and their levels can rise during times of high stress like drought, fire or microbial infection (Cock, 2011). Many of these compounds display antimicrobial, antioxidant, cytotoxic and other medicinally useful activities (Cock, 2011). There is a range of phytochemical constituents, which can be divided into three main chemically distinct classes, namely, terpenes, phenolics and nitrogen-containing compounds (alkaloids), that contribute to these pharmacological effects (Cock, 2011).

NPs have been and remain a prime source of NCEs, leading to the development of a myriad of novel drugs to combat a diverse range of diseases (Clardy and Walsh, 2004; Newman and Cragg, 2007). Bioprospecting can result in whole new classes of materials that even the best chemists could probably never even imagine. Nature has produced chemical combinations for millennia, and so the amount of chemical diversity within natural compounds is extensive (Macilwain, 1998; Petsko, 1996; Tulp and Bohlin, 2004; Verdine, 1996). And while the ability to synthesise new proteins is growing exponentially, our current understanding of structural biology is still inadequate to allow us to produce the kinds of molecules available in nature. Hence, NPs provide a virtually limitless source of novel and complex chemical structures that probably would never otherwise have been the subject of a beginning synthetic programme (Fabricant and Farnsworth, 2001). Moreover, the chemical properties of NPs, such as lipophilicity and binding affinities for specific receptors, are generally superior to chemically synthesised compounds and less likely to cause detrimental interactions within the complex microenvironment of a cell (Dobson, 2004; Feher and Schmidt, 2003).

Historically, NPs have been a principal source of new chemotherapeutic agents, and many successful drugs were initially synthesised to mimic the action of molecules found in nature (Feher and Schmidt, 2003). In fact, more than half of all drugs in current clinical use have an NP origin (Cragg et al., 1997). Furthermore, in the past century, drugs that trace their heritage to NPs have more than doubled the average lifespan of humans and ameliorated non-fatal, but still debilitating, conditions such as chronic pain and depression (Verdine, 1996). Of the over 1000 NCEs approved as drugs between 1981 and 2002 by the U.S. Food and Drug Administration (FDA), 5% were secondary metabolites and a further 23% were derived from NPs (Clardy and Walsh, 2004). The proportion was even higher if compounds 'inspired by' NPs are included (Li and Vederas, 2009; Newman and Cragg, 2007). While the expansion of synthetic medicinal chemistry during the 1990s caused the proportion of new drugs based on NPs to drop from 80% in 1990 to only about 50% by the end of the decade, 13 drugs derived from NPs were still approved by the FDA between 2005 and 2007, with 5 of those being the first members of new classes (Li and Vederas, 2009).

NPs can have a plant, animal or microbiological origin and be found anywhere in the world. As chemical diversity is a function of biodiversity, it is often assumed that tropical regions, where about two-thirds of biodiversity exists (Pimm and Raven, 2000), will provide us with a plethora of medicines (Tulp and Bohlin, 2002). However, Australia is one of 17 'megadiverse' regions, among naturally rich countries such as Brazil and the Congo. The heathlands and woodlands of South-western Australia comprise one of 34 worldwide biodiversity hotspots. Australia possesses more than 80 globally unique families of plants and animals (Morton and Sheppard, 2014) and contains an estimated 7.9% of all the world's accepted described plant species (including algae) (Chapman, 2009).

Plants as Sources of Drugs

Prominent pharmacognoscists and ethnobotanists, such as Norman Farnsworth and James Duke, believed that 'somewhere in the plant kingdom there is a remedy for everything' (Duke, 1990). More than 7000 compounds used in Western medicine are derived from plants (Clapp and Crook, 2002). It is well established that plants have been a prime source of highly effective conventional drugs (Ansah and Gooderham, 2002; Cordell et al., 1991; Newman and Cragg, 2005). In fact, more than half of all pharmaceuticals are derived wholly or partly from plants (Melnick, 2006). It is clear that, despite various challenges encountered in plant-based drug discovery, NPs isolated from plants will continue to be an essential component in the search for new medicines (Jachak and Saklani, 2007). There are distinct advantages to using plants as the starting point in any drug development programme. In most cases, plants are a renewable source of starting material and any bioactive compounds obtained from plants that have been used long-term by people might be expected to have low human toxicity (Fabricant and Farnsworth, 2001).

After insects, plants represent the second largest source of global biodiversity (15%) (Tan et al., 2006). Estimates of the number of higher plant species on Earth range from 200,000 to 1,000,000 (Pimm et al., 1995) and only a small fraction have been examined for medicinal value. It therefore seems logical that an abundance of drugs from plant sources are yet to be discovered. Approaches to selecting plants to be tested for new bioactive compounds range from random selection to more guided selection strategies such as the chemotaxonomic approach and the ethnopharmacological approach (Cordell et al., 1991; Shoemaker et al., 2005; Soejarto, 1996).

Ethnopharmacology

One undeniably valid approach to discovering drugs from plants is ethnopharmacology. Ethnomedicine has been broadly defined as the use of plants by humans as medicines, but this use could be more accurately called ethnobotanic medicine (Fabricant and Farnsworth, 2001). Northridge (2002) defines ethnomedicine as the folk medicines of particular ethnic groups and asserts it is dependent on location. All over the world, indigenous cultures traditionally used the plants available to them in their local environments to treat diseases and promote health and these folk remedies were generally passed down orally (Northridge, 2002). Fossil records indicate that people have used plants as medicines for at least 60,000 years (Kong et al., 2003) and almost 65% of the world's population continues to use plants as their primary mode of healthcare (Dufault et al., 2001; Fabricant and Farnsworth, 2001).

Ethnopharmacology is a highly diversified approach to drug discovery involving the observation, description and experimental investigation of indigenous drugs and their biological activities (Fabricant and Farnsworth, 2001). Intuitively, the rate of drug discovery through the use of ethnopharmacological data would be expected to be much greater than with random collection. Indeed, Chapuis et al. (1988) pre-selected plant species based on traditional medicinal use and demonstrated a relatively high proportion showed promising activity. Spjut and Perdue (1976) revealed in a retro-spective (to that time) analysis of the early NCI programme in which many species of plants were tested for antitumour properties that the percentage of active leads based on ethnomedical use was substantially above that based on either random screening or taxonomy.

Fabricant and Farnsworth (2001) pointed out that any of the 250,000 higher plant species on earth could conceivably produce a new drug. However, given the virtually limitless introduction of novel mechanism-based bioassays, there is a very low probability of collecting all 250,000 species and screening them for more than one biological activity. Thus, researchers should judiciously choose species most likely to produce useful activity and the biological targets must represent the activi-ties best correlated with the rationale for plant selection. Hence, selection of plants based on long-term human use in conjunction with appropriate biologic assays that correlate with the ethnomedical uses seems to be most appropriate (Fabricant and Farnsworth, 2001).

Challenges of Ethnopharmacology

The search for new drugs from terrestrial plants is a complex process that demands the involvement of not only scientific expertise, but also proficiency in a broad spectrum of human endeavours including diplomacy, international laws and legal understandings, social sciences, politics, anthropology and basic common sense. Equally important is the fact that such endeavours must be governed by inter-national bureaucratic and regulatory procedures (Tan et al., 2006).

Ethnopharmacology has been criticised for taking advantage of the indigenous cultures who provided the original knowledge about plants they have traditionally used as medicines (Hamilton, 2006; Shiva, 1997; Stone, 2008; Verma, 2002). While these criticisms may have some valid historical basis (Alter, 2000; Gollin, 1999; Philip, 2001; Wallace, 1996), new regulations and guidelines for bioprospecting and NP research are now in place to protect nations and indigenous cultures from biopi-racy (Gollin, 1999). However, while new legislation concerning benefit sharing has been introduced in many countries, there is widespread concern that the laws are not stringent enough (Roehrs, 2007). Despite the new regulations, often only an extremely small percentage of a company's huge profits are given to those who have nurtured the traditional knowledge from which the medicine is derived (Chacko and Sambuc, 2003).

Yet another factor alienating 'Big Pharma' from the ethnopharmacological approach is cost. Pharmaceutical companies generally prefer synthetic compounds for proprietary economic reasons. This is understandable given it costs millions of dollars to develop a new drug. When fees and royalties have to be paid to host coun-tries, the expense becomes even higher. Moreover, the company has only 10 years to recoup its investment before a patent expires and generic manufacturers can copy the design. There is no incentive for pharmaceutical companies to determine which alternative phytochemical is best, because plants are not patentable. Indeed, even if an NP works better, they often prefer the semi-synthetic compound as it guarantees higher profits (Dufault et al., 2001).

Another problem associated with any drug discovery programme based on plants, either chosen on the basis of ethnomedical data or randomly selected, is that, as biological systems, plants display inherent individual variability in their chemistry and can therefore display varying degrees of bioactivity (Fabricant and Farnsworth, 2001). In some cases, variable activities between plant collections are very likely due to incorrect species identification. This can be especially significant when the ethnopharmacological strategy is employed as locals often only know plants by their vernacular names and these can change from region to region (Guarrera and Lucia, 2007). In Australia, there can be many names for the same species between different Aboriginal languages and even within one language. For example, *Euphorbia drummondii* has eight different names within a radius of about 200 km of Alice Springs. Within an area less than 600,000 km^2, the same plant is known variously by 17 different Aboriginal names. To complicate matters further, different species may have the same Aboriginal name. For example, the Alyawarr plant name, 'Amikwel' can refer to either *Euphorbia tannensis* or *Sarcostemma australe*. In the Pitjantjatjara language, 'Ipi-ipi' refers to *E. drummondii*, *E. tannensis*, *E. eremophila* and any plant with milky sap in general (Latz, 1995). Therefore, to ensure plants are correctly identified each time they are collected, it is vital that collections are always documented with representative voucher specimens and that these are preserved for independent verification (Farnsworth and Bingel, 1977; Hildreth et al., 2007; Soejarto, 1996).

While collecting plant samples randomly in a specific geographic area can be done simply and quickly, collecting plants on the basis of ethnomedical knowledge is a much more time-consuming process. Not only does it require considerable preliminary planning to determine where each plant grows and how plentiful it is, particular arrangements must also be made to collect the plants, such as acquiring permits and finding local botanists familiar with the flora of the region (Fabricant and Farnsworth, 2001). Moreover, many indigenous cultures, including Australian Aboriginal communities, operate on a slower timeline compared to the faster pace of modern society. Indeed, researchers are often taken aback by the almost complete lack of awareness of time as a concept by Aboriginal people (Zur, 2007). This type of seemingly timeless culture actually operates on what has been coined 'polychronic time' (Hall and Reed Hall, 2001). In extreme polychronic cultures, events take place as solitary entities, separate from each other, and in no way connected by an ongoing timeline as many Westerners understand it. In such cultures, even when it appears that everything is at a standstill, a great deal is probably going on behind the scenes. Scheduling cannot happen until meetings have taken place to permit essential discussions. Hence, the amount of time, especially the amount of lead time, required in forging collaborative relationships is quite protracted. Therefore, it is vital that researchers are always aware of the local 'time system' so that they can make the necessary allowances (Hall and Reed Hall, 2001). Compounding the research obstacles associated with the polychronicity of Aboriginal society is their notion of equality in knowledge sharing. Relationships must be established before knowledge is shared. The key word is 'shared', so there is an expectation of a two-way exchange of information. This is facilitated by face-to-face communication, which is also desirable for cultural reasons of trust and respect. The amount of time necessary to build this trust and respect is naturally not up to the researcher. Therefore, multiple trips to remote communities are required and this obviously consumes yet more time (Zur, 2007). All this means that Aboriginal-related research takes much longer than a Western person might expect. The obligatory abandonment of usual scheduling procedures can be extremely frustrating, especially for drug companies which are generally driven by profit and therefore pathologically attached to rigid timetables.

Other cross-cultural differences can also be a barrier to collaborative ethnopharmacological research efforts. When people unwittingly apply their own rules to another culture, critical steps can be omitted and effective research rendered unachievable. If two people from different cultures meet, they can have difficulty relating, because they are not 'in sync'. This is significant because synchrony, the subtle ability to move together, is imperative for all collaborations. Therefore, respecting cross-cultural differences by paying attention to aspects of 'the silent language' is paramount (Hall and Reed Hall, 2001). For example, the processes of observation, recording and other typical Western means of generating data are often in direct opposition to the way knowledge is traditionally shared in Aboriginal society. Whereas physical possessions have been traditionally valueless to indigenous Australians, knowledge has always been treated as a significant possession. This means researchers cannot simply sit down and begin asking questions or they risk exhibiting rudeness severe enough to ruin their chances of ever getting close enough to learn (Zur, 2007). Any reluctance to share traditional knowledge with outsiders may be due to cultural reasons or mistrust regarding the way that the information will be used. A lack of building appropriate trust relationships and respect for the worldview of Aboriginal people from researchers can potentially cause an unwillingness to disclose knowledge (Oliver, 2013). The concept of equality in knowledge sharing means that, in return for traditional knowledge, researchers must be able to offer something back to the community that is seen as valuable. However, it is important to avoid neocolonialism, which could occur if the researcher attempts to impose West-centric structures and ideas in a completely inappropriate context which in no way meets the needs of the people sharing their knowledge. Therefore, an essential first step is to communicate effectively and respectfully how the results of one's research may help the people involved (Zur, 2007).

Besides disparities in time perception and the concept of sharing, other cross-cultural differences may hinder information exchange. For example, where there is a language barrier, a skilful interpreter might be required. However, it is vital that the interpreter chosen must have an appropriate awareness of cultural nuances and obey the unwritten laws of custom so as not to offend the local people (Hall and Reed Hall, 2001). While we have gleaned some of what our ancestors knew from current and ancient texts, valuable data have not always been recorded (Lewis and Elvin-Lewis, 1977). This is especially true of traditional indigenous cultures, including Australian Aboriginals, who did not have a written language. Additionally, some of the knowledge is highly sacred and is often kept secret, even from members of the particular culture (Fabricant and Farnsworth, 2001; Low, 1990; Turton, 1997). It therefore makes sense to study the practices of indigenous peoples in case they are lost forever, either through neglect or because the vegetation has been irrevocably changed. Today's researchers, with broader insight and scientific expertise, have much greater scope to utilise this information than any previous generation (Lewis and Elvin-Lewis, 1977). Some might consider it tantamount to gross negligence if we did not seize the opportunity before it is too late.

OTHER COMMERCIAL PROSPECTS

Aboriginal people have historically shared their traditional knowledge generously. Indeed, it was their open revelation of details of edible seeds, fruits and roots, water sources, and how best to catch animals that convinced sceptical outback settlers of the authenticity of Aboriginal claims to tribal lands (Isaacs, 1987). However, as a consequence of relying on oral medicinal lore and having no written language, as well as young Aboriginals moving away from traditional lands and showing a declining interest in their heritage, this well of information is in serious danger of drying up. For this reason, a project aimed at determining the therapeutic effectiveness of Aboriginal traditional medicines was undertaken in the mid-1980s. The importance of this endeavour was recognised by both the Northern Territory and federal governments who jointly supported the project as part of a bicentennial commemorative programme (Stack, 1989). It resulted in the publication of an Aboriginal pharmacopoeia, which linked traditional Aboriginal medicine with modern science (ACNTA, 1988). Over 40 Aboriginal communities shared their traditional ethnomedical knowledge about specific bush medicines which is detailed in the internationally recognised document, along with a botanical description for each species and the chemical composition and therapeutic activity known to that time. More recently, the web-based Customary Medicinal Knowledgebase (http://biolinfo.org/cmkb/index.php) has been developed as an integrated multidisciplinary resource to document, conserve and disseminate the valuable Aboriginal traditional knowledge that is scattered throughout the published literature and amongst various Aboriginal communities (Gaikwad et al., 2008).

Provided that local guides can be persuaded to help, this procedure could be repeated for other Aboriginal communities living in different regions of Australia. In this way, not only will Aboriginal traditional knowledge be recorded for preservation, but it will also hopefully be scientifically validated too. By using ethnopharmacological knowledge from other Australian landscapes, the chances of discovering a novel therapeutically useful agent will be multiplied.

Even if a novel compound was not identified, the fact that extracts of the medicinal plant were bioactive would still be significant. In the end, if bioactivity is not due to an NCE that can be exploited, then maybe it does not really matter what the active principle(s) are. Perhaps the real point is that, regardless of what compound(s) are responsible, something in the plant, acting alone or in combination, is bioactive. It seems likely that the various active constituents of extracts have to work as a whole, not in isolation, to be most effective. This would support the idea of the usefulness of traditional herbal medicines and validate Aboriginal traditional knowledge, reconciling traditional Aboriginal knowledge with Western medicinal ideals. While many more experiments would be required before any drug could be developed, preliminary data may pique the interest of a pharmaceutical company willing to invest in such a project. In such a case, the Aboriginal communities would have to be considered as major stakeholders of the enterprise. If not, marketable products based on some plants screened may still be feasible. As mentioned in the introduction, there is a growing demand for medicines based on T/CAM, so the communities may choose to develop extracts of plants as alternative remedies for treatment of particular ailments and supply an industry partner with the required raw materials.

An excellent example of a successful international company founded on Aboriginal ethnomedical knowledge is Mount Romance, a manufacturer of a range of beauty products based on the sandalwood tree, *Santalum spicatum*, and located in Albany, Western Australia (Mount Romance. http://www.mtromance.com.au/). A groundbreaking partnership between Mt Romance, Aveda (a U.S.-based multinational cosmetics corporation) and the Kutkabubba Aboriginal community (represented by the Songman Circle of Wisdom) was launched in 2004 at Murdoch University in Perth, representing the world's first case of indigenous intellectual accreditation. According to this protocol, Aveda and Mt Romance each donated $50,000 to the Kutkabubba community for using the land and knowledge of the Aboriginal people. Ongoing funds are used by the community for any purpose they desire. Under the accreditation protocol, the partnership provides a new approach to protecting indigenous knowledge, which is very different from the current patenting regime for intellectual property. The Indigenous Accreditation is a voluntary undertaking under a sustainability framework, allowing a holistic approach to indigenous knowledge (Marinova and Raven, 2006). As such, it represents a good model for partnerships between indigenous communities and other groups who wish to apply their traditional knowledge or employ their skills for commercial gain.

Another option is for Aboriginal communities themselves to set up a cottage industry based on skin creams or lotions to be sold directly to the public. Various examples of such cottage-style industry products traditionally used by Aboriginals exist. Many, such as the herbal remedy Gumbi(y) Gumbi(y) (*Pittosporum phylliraeoides*) and emu oil, are sold at local markets or via various internet sites and can be found simply by doing an online search (Essentially Bliss, http://www.essentiallybliss.com.au; Gumby Gumby, http://www.gumbygumby.com/). Similarly, the Scotdesco Aboriginal community operates several enterprising schemes, including running Aboriginal cultural awareness presentations and Wirangu language and cultural experience camps. In addition, they sell books on the Wirangu language, various arts and crafts, and jars of Gujaru, a natural bush remedy to reduce pain, relieve breathing difficulties and dry itchy skin or rashes (Scotdesco Aboriginal Community, http://www.scotdesco.com/scotdesco/about.htm). Provided that further tests indicate no human toxicity, other herbal remedies based on any of the plants that showed bioactivity could be similarly developed and marketed as traditional bush medicines with the additional selling point that the effects have been scientifically validated. However, even if it proves to be unviable to profit financially from this knowledge, it is hoped that Aboriginal people who shared their traditional knowledge of medicinal plants will still feel empowered by the realisation that it has been scientifically veri-fied and therefore seen to be valuable by the Western world.

TRADITIONAL KNOWLEDGE AND WESTERN MEDICINE

The first point that must be recognised is that traditional methods of making herbal medicines, treatment of the extract prior to administration, the effects of other substances used in herbal mixtures and the dose used should all be taken into account when scientifically evaluating the efficacy or mechanism of action of traditional med-icines (Houghton et al., 2007). For example, sometimes, a traditional medicine is prescribed for a substantial length of time, often with no significant improvement expected for weeks or even months. This type of prolonged dosage is usually not con-sidered in new drug-screening programmes, largely due to a lack of awareness on the part of the researchers but also because of practical reasons. In vitro technologies are not suitable for assessing the effects of repeated interactions with tissues or molecular targets, which also requires large quantities of test materials. Thus, traditional claims of efficacy may be falsely rejected (Etkin and Elisabetsky, 2005).

Second, and related to this, is that traditional medical practices are generally holistic in nature. For example, many traditional remedies have been shrouded in religion and even magic over the centuries (Dubick, 1986). Religion and mysticism can often play a big part in the lives of indigenous cultures and their beliefs, influenc-ing their medical treatments much more commonly than in conventional, Western therapies, and this is certainly the case for traditional Aboriginal medicine (Maher, 1999). While Western medicine often seems to underestimate or ignore the power of the human spirit (Cassell, 1998), in a 1996 survey of 296 practicing U.S. physicians, 99% were convinced that religious beliefs can heal, and 75% believed that prayers of others could promote a patient's recovery (Sloan et al., 1999). Both systems of health-care have their place and should be free to operate according to their own medical standards (Fan and Holliday, 2007). However, if the goal of ethnopharmacological research is to discover a new drug, then evidence-based, placebo-controlled experi-ments are an absolute requirement.

Additionally, it is important to realise that the continued knowledge of, and use of, these resources requires not only their recognition as local knowledge, but also their multidisciplinary study. Obviously, this is only possible if the traditional keep-ers of this knowledge have a say in its future use and benefit somehow from such research and development (Heinrich and Bremner, 2006). It can be a delicate and time-consuming process to earn the trust of Aboriginal community leaders so as to enlist their cooperation in identifying plants with traditional medicinal value.

It has been suggested that because of the holistic approach of traditional medicine, a holistic approach to study drug activity of these medicines seems more appropriate than a reductionist approach (Verpoorte et al., 2006). Etkin and Elisabetsky (2005) argued for a transdisciplinary approach to ethnopharmacology, spanning both the biological and social sciences, creating a dynamic link that encourages dialogue and collaboration. Currently, ethnopharmacology is heavily focussed on the biology and pharmacology of the (mostly) botanical sources of potential drugs, only sometimes touching on areas like cultural anthropology (Gertsch, 2009). A truly integrated approach would aid in facing the particular challenges of ethnopharmacology.

CONCLUSIONS

Australia is a vast continent, covering a diverse range of botanical environments, including tropical coast, rainforest, open scrub, wet sclerophyll forest, woodland, desert, temperate riverine and alpine areas. Aboriginal people once lived in all these zones and continue to live in most, utilising the natural foods and medicines unique to Australia (Isaacs, 1987). Aboriginal Australians maintain the oldest continuous culture on earth, but their wealth of phytochemical knowledge is largely unknown to the Western world. To date, only a limited number of plants used as traditional medicines by Australian Aboriginal people have been investigated in order to establish the biologically active constituents present (Locher and Currie, 2010). There is still great scope for further study in this area. Many of the ethnomedicines could potentially be used as T/CAMs or even provide leads for new drugs. However, simply scientifically verifying the medicinal value of the extracts of plants traditionally used as medicines by Aboriginal people has its own merit. The validation of the bioactivity of their medicines will hopefully give the indigenous communities who provided the traditional knowledge a sense of empowerment and can only add to their rich heritage. It might also inspire other Aboriginal communities to share their traditional knowledge of plant medicines to use as alternative therapies or to aid in the never-ending search for new compounds to combat diseases.

ACKNOWLEDGEMENT

The author's original research was undertaken with some financial support of the Australian Government Cooperative Research Centres Program through the Desert Knowledge Cooperative Research Centre (DKCRC) and its parent organisation Ninti One Limited (http://www.nintione.com.au/). The views expressed here do not, however, necessarily reflect the views of all Ninti One Limited partners or participant organisations of the DKCRC.

REFERENCES

ACNTA, 1988. *Traditional Bush Medicines: An Aboriginal Pharmacopoeia*. Greenhouse Publications, Melbourne, Victoria, Australia.

Afifi, W.M., Ragab, E.H., Mohammed, A.I., El-Hela, A.A., 2014. Chemical constituents and biological activity of *Ficus platypoda* (Miq.) leaves. *Journal of Biomedical and Pharmaceutical Research* 3, 21–37.

Afzal, M., Mehdi, F.S., Abbasi, F.M. et al., 2011. Efficacy of *Avicennia marina* (Forsk.) Vierh. leaves extracts against some atmospheric fungi. *African Journal of Biotechnology* 10, 10790–10794.

Ahmed, S., Sultana, M., Hasan, M.M.U., Azhar, I., 2011. Analgesic and antiemetic activity of *Cleome viscosa* L. *Pakistani Journal of Botany* 43, 119–122.

Akiyama, H., Fujii, K., Yamasaki, O., Oono, T., Iwatsuki, K., 2001. Antibacterial action of several tannins against *Staphylococcus aureus*. *Journal of Antimicrobial Chemotherapy* 48, 487–491.

Alam, F., Shahriar, M., Bhuiyan, M.A., 2014. In vivo phamacological investigations of bark extracts of *Carissa carandas*. *International Journal of Pharmacy and Pharmaceutical Sciences* 6, 180–185.

Alipieva, K., Korkina, L., Orhan, I.E., Georgiev, M.I., 2014. Verbascoside – A review of its occurrence, (bio)synthesis and pharmacological significance. *Biotechnology Advances* 32, 1065–1076.

Alter, J.M., 2000. International biopiracy versus the value of local knowledge. *Capitalism Nature Socialism* 11, 59–66.

Amabeoku, G.J., Eagles, P., Scott, G., Mayeng, I., Springfield, E., 2001. Analgesic and antipyretic effects of *Dodonaea angustifolia* and *Salvia africana-lutea*. *Journal of Ethnopharmacology* 75, 117–124.

Ansah, C., Gooderham, N.J., 2002. The popular herbal antimalarial, extract of *Cryptolepis sanguinolenta*, is potently cytotoxic. *Toxicological Sciences* 70, 245–251.

Astin, J.A., 1998. Why patients use alternative medicine: Results of a national study. *Journal of the American Medical Association* 279, 1548–1553.

Ávila-Peña, D., Peña, N., Quintero, L., Suárez-Roca, H., 2007. Antinociceptive activity of *Syzygium jambos* leaves extract on rats. *Journal of Ethnopharmacology* 112, 380–385.

Baddeley, G.V., Bealing, A.J., Jefferies, P.R., Retallack, R.W., 1964a. The chemistry of the Euphorbiaceae-VI: A triterpene from *Beyeria leschenaultii*. *Australian Journal of Chemistry* 17, 908–914.

Baddeley, G.V., Jefferies, P.R., Retallack, R.W., 1964b. Chemistry of the Euphorbiaceae-IX: Further constituents of *Beyeria leschenaultii*. *Tetrahedron* 20, 1983–1985.

Bandaranayake, W.M., 1998. Traditional and medicinal uses of mangroves. *Mangroves and Salt Marshes* 2, 133–148.

Bandaranayake, W.M., 2002. Bioactivities, bioactive compounds and chemical constituents of mangrove plants. *Wetlands Ecology and Management* 10, 421–452.

Bawankule, D.U., Chattopadhyay, S.K., Pal, A., Saxena, K., Yadav, S., Faridi, U., Darokar, M.P., Gupta, A.K., Khanuja, S.P., 2008. Modulation of inflammatory mediators by coumarinolignoids from *Cleome viscosa* in female swiss albino mice. *Inflammopharmacology* 16, 272–277.

Bhaskar, V.H., Balakrishnan, N., 2009. Analgesic, anti-inflammatory and antipyretic activities of *Pergularia daemia* and *Carissa carandas*. *DARU Journal of Pharmaceutical Sciences* 17, 168–174.

Bodeker, G., Kronenberg, F., 2002. A public health agenda for traditional, complementary, and alternative medicine. *American Journal of Public Health* 92, 1582–1591.

Campbell, J., 1983. Smallpox in Aboriginal Australia, 1829–31. *Historical Studies* 20, 536–556.

Campbell, J., 2007. *Invisible Invaders: Smallpox and Other Diseases in Aboriginal Australia, 1780–1880*. Melbourne University Press, Carlton, Victoria, Australia, pp. xiv, 266. Available at: http://search.informit.com.au/documentSummary;dn=248035834412189; res=IELIND EISBN: 0522849393 (cited 09 July 2015).

Carrick, S., Clapham, K., Paul, C., Plant, A., Redman, S., 1996. *Breast Cancer and Aboriginal and Torres Strait Islander Women*. NHMRC National Breast Cancer Centre, Canberra, Australian Capital Territory, Australia.

Carson, C.F., Hammer, K.A., Riley, T.V., 2006. *Melaleuca alternifolia* (Tea Tree) oil: A review of antimicrobial and other medicinal properties. *Clinical Microbiology Reviews* 19, 50–62.

Cassell, E.J., 1998. The nature of suffering and the goals of medicine. *Loss, Grief and Care* 8, 129–142.

CCWA, 2001. Annual report 2000/01. Department of Mineral and Petroleum Resources, Perth, Western Australia, Australia.

Chacko, S., Sambuc, H.P., 2003. Blockbusters, traditional knowledge and intellectual property. *Indigenous Law Bulletin* 5, 12–13.

Chandrasekaran, M., Kannathasan, K., Venkatesalu, V., Prabhakar, K., 2009. Antibacterial activity of some salt marsh halophytes and mangrove plants against methicillin resistant *Staphylococcus aureus*. *World Journal of Microbiology and Biotechnology* 25, 155–160.

Chang, S.T., Tung, Y.T., 2013. Use of *Acacia* extracts and their compounds on inhibition of xanthine oxidase. United States Patent. Patent No. 8,877,260 B2.

Chapman, A.D., 2009. *Numbers of Living Species in Australia and the World*. Australian Government, Canberra, Australian Capital Territory, Australia.

Chapuis, J.C., Sordat, B., Hostettmann, K., 1988. Screening for cytotoxic activity of plants used in traditional medicine. *Journal of Ethnopharmacology* 23, 273–284.

Chawla, R., Kumar, S., Sharma, A., 2012. The genus *Clematis* (Ranunculaceae): Chemical and pharmacological perspectives. *Journal of Ethnopharmacology* 143, 116–150.

Chung, L.Y., Soo, W.K., Chan, K.Y., Mustafa, M.R., Goh, S.H., Imiyabir, Z., 2009. Lipoxygenase inhibiting activity of some Malaysian plants. *Pharmaceutical Biology* 47, 1142–1148.

Çitoğlu, G., Tanker, M., Gümüşel, B., 1998. Anti inflammatory effects of lycorine and haemanthidine. *Phytotherapy Research* 12, 205–206.

Çitoğlu, G.S., Acıkara, Ö.B., Yılmaz, B.S., Özbek, H., 2012. Evaluation of analgesic, antiinflammatory and hepatoprotective effects of lycorine from *Sternbergia fisheriana* (Herbert) Rupr. *Fitoterapia* 83, 81–87.

Clapp, R.A., Crook, C., 2002. Drowning in the magic well: Shaman Pharmaceuticals and the elusive value of traditional knowledge. *The Journal of Environment Development* 11, 79–102.

Clardy, J., Walsh, C., 2004. Lessons from natural molecules. *Nature* 432, 829–837.

Clarke, P.A., 2007. *Aboriginal People and Their Plants*. Rosenberg Publishing, Dural Delivery Centre, Dural, New South Wales, Australia.

Clarke, P.A., 2008. Aboriginal healing practices and Australian bush medicine. *Journal of the Anthropological Society of South Australia* 33, 3–38.

Cleland, J.B., 1931. Plants, including fungi, poisonous or otherwise injurious to Man in Australia. Series III. *The Medical Journal of Australia* ii(25), 775.

Cock, I.E., 2011. *Medicinal and Aromatic Plants – Australia, Ethnopharmacology Section, Biological, Physiological and Health Sciences, Encyclopedia of Life Support Systems (EOLSS)*. EOLSS, Oxford, U.K.

Coffeen, U., Lopez-Avila, A., Ortega-Legaspi, J.M., Angel, R.D., Lopez-Munoz, F.J., Pellicer, F., 2008. Dopamine receptors in the anterior insular cortex modulate long-term nociception in the rat. *European Journal of Pain* 12, 535–543.

Colegate, S.M., Molyneux, R.J., 2008. An introduction and overview. In: Colegate, S.M., Molyneux, R.J. (eds.), *Bioactive Natural Products: Detection, Isolation and Structural Determination*. CRC Press, Boca Raton, FL, pp. 1–10.

Cordell, G.A., Beecher, C.W.W., Pezzuto, J.M., 1991. Can ethnopharmacology contribute to the development of new anticancer drugs? *Journal of Ethnopharmacology* 32, 117–133.

Cowan, M.M., 1999. Plant products as antimicrobial agents. *Clinical Microbiology Reviews* 12, 564–582.

Cragg, G.M., Newman, D.J., Snader, K.M., 1997. Natural products in drug discovery and development. *Journal of Natural Products* 60, 52–60.

Crago, J., 2011. *Nature's Helping Hand – Scaevola spinescens, History and Use in Western Australia: The Maroon Bush Story*. Aussie Outback Books, Mt Helena, Western Australia, Australia.

Cribb, A.B., Cribb, J.W., 1981. *Wild Medicine in Australia*. Collins, Sydney, New South Wales, Australia.

Croft, K.D., Ghisalberti, E.L., Jefferies, P.R., 1978. Biosynthesis of ent-kauranes: Evidence for a C-17, C-16 hydrogen 1,2-shift. *Phytochemistry* 17, 695–699.

Cunningham, J., Rumbold, A.R., Zhang, X., Condon, J.R., 2008. Incidence, aetiology, and outcomes of cancer in Indigenous peoples in Australia. *The Lancet Oncology* 9, 585–595.

Curir, P., Beruto, M., Dolci, M., 1995. *Eucalyptus* species: In vitro culture and production of essential oils and other secondary metabolites. In: Bajaj, Y.P.S. (ed.), *Biotechnology in Agriculture and Forestry*, Volume 33: Medicinal and Aromatic Plants VIII. Springer-Verlag, Berlin, Germany, pp. 194–214.

Daw, B., Walley, T., Keighery, G., 1997. *Bush Tucker Plants of the South-West*. Department of Conservation and Land Management, Perth, Western Australia, Australia.

Debboun, M., Frances, S.P., Strickman, D. (eds.), 2015. *Insect Repellents Handbook*, 2nd ed. CRC Press, Boca Raton, FL.

Dobson, C.M., 2004. Chemical space and biology. *Nature* 432, 824–828.

Dobson, V.P., 2007. *Arrernte Traditional Healing*. IAD Press, Alice Springs, Northern Territory, Australia.

dos Santos, M.D., Almeida, M.C., Lopes, N.P., de Souza, G.E.P., 2006. Evaluation of the anti-inflammatory, analgesic and antipyretic activities of the natural polyphenol chlorogenic acid. *Biological and Pharmaceutical Bulletin* 29, 2236–2240.

Dubick, M.A., 1986. Historical perspectives on the use of herbal preparations to promote health. *Journal of Nutrition* 116, 1348–1354.

Dufault, R.J., Hassell, R., Rushing, J.W., McCutcheon, G., Shepard, M., Keinath, A., 2001. Revival of herbalism and its roots in medicine. *Journal of Agromedicine* 7, 21–29.

Duke, J.A., 1990. Promising phytomedicinals. In: Janick, J., Simon, J.E. (eds.), *Advances in New Crops*. Timber Press, Portland, OR, pp. 491–498.

Elisabetsky, E., Costa-Campos, L., 2006. The alkaloid alstonine: A review of its pharmacological properties. *Evidence-Based Complementary and Alternative Medicine* 3, 39–48.

Erdelmeier, C.A., Hauer, H., Sticher, O., Rali, T., 1994. 10-Deoxysecogalioside: A new iridoid glycoside from *Timonius timon*. *Planta Medica* 60, 484–485.

Etkin, N.L., Elisabetsky, E., 2005. Seeking a transdisciplinary and culturally germane science: The future of ethnopharmacology. *Journal of Ethnopharmacology* 100, 23–26.

Everist, S.L., 1974. *Poisonous Plants of Australia*. Angus & Robertson, Sydney, New South Wales, Australia.

Fabricant, D.S., Farnsworth, N.R., 2001. The value of plants used in traditional medicine for drug discovery. *Environmental Health Perspectives* 109(Suppl. 1), 69–75.

Fan, R., Holliday, I., 2007. Which medicine? Whose standard? Critical reflections on medical integration in China. *Journal of Medical Ethics* 3, 454–461.

Farivar-Mohseni, H., Kandzari, S.J., Zaslau, S., Riggs, D.R., Jackson, B.J., McFadden, D.W., 2004. Synergistic effects of COX-1 and -2 inhibition on bladder and prostate cancer in vitro. *American Journal of Surgery* 188, 505–510.

Farnsworth, N.R., Bingel, A.S., 1977. Problems and prospects of discovering new drugs from higher plants by pharmacological screening. In: Wagner, H., Wolff, P. (eds.), *New Natural Products and Plant Drugs with Pharmacological, Biological or Therapeutical Activity*. Springer-Verlag, Berlin, Germany, pp. 1–22.

Feher, M., Schmidt, J.M., 2003. Property distributions: Differences between drugs, natural products, and molecules from combinatorial chemistry. *Journal of Chemical Information and Modeling* 43, 218–227.

Fennell, D., Liberato, A.S.Q., and Zsembik, B., 2009. Definitions and patterns of CAM use by the lay public. *Complementary Therapies in Medicine* 17, 71–77.

Fletcher, M.T., McKenzie, R.A., Blaney, B.J., Reichmann, K.G., 2008. Pyrrolizidine alkaloids in *Crotalaria* taxa from Northern Australia: Risk to grazing livestock. *Journal of Agricultural and Food Chemistry* 57, 311–319.

Gaikwad, J., Khanna, V., Vemulpad, S., Jamie, J., Kohen, J., Shoba Ranganathan, S., 2008. CMKb: A web-based prototype for integrating Australian Aboriginal customary medicinal plant knowledge. *BMC Bioinformatics* 9(Suppl. 12), S25.

Galipalli, S., Patel, N.K., Prasanna, K., Bhutani, K.K., 2015. Activity-guided investigation of *Carissa carandas* (L.) roots for anti-inflammatory constituents. *Natural Product Research* 29(17), 1670–1672.

Gandomani, M.Z., Molaali, E.F., Gandomani, Z.Z., Madani, H., Seyed, J.M., 2012. Evaluation of anti-inflammatory effect of hydroalcoholic extract of mangrove (*Avicennia marina*) leaves in male rats. *Medical Journal of Tabriz University of Medical Sciences* 34, 80–85.

Geeta, V.D.M., Kedlaya, R., Deepa, S., Ballal, M., 2001. Activity of *Ocimum sanctum* (the traditional Indian medicinal plant) against the enteric pathogens. *Indian Journal of Medical Sciences* 55, 434–438.

Gertsch, J., 2009. How scientific is the science in ethnopharmacology? Historical perspectives and epistemological problems. *Journal of Ethnopharmacology* 122, 177–183.

Ghisalberti, E.L., 1994. The ethnopharmacology and phytochemistry of *Eremophila* species (Myoporaceae). *Journal of Ethnopharmacology* 44, 1–9.

Ghisalberti, E.L., 1995. *Eremophila* species (Poverty bush; Emu bush): In vitro culture and the production of verbascoside. In: Bajaj, Y.P.S. (ed.), *Biotechnology in Agriculture and Forestry*, Vol. 33: Medicinal and Aromatic Plants VIII. Springer-Verlag, Berlin, Germany, pp. 176–193.

Ghisalberti, E.L., 1998a. Ethnopharmacology and phytochemistry of *Dodonaea* species. *Fitoterapia* 69, 99–113.

Ghisalberti, E.L., 1998b. Biological and pharmacological activity of naturally occurring iridoids and secoiridoids. *Phytomedicine* 5, 147–163.

Ghisalberti, E.L., 2004. The Goodeniaceae. *Fitoterapia* 75, 429–446.

Ghisalberti, E.L., Jefferies, P.R., 1968. The chemistry of the Euphorbiaceae. XVIII. Modified Beyer-15-enes from a *Beyeria leschenaultii* Var. *Australian Journal of Chemistry* 21, 439–457.

Godhwani, S., Godhwani, J.L., Vyas, D.S., 1987. *Ocimum sanctum*: An experimental study evaluating its anti-inflammatory, analgesic and antipyretic activity in animals. *Journal of Ethnopharmacology* 21, 153–163.

Gollin, M.A., 1999. New rules for natural product research. *Nature Biotechnology* 17, 921–922.

Goyal, P.K., 2012. Phytochemical and pharmacological properties of the genus *Grewia*: A review. *International Journal of Pharmacy and Pharmaceutical Sciences* 4(S4), 72–78.

Grice, I.D., Garhnam, B., Pierens, G., Rogers, K., Tindal, D., Griffiths, L.R., 2003. Isolation of two phenylethanoid glycosides from *Eremophila gilesii*. *Journal of Ethnopharmacology* 86, 123–125.

Grice, I.D., Rogers, K.L., Griffiths, L.R., 2011. Isolation of bioactive compounds that relate to the anti-platelet activity of *Cymbopogon ambiguus*. *Evidence-Based Complementary and Alternative Medicine* 2011, 1–8.

Grieve, M., 1931. *A Modern Herbal*, Vol. 1. Hafner Publishing, Darien, CT.

Guarrera, P.M., Lucia, L.M., 2007. Ethnobotanical remarks on central and southern Italy. *Journal of Ethnobiology and Ethnomedicine* 3, 23.

Hall, E.T., Reed, H.M., 2001. Key concepts: Underlying structures of culture. In: Albrecht, M.H. (ed.), *International HRM: Managing Diversity in the Workplace*. Blackwell, Oxford, U.K., pp. 24–40.

Hamilton, C., 2006. Biodiversity, biopiracy and benefits: What allegations of biopiracy tell us about intellectual property. *Developing World Bioethics* 6, 158–173.

Hampson, P., Chahal, H., Khanim, F., Hayden, R., Mulder, A., Assi, L.K., Bunce, C.M., Lord, J.M., 2005. PEP005, a selective small-molecule activator of protein kinase C, has potent antileukemic activity mediated via the delta isoform of PKC. *Blood* 106, 1362–1368.

Hao, D.C., Gu, X.J., Xiao, P.G., Peng, Y., 2013. Chemical and biological research of *Clematis* medicinal resources. *Chinese Science Bulletin* 58, 1120–1129.

Haslam, E., 1996. Natural polyphenols (vegetable tannins) as drugs: Possible modes of action. *Journal of Natural Products* 59, 205–215.

Heinrich, M., Bremner, P., 2006. Ethnobotany and ethnopharmacy – Their role for anti-cancer drug development. *Current Drug Targets* 7, 239–245.

Hettiarachchi, D.S., Locher, C., Longmore, R.B., 2011. Antibacterial compounds from the root of the indigenous Australian medicinal plant *Carissa lanceolata* R.Br. *Natural Product Research* 25, 1388–1395.

Hildreth, J., Hrabeta-Robinson, E., Applequist, W., Betz, J., Miller, J., 2007. Standard operating procedure for the collection and preparation of voucher plant specimens for use in the nutraceutical industry. *Analytical and Bioanalytical Chemistry* 389, 13–17.

Hornfeldt, C.S., Collins, J.E., 1990. Toxicity of nightshade berries (*Solanum dulcamara*) in mice. *Clinical Toxicology* 28, 185–192.

Houghton, P.J., Howes, M.J., Lee, C.C., Steventon, G., 2007. Uses and abuses of in vitro tests in ethnopharmacology: Visualizing an elephant. *Journal of Ethnopharmacology* 110, 391–400.

Hungerford, N.L., Sands, D.P.A., Kitching, W., 1998. Isolation and structure of some constituents of the Australian medicinal plant *Tinospora smilacina* ('Snakevine'). *Australian Journal of Chemistry* 51, 1103–1112.

Hurst, D., 2009. Old wives' tale to fight skin cancer. *Brisbane Times*, 18th May. http://www.brisbanetimes.com.au/queensland/old-wives-tale-to-fight-skin-cancer-20090518-bca6.html (accessed 3 April 2010).

Isaacs, J., 1987. *Bush Food: Aboriginal Food and Herbal Medicine*. Weldons Pty Ltd, Sydney, New South Wales, Australia.

Isacchi, B., Iacopi, R., Bergonzi, M.C., Ghelardini, C., Galeotti, N., Norcini, M., Vivoli, E., Vincieri, F.F., Bilia, A.R., 2011. Antihyperalgesic activity of verbascoside in two models of neuropathic pain. *Journal of Pharmacy and Pharmacology* 63, 594–601.

Jachak, S.M., Saklani, A., 2007. Challenges and opportunities in drug discovery from plants. *Current Science* 92, 1251–1257.

Jassbi, A.R., 2006. Chemistry and biological activity of secondary metabolites in *Euphorbia* from Iran. *Phytochemistry* 67, 1977–1984.

Jia, Q., Nichols, T.C., Rhoden, E., Waite, S., 2006. Isolation of a dual COX-2 and 5-lipoxygenase inhibitor from *Acacia*. United States Patent. Patent No. 7,108,868.

Jia, Q., Nichols, T.C., Rhoden, E.E., Waite, S., 2012. Isolation of a dual COX-2 and 5-lipoxygenase inhibitor from *Acacia*. United States Patent. Patent No. 8,124,134 B2.

Jia, Q., Nichols, T.C., Rhoden, E.E., Waite, S., 2013. Isolation of a dual COX-2 and 5-lipoxygenase inhibitor from *Acacia*. United States Patent. Patent No. 8,568,799.

Kandati, V., Govardhan, P., Reddy, C.S., Nath, A.R., Reddy, R.R., 2012. In vitro and in vivo anti-inflammatory activity of *Syzygium alternifolium*. *Journal of Medicinal Plants Research* 6, 4995–5001.

Kelm, M.A., Nair, M.G., Strasburg, G.M., DeWitt, D.L., 2000. Antioxidant and cyclooxygenase inhibitory phenolic compounds from *Ocimum sanctum* Linn. *Phytomedicine* 7, 7–13.

Kerr, P.G., 1999. *Scaevola spinescens* R.Br. (Goodeniaceae) phytochemical and biological screening studies of a West Australian medicinal plant. PhD thesis, School of Pharmacy, Curtin University of Technology, Bentley, Western Australia, Australia.

Kerr, P.G., Longmore, R.B., Betts, T.J., 1996. Myricadiol and other taraxerenes from *Scaevola spinescens*. *Planta Medica* 62, 519–522.

Khafagi, I., Gab-Alla, A., Salama, W., Fouda, M., 2003. Biological activities and phytochemical constituents of the gray mangrove *Avicennia marina* (Forssk.) Vierh. *Egyptian Journal of Biology* 5, 62–69.

Khan, I.A., Sticher, O., Rali, T., 1993. New triterpenes from the leaves of *Timonius timon*. *Journal of Natural Products* 56, 2163–2165.

Khanna, N., Bhatia, J., 2003. Antinociceptive action of *Ocimum sanctum* (Tulsi) in mice: Possible mechanisms involved. *Journal of Ethnopharmacology* 88, 293–296.

Kong, J.M., Goh, N.K., Chia, L.S., Chia, T.F., 2003. Recent advances in traditional plant drugs and orchids. *Acta Pharmacologica Sinica* 24, 7–21.

Küpeli, E., Harput, U.S., Varel, M., Yesilada, E., Saracoglu, I., 2005. Bioassay-guided isolation of iridoid glucosides with antinociceptive and anti-inflammatory activities from *Veronica anagallis-aquatica* L. *Journal of Ethnopharmacology* 102, 170–176.

Lassak, E.V., McCarthy, T., 2001. *Australian Medicinal Plants*. Reed New Holland, Sydney, New South Wales, Australia.

Lassak, E.V., Southwell, I., 1972. The coumarins from the resin of *Euodia vitiflora*. *Australian Journal of Chemistry* 25, 2491–2496.

Latz, P.K., 1995. *Bushfires and Bushtucker: Aboriginal Plant Use in Central Australia*. IAD Press, Alice Springs, Northern Territory, Australia.

Lawai, D., Yunusa, D., 2013. *Dodonea viscosa* Linn: Its medicinal, pharmacological and phytochemical properties. *International Journal of Innovation and Applied Studies* 2, 477–483.

Lee, I.S., Jin, W.Y., Zhang, X., Hung, T.M., Song, K.S., Seong, Y.H., Bae, K., 2006. Cytotoxic and COX-2 inhibitory constituents from the aerial parts of *Aralia cordata*. *Archives of Pharmacal Research* 29, 548–555.

Lewis, W.H., Elvin-Lewis, M.P.F., 1977. *Medical Botany: Plants Affecting Man's Health*. John Wiley & Sons, New York.

Li, J.W.H., Vederas, J.C., 2009. Drug discovery and natural products: End of an era or an endless frontier? *Science* 325, 161–165.

Li, R.W., Lin, G.D., Leach, D.N., Waterman, P.G., Myers, S.P., 2006. Inhibition of COXs and 5-LOX and activation of PPARs by Australian *Clematis* species (Ranunculaceae). *Journal of Ethnopharmacology* 104, 138–143.

Li, R.W., Myers, S.P., Leach, D.N., Lin, G.D., Leach, G., 2003. A cross-cultural study: Anti-inflammatory activity of Australian and Chinese plants. *Journal of Ethnopharmacology* 85, 25–32.

Lim, T.K., 2012. *Syzygium suborbiculare*. In: Lim, T.K. (ed.), *Edible Medicinal and Non-Medicinal Plants*, Vol. 3: Fruits. Springer, Dordrecht, the Netherlands, pp. 789–790.

Lindsay, E.A., Berry, Y., Jamie, J.F., Bremner, J.B., 2000. Antibacterial compounds from *Carissa lanceolata* R.Br. *Phytochemistry* 55, 403–406.

Liu, Q., Harrington, D., Kohen, J.L., Vemulpad, S., Jamie, J.F., 2006. Bactericidal and cyclooxygenase inhibitory diterpenes from *Eremophila sturtii*. *Phytochemistry* 67, 1256–1261.

Locher, C., Currie, L., 2010. Revisiting kinos – An Australian perspective. *Journal of Ethnopharmacology* 128, 259–267.

Locher, C., Semple, S.J., Simpson, B.S., 2013. Traditional Australian Aboriginal medicinal plants: An untapped resource for novel therapeutic compounds? *Future Medicinal Chemistry* 5, 733–736.

Lord, J.M., 1987. The use of cytotoxic plant lectins in cancer therapy. *Plant Physiology* 85, 1–3.

Low, T., 1990. *Bush Medicine: A Pharmacopoeia of Natural Remedies*. Collins/Angus & Robertson Publishers, Sydney, New South Wales, Australia.

Luis, A., Gil, N., Amaral, M.E., Duarte, A.P., 2012. Antioxidant activities of extracts from *Acacia melanoxylon*, *Acacia dealbata* and *Olea europaea* and alkaloids estimation. *International Journal of Pharmacy and Pharmaceutical Sciences* 4(S1), 225–231.

Macilwain, C., 1998. When rhetoric hits reality in debate on bioprospecting. *Nature* 392, 535–537, 539–540.

Macleod, J.K., Rasmussen, H.B., 1999. A hydroxy-β-caryophyllene from *Pterocaulon serrulatum*. *Phytochemistry* 50, 105–108.

Maher, P., 1999. A review of 'traditional' aboriginal health beliefs. *Australian Journal of Rural Health* 7, 229–236.

Mahomoodally, M.F., Gurib-Fakim, A., Subratty, A.H., 2010. Screening for alternative antibiotics: An investigation into the antimicrobial activities of medicinal food plants of Mauritius. *Journal of Food Science* 75(3), M173–M177.

Mali, R.G., 2010. *Cleome viscosa* (wild mustard): A review on ethnobotany, phytochemistry, and pharmacology. *Pharmaceutical Biology* 48, 105–112.

Maria de Souza, M., Madeira, A., Berti, C., Krogh, R., Yunes, R.A., Cechinel-Filho, V., 2000. Antinociceptive properties of the methanolic extract obtained from *Ipomoea pes-caprae* (L.) R. Br. *Journal of Ethnopharmacology* 69, 85–90.

Marinova, D., Raven, M., 2006. Indigenous knowledge and intellectual property: A sustainability agenda. *Journal of Economic Surveys* 20, 587–605.

Marzoug, H.N.B., Romdhane, M., Lebrihi, A., Mathieu, F., Couderc, F., Abderraba, M., Khouja, M.L., Bouajila, J., 2011. *Eucalyptus oleosa* essential oils: Chemical composition and antimicrobial and antioxidant activities of the oils from different plant parts (stems, leaves, flowers and fruits). *Molecules* 16, 1695–1709.

McCarthy, J., Ratcliffe, A., 2009. *A Field Guide to the Edible, Medicinal and Useful Plants of Australia*, Vol. 1. Alchemy Publishing, Mount Burrell, New South Wales, Australia.

McClatchey, W., Stevens, J., 2001. An overview of recent developments in bioprospecting and pharmaceutical development. In: Saxena, P.K. (ed.), *Development of Plant-Based Medicines: Conservation, Efficacy and Safety.* Kluwer Academic Publishers, Dordrecht, the Netherlands, pp. 17–46.

McCormick, J.L., McKee, T.C., Cardellina, J.H., Boyd, M.R., 1996. HIV inhibitory natural products. 26. Quinoline alkaloids from *Euodia roxburghiana. Journal of Natural Products* 59, 469–471.

McDonald, H., 2006. Australian Aboriginal traditional healing practices. In: Dunning, T. (ed.), *Complementary Therapies and the Management of Diabetes and Vascular Disease.* Wiley, Chichester, U.K., pp. 272–290.

McFadden, D.W., Riggs, D.R., Jackson, B.J., Cunningham, C., 2006. Additive effects of COX-1 and COX-2 inhibition on breast cancer in vitro. *International Journal of Oncology* 29, 1019–1023.

McGrath, P., Holewa, H., Ogilvie, K., Rayner, R., Patton, M.A., 2006. Insights on Aboriginal peoples' views of cancer in Australia. *Contemporary Nurse* 22, 240–254.

Meilak, M., Palombo, E.A., 2008. Anti-mycobacterial activity of extracts derived from Australian medicinal plants. *Research Journal of Microbiology* 3, 535–538.

Melnick, S.J., 2006. Developmental therapeutics: Review of biologically based CAM therapies for potential application in children with cancer: Part I. *Journal of Pediatric Hematology/Oncology* 28, 221–230.

Melo, G.O.de., Malvar, D.do.C., Vanderlinde, F.A., Rocha, F.F., Pires, P.A., Costa, E.A., Matos, L.G.de., Kaiser, C.R., Costa, S.S., 2009. Antinociceptive and anti-inflammatory kaempferol glycosides from *Sedum dendroideum. Journal of Ethnopharmacology* 124, 228–232.

Mijajlovic, S., Smith, J., Watson, K., Parsons, P., Lloyd Jones, G., 2006. Traditional Australian medicinal plants: Screening for activity against human cancer cell lines. *Journal of the Australian Traditional-Medicine Society* 12, 129–132.

Mishra, P., Mishra, S., 2011. Study of the antibacterial activity of *Ocimum sanctum* extract against Gram positive and Gram negative bacteria. *American Journal of Food Technology* 6, 336–341.

Mohseni, H., Zaslau, S., McFadden, D., Riggs, D.R., Jackson, B.J., Kandzari, S., 2004. COX-2 inhibition demonstrates potent anti-proliferative effects on bladder cancer in vitro. *Journal of Surgical Research* 119, 138–142.

Mollika, S., Islam, N., Parvin, N., Kabir, A., Sayem, M.W., Luthfunnesa Md, Saha, R., 2014. Evaluation of analgesic, anti-inflammatory and CNS activities of the methanolic extract of *Syzygium samarangense* leave. *Global Journal of Pharmacology* 8, 39–46.

Morton, J.F., 1990. Mucilaginous plants and their uses in medicine. *Journal of Ethnopharmacology* 29, 245–266.

Morton, S., Sheppard, A., 2014. *Biodiversity: Science and Solutions for Australia*. CSIRO, Collingwood, Victoria, Australia.

Mukherjee, P.K., Saha, K., Murugesan, T., Mandal, M., Pal, Saha, B.P., 1998. Screening of anti-diarrhoeal profile of some plant extracts of a specific region of West Bengal, India. *Journal of Ethnopharmacology* 60, 85–89.

Mulyaningsih, S., Sporer, F., Reichling, J., Wink, M., 2011. Antibacterial activity of essential oils from *Eucalyptus* and of selected components against multidrug-resistant bacterial pathogens. *Pharmaceutical Biology* 49, 893–899.

Muruganandan, S., Srinivasan, K., Chandra, S., Tandan, S.K., Lal, J., Raviprakash, V., 2001. Anti-inflammatory activity of *Syzygium cumini* bark. *Fitoterapia* 72, 369–375.

Newman, D.J., Cragg, G.M., 2005. The discovery of anticancer drugs from natural sources. In: Zhang, L., Demain, A.L. (eds.), *Natural Products: Drug Discovery and Therapeutic Medicine*. Humana Press, Totowa, NJ, pp. 129–168.

Newman, D.J., Cragg, G.M., 2007. Natural products as sources of new drugs over the last 25 years. *Journal of Natural Products* 70, 461–477.

Ngaanyatjarra, Pitjantjatjara Yankunytjatjara Women's Council Aboriginal Corporation, 2013. *Traditional Healers of Central Australia: Ngangkari*. Magabala Books, Broome, Western Australia, Australia.

Ngulde, S.I., Sandabe, U.K., Barkindo, A.A., Tijjani, M.B., Hussaini, I.M., 2013. Antinociceptive and anti-inflammatory activities of the ethanol extract of *Carissa edulis* Vahl. root bark in rats and mice. *International Journal of Modern Biology and Medicine* 4, 85–95.

Northridge, M.E., 2002. Integrating ethnomedicine into public health. *American Journal of Public Health* 92, 1561.

O'Connell, J.F., Latz, P.K., Barnett, P., 1983. Traditional and modern plant use among the Alyawara of Central Australia. *Economic Botany* 37, 80–109.

O'Donnell, F., Ramachandran, V.N., Smyth, T.J.P., Smyth, W.F., Brooks, P., 2009. An investigation of bioactive phytochemicals in the leaves of *Melicope vitiflora* by electrospray ionisation ion trap mass spectrometry. *Analytica Chimica Acta* 634, 115–120.

Ogbourne, S.M., Hampson, P., Lord, J.M., Parsons, P., De Witte, P.A., Suhrbier, A., 2007. Proceedings of the First International Conference on PEP005. *Anticancer Drugs* 18, 357–362.

Ogbourne, S.M., Suhrbier, A., Jones, B. et al., 2004. Antitumor activity of 3-ingenyl angelate: Plasma membrane and mitochondrial disruption and necrotic cell death. *Cancer Research* 64, 2833–2839.

Olajide, O.A., Awe, S.O., Makinde, J.M., Ekhelar, A.I., Olusola, A., Morebise, O., Okpako, D.T., 2000. Studies on the anti-inflammatory, antipyretic and analgesic properties of *Alstonia boonei* stem bark. *Journal of Ethnopharmacology* 71, 179–186.

Oliver, S.J., 2013. The role of traditional medicine practice in primary health care within Aboriginal Australia: A review of the literature. *Journal of Ethnobiology and Ethnomedicine* 9, 46.

Packer, J., Brouwer, N., Harrington, D., Gaikwad, J., Heron, R., Yaegl Community Elders, Ranganathan, S., Vemulpad, S., Jamie, J., 2012. An ethnobotanical study of medicinal plants used by the Yaegl Aboriginal community in northern New South Wales, Australia. *Journal of Ethnopharmacology* 139, 244–255.

Palombo, E.A., Semple, S., 2001. Antibacterial activity of traditional Australian medicinal plants. *Journal of Ethnopharmacology* 77, 151–157.

Panahi, Y., Akhavan, A., Sahebkar, A., Hosseini, S.M., Taghizadeh, M., Akbari, H., Sharif, M.R., Imani, S., 2014. Investigation of the effectiveness of *Syzygium aromaticum*, *Lavandula angustifolia* and *Geranium robertianum* essential oils in the treatment of acute external otitis: A comparative trial with ciprofloxacin. *Journal of Microbiology, Immunology and Infection* 47, 211–216.

Parimaladevi, B., Boominathan, R., Mandal, S.C., 2002. Evaluation of anti-diarrheal activity of *Cleome viscosa* L. extract in rats. *Phytomedicine* 9, 739–742.

Parimaladevi, B., Boominathan, R., Mandal, S.C., 2003a. Studies on analgesic activity of *Cleome viscosa* in mice. *Fitoterapia* 74, 262–266.

Parimaladevi, B., Boominathan, R., Mandal, S.C., 2003b. Evaluation of antipyretic potential of *Cleome viscosa* Linn. (Capparidaceae) extract in rats. *Journal of Ethnopharmacology* 87, 11–13.

Parris, W.C.V., Smith, H.S., 2003. Alternative pain medicine. *Pain Practice* 3, 105–116.

Patel, S., 2013. Food, pharmaceutical and industrial potential of *Carissa* genus: An overview. *Reviews in Environmental Science and Bio/Technology* 12, 201–208.

Pattamadilok, D., Pengsuparp, T., Phummiratch, D., Ongpipattanakul, B., Meksuriyen, D., Kawanishi, K., Norito Kaneda, N., Suttisri, R., 2008. Canarosine: A new guanidine alkaloid from *Canavalia rosea* with inhibitory activity on dopamine D_1 receptors. *Journal of Asian Natural Products Research* 10, 915–918.

Patwardhan, B., Patwardhan, A., 2006. Traditional medicine for global health. Public health and intellectual property aspects. *Technology Monitor Special Feature: Traditional Medicine: S&T Advancement*, 17–22 November–December 2006.

Payne, T.G., Jefferies, P.R., 1973. The chemistry of *Dodonaea* spp-IV. Diterpene and flavonoid components of *D. attenuata*. *Tetrahedron* 29, 2575–2583.

Pearn, J., 2005. The world's longest surviving paediatric practices: Some themes of Aboriginal medical ethnobotany in Australia. *Journal of Paediatrics and Child Health* 41, 284–290.

Pengelly, A., 1999. *Dodonea viscosa*. *Australian Journal of Medical Herbalism* 11, 11–14.

Pennacchio, M., Alexander, E., Ghisalberti, E.L., Richmond, G.S., 1995. Cardioactive effects of *Eremophila alternifolia* extracts. *Journal of Ethnopharmacology* 47, 91–95.

Pennacchio, M., Alexander, E., Syah, Y.M., Ghisalberti, E.L., 1996a. The effect of verbascoside on cyclic 3′,5′-adenosine monophosphate levels in isolated rat heart. *European Journal of Pharmacology* 305, 169–171.

Pennacchio, M., Ghisalberti, E.L., 2000. Indigenous knowledge and pharmaceuticals. *Journal of Australian Studies* 24, 173–175.

Pennacchio, M., Kemp, A.S., Taylor, R.P., Wickens, K.M., Kienow, L., 2005. Interesting biological activities from plants traditionally used by Native Australians. *Journal of Ethnopharmacology* 96, 597–601.

Pennacchio, M., Syah, Y.M., Ghisalberti, E.L., Alexander, E., 1996b. Cardioactive compounds from *Eremophila* species. *Journal of Ethnopharmacology* 53, 21–27.

Petsko, G.A., 1996. For medicinal purposes. *Nature* 384(6604 Suppl.), 7–9.

Philip, K., 2001. Seeds of neo-colonialism? Reflections on ecological politics in the new world order. *Capitalism Nature Socialism* 12, 3–47.

Pilbeam, D.J., Lyon-Joyce, A.J., Bell, E.A., 1983. Occurrence of the pyrrolizidine alkaloid monocrotaline in *Crotalaria* seeds. *Journal of Natural Products* 46, 601–605.

Pimm, S.L., Raven, P., 2000. Biodiversity: Extinction by numbers. *Nature* 403(6772), 843–845.

Pimm, S.L., Russell, G.J., Gittleman, J.L., Brooks, T.M., 1995. The future of biodiversity. *Science* 269, 347–350.

Pongprayoon, U., Bohlin, L., Wasuwat, S., 1991. Neutralization of toxic effects of different crude jellyfish venoms by an extract of *Ipomoea pes-caprae* (L.) R. Br. *Journal of Ethnopharmacology* 35, 65–69.

Poompozhil, S., Kumarasamy, D., 2014. Studies on phytochemical constituents of some selected mangroves. *Journal of Academia and Industrial Research* 2, 590–592.

Quintans, J.S.S., Brito, R.G., Pedro Gregório, V. et al., 2014. Antinociceptive activity of *Syzygium cumini* leaves ethanol extract on orofacial nociception protocols in rodents. *Pharmaceutical Biology* 52, 762–766.

Rahman, M.A., Toi, M., 2003. Anti-angiogenic therapy in breast cancer. *Biomedicine and Pharmacotherapy* 57, 463–470.

Rajendran, P., Rengarajan, T., Nandakumar, N., Palaniswami, R., Nishigaki, Y., Nishigaki, I., 2014. Kaempferol, a potential cytostatic and cure for inflammatory disorders. *European Journal of Medicinal Chemistry* 86, 103–112.

Rao, R.J., Kumar, U.S., Reddy, S.V., Ashok, K.T., Rao, J.M., 2005. Antioxidants and a new germacrane sesquiterpene from *Carissa spinarum*. *Natural Product Research* 19, 763–769.

Reid, E.J., Betts, T.J., 1979. Records of Western Australian plants used by Aboriginals as medicinal agents. *Planta Medica* 36, 164–173.

Revathi, P., Senthinath, T.J., Thirumalaikolundusubramanian, P., Prabhu, N., 2014. An overview of antidiabetic profile of mangrove plants. *International Journal of Pharmacy and Pharmaceutical Sciences* 6, 1–5.

Rise, C.L., Prabhu, V.V., Guruvayoorappan, C., 2012. Effect of marine mangrove *Avicennia marina* (Forssk.) Vierh against acetic acid-induced ulcerative colitis in experimental mice. *Journal of Environmental Pathology, Toxicology and Oncology* 31, 179–192.

Roeder, E., Wiedenfeld, H., 2009. Pyrrolizidine alkaloids in medicinal plants of Mongolia, Nepal and Tibet. *Pharmazie* 64, 699–716.

Roehrs, P., 2007. *NGOs' Perspectives on Piracy and Protection.* Centre for Applied Studies in International Negotiations, Geneva, Switzerland.

Rogers, K.L., Fong, W.F., Redburn, J., Griffiths, L.R., 2002. Fluorescence detection of plant extracts that affect neuronal voltage-gated Ca^{2+} channels. *European Journal of Pharmaceutical Sciences* 15, 321–330.

Rogers, K.L., Grice, I.D., Griffiths, L.R., 2000. Inhibition of platelet aggregation and 5-HT release by extracts of Australian plants used traditionally as headache treatments. *European Journal of Pharmacological Sciences* 9, 355–363.

Rogers, K.L., Grice, I.D., Griffiths, L.R., 2001. Modulation of in vitro platelet 5-HT release by species of *Erythrina* and *Cymbopogon*. *Life Sciences* 69, 1817–1829.

Roome, T., Dar, A., Ali, S., Naqvi, S., Choudhary, M.I., 2008a. A study on antioxidant, free radical scavenging, anti-inflammatory and hepatoprotective actions of *Aegiceras corniculatum* (stem) extracts. *Journal of Ethnopharmacology* 118, 514–521.

Roome, T., Dar, A., Naqvi, S., Ali, S., Choudhary, M.I., 2008b. *Aegiceras corniculatum* extract suppresses initial and late phases of inflammation in rat paw and attenuates the production of eicosanoids in rat neutrophils and human platelets. *Journal of Ethnopharmacology* 120, 248–254.

Roome, T., Dar, A., Naqvi, S., Choudhary, M.I., 2011. Evaluation of antinociceptive effect of *Aegiceras corniculatum* stems extracts and its possible mechanism of action in rodents. *Journal of Ethnopharmacology* 135, 351–358.

Sadgrove, N.J., Hitchcock, M., Watson, K., Lloyd Jones, G., 2013. Chemical and biological characterization of novel essential oils from *Eremophila bignoniiflora* (F. Muell) (Myoporaceae): A traditional Aboriginal Australian bush medicine. *Phytotherapy Research* 27, 1508–1516.

Saleem, M., 2009. Lupeol, a novel anti-inflammatory and anti-cancer dietary triterpene. *Cancer Letters* 285, 109–115.

Salinas-Sanchez, D.O., Herrera-Ruiz, M., Perez, S., Jimenez-Ferrer, E., Zamilpa, A., 2012. Anti-inflammatory activity of hautriwaic acid isolated from *Dodonaea viscosa* leaves. *Molecules* 17, 4292–4299.

Salminen, A., Lehtonen, M., Suuronen, T., Kaarniranta, K., Huuskonen, J., 2008. Terpenoids: Natural inhibitors of NF-[kappa]B signaling with anti-inflammatory and anticancer potential. *Cellular and Molecular Life Sciences* 65, 2979–2999.

Savigni, D., 2010. Evaluation of the anticancer potential of plants used as traditional medicines by Aboriginal people. PhD thesis, Physiology, University of Western Australia, Perth, Western Australia, Australia.

Scalbert, A., 1991. Antimicrobial properties of tannins. *Phytochemistry* 30, 3875–3883.

Scarborough, J., 1978. Theophrastus on herbals and herbal remedies. *Journal of the History of Biology* 11, 353–385.

Scott, I.U., Karp, C.L., 1996. *Euphorbia* sap keratopathy: Four cases and a possible pathogenic mechanism. *British Journal of Ophthalmology* 80, 823–826.

Semple, S.J., Nobbs, S.F., Pyke, S.M., Reynolds, G.D., Flower, R.L., 1999. Antiviral flavonoid from *Pterocaulon sphacelatum*, an Australian Aboriginal medicine. *Journal of Ethnopharmacology* 68, 283–288.

Semple, S.J., Reynolds, G.D., O'Leary, M.C., Flower, R.L.P., 1998. Screening of Australian medicinal plants for antiviral activity. *Journal of Ethnopharmacology* 60, 163–172.

Shiva, V., 1997. *Biopiracy: The Plunder of Nature and Knowledge*. South End Press, Cambridge, MA.

Shoemaker, M., Hamilton, B., Dairkee, S.H., Cohen, I., Campbell, M.J., 2005. In vitro anticancer activity of twelve Chinese medicinal herbs. *Phytotherapy Research* 19, 649–651.

Siller, G., Gebauer, K., Welburn, P., Katsamas, J., Ogbourne, S.M., 2009. PEP005 (ingenol mebutate) gel, a novel agent for the treatment of actinic keratosis: Results of a randomized, double-blind, vehicle-controlled, multicentre, phase IIa study. *Australasian Journal of Dermatology* 50, 16–22.

Simes, J.J.H., Tracey, J.G., Webb, L.J., Dunstan, W.J., 1959. An Australian phytochemical survey III. Saponins in eastern Australian flowering plants. Bulletin No. 281. CSIRO, Melbourne, Victoria, Australia.

Simpson, B., Claudie, D., Gerber, J., Pyke, S., Wang, J., McKinnon, R., Semple, S., 2011. In vivo activity of benzoyl ester clerodane diterpenoid derivatives from *Dodonaea polyandra*. *Journal of Natural Products* 74, 650–657.

Simpson, B., Claudie, D., Smith, N., Wang, J., McKinnon, R., Semple, S., 2010. Evaluation of the anti-inflammatory properties of *Dodonaea polyandra*, a Kaanju traditional medicine. *Journal of Ethnopharmacology* 132, 340–343.

Singab, A.N., Youssef, F.S., Ashour, M.L., Wink, M., 2013. The genus *Eremophila* (Scrophulariaceae): An ethnobotanical, biological and phytochemical review. *The Journal of Pharmacy and Pharmacology* 65, 1239–1279.

Singh, A.K., Lohani, M., Singh, U.P., 2010. Analgesic activity of methanolic extract of *Aegiceras corniculatam* Linn. *International Journal of PharmTech Research* 2, 1058–1060.

Singh, B., Bhat, T.K., Singh, B., 2003. Potential therapeutic applications of some antinutritional plant secondary metabolites. *Journal of Agricultural and Food Chemistry* 51, 5579–5597.

Singh, S., Majumdar, D.K., 1999. Effect of *Ocimum sanctum* fixed oil on vascular permeability and leucocytes migration. *Indian Journal of Experimental Biology* 37, 1136–1138.

Singh, S., Taneja, M., Majumdar, D.K., 2007. Biological activities of *Ocimum sanctum* L. fixed oil-an overview. *Indian Journal of Experimental Biology* 45, 403–412.

Sloan, R.P., Bagiella, E., Powell, T., 1999. Religion, spirituality, and medicine. *The Lancet* 353, 664–667.

Smith, J.E., Tucker, D., Watson, K., Lloyd Jones, G., 2007. Identification of antibacterial constituents from the indigenous Australian medicinal plant *Eremophila duttonii* F. Muell. (Myoporaceae). *Journal of Ethnopharmacology* 112, 386–393.

Smith, N.M., 1991. Ethnobotanical field notes from the Northern Territory, Australia. *Journal of the Adelaide Botanic Gardens* 14, 1–65.

Smyth, T., Ramachandran, V.N., Smyth, W.F., 2009. A study of the antimicrobial activity of selected naturally occurring and synthetic coumarins. *International Journal of Antimicrobial Agents* 33, 421–426.

Smyth, W.F., Smyth, T.J.P., Ramachandran, V.N., O'Donnell, F., Brooks, P., 2012. Dereplication of phytochemicals in plants by LC-ESI-MS and ESI-MSn. *TrAC Trends in Analytical Chemistry* 33, 46–54.

Soejarto, D.D., 1996. Biodiversity prospecting and benefit-sharing: Perspectives from the field. *Journal of Ethnopharmacology* 51, 1–15.

Soetardjo, S. Jr., Jong, P.C., Ahmad, M.N., Lachimanan, Y.L., Sreenivasan, S., 2007. Chemical composition and biological activity of the *Centipeda minima* (Asteraceae). *Malaysian Journal of Nutrition* 13, 81–87.

Sosa-Sequera, M.C., Suárez, O., Daló, N.L., 2010. Kaurenic acid: An in vivo experimental study of its anti-inflammatory and antipyretic effects. *Indian Journal of Pharmacology* 42, 293–296.

Southwell, I.A., Tucker, D.J., 1993. Protoanemonin in Australian *Clematis*. *Phytochemistry* 33, 1099–1102.

Spjut, R.W., Perdue, R.E. Jr., 1976. Plant folklore: A tool for predicting sources of antitumor activity? *Cancer Treatment Reports* 60, 979–985.

Stack, E.M., 1989. *Aboriginal Pharmacopoeia*. The Third Eric Johnston Lecture. Northern Territory Library Service, Darwin, Northern Territory.

Stone, R., 2008. Intellectual property: Chinese province crafts pioneering law to thwart biopiracy. *Science* 320, 732–733.

Strohl, W.R., 2000. The role of natural products in a modern drug discovery program. *Drug Discovery Today* 5, 39–41.

Sudhakar, M., Rao, C.V., Rao, P.M., Raju, D.B., 2006. Evaluation of antimicrobial activity of *Cleome viscosa* and *Gmelina asiatica*. *Fitoterapia* 77, 47–49.

Sweeney, A.P., Wyllie, S.G., Shalliker, R.A., Markham, J.L., 2001. Xanthine oxidase inhibitory activity of selected Australian native plants. *Journal of Ethnopharmacology* 75, 273–277.

Takasaki, M., Konoshima, T., Tokuda, H., Masuda, K., Arai, Y., Shiojima, K., Ageta, H., 1999. Anti-carcinogenic activity of Taraxacum plant. II. *Biological and Pharmaceutical Bulletin* 22, 606–610.

Tan, G., Gyllenhaal, C., Soejarto, D.D., 2006. Biodiversity as a source of anticancer drugs. *Current Drug Targets* 7, 265–277.

Tanaka, N., Yuhara, H., Wada, H., Murakami, T., Cambie, R.C., Braggins, J.E., 1993. Phenolic constituents of *Pteridium esculentum*. *Phytochemistry* 32, 1037–1039.

Taylor, H.R., 2001. Trachoma in Australia. *Medical Journal of Australia* 175, 371–372.

Taylor, R.S., Towers, G.H., 1998. Antibacterial constituents of the Nepalese medicinal herb, *Centipeda minima*. *Phytochemistry* 47, 631–634.

Teffo, L.S., Aderogba, M.A., Eloff, J.N., 2010. Antibacterial and antioxidant activities of four kaempferol methyl ethers isolated from *Dodonaea viscosa* Jacq. var. angustifolia leaf extracts. *South African Journal of Botany* 76, 25–29.

Toledo-Pereyra, L.H., 1973. Galen's contribution to surgery. *Journal of the History of Medicine and Allied Sciences* 28, 357–375.

Torel, J., Cillard, J., Cillard, P., 1986. Antioxidant activity of flavonoids and reactivity with peroxy radical. *Phytochemistry* 25, 383–385.

Trachtenberg, D., 2002. Alternative therapies and public health: Crisis or opportunity? *American Journal of Public Health* 92, 1566–1567.

Tulp, M., Bohlin, L., 2002. Functional versus chemical diversity: Is biodiversity important for drug discovery? *Trends in Pharmacological Sciences* 23, 225–231.

Tulp, M., Bohlin, L., 2004. Unconventional natural sources for future drug discovery. *Drug Discovery Today* 9, 450–458.

Turton, C.L.R., 1997. Ways of knowing about health: An Aboriginal perspective. *Advances in Nursing Science* 19, 28–36.

Verdine, G.L., 1996. The combinatorial chemistry of nature. *Nature* 384(6604 Suppl.), 11–13.

Verma, I.M., 2002. Biopiracy: Distrust widens the rich-poor divide. *Molecular Therapy* 5, 95.

Verpoorte, R., Kim, H.K., Choi, Y.H., 2006. Plants as sources of medicines. In: Bogers, R.J., Craker, L.E., Lange, D. (eds.), *New Perspectives in Medicinal and Aromatic Plants: Agricultural, Commercial, Ecological, Legal, Pharmacological and Social Aspects.* Springer, Dordrecht, the Netherlands, pp. 261–273.

Vieira, D., Padoani, C., Soares, J.S., Adriano, J., Filho, V.C., de Souza, M.M., Bresolin, T.M.B., Couto, A.G., 2013. Development of hydroethanolic extract of *Ipomoea pes-caprae* using factorial design followed by antinociceptive and antiinflammatory evaluation. *Brazilian Journal of Pharmacognosy* 23, 72–78.

von Martius, S., Hammer, K.A., Locher, C., 2012. Chemical characteristics and antimicrobial effects of some *Eucalyptus* kinos. *Journal of Ethnopharmacology* 144, 293–299.

Vuong, Q.V., Hirun, S., Phillips, P.A., Chuen, T.L.K., Bowyer, M.C., Goldsmith, C.D., Scarlett, C.J., 2014. Fruit-derived phenolic compounds and pancreatic cancer: Perspectives from Australian native fruits. *Journal of Ethnopharmacology* 152, 227–242.

Wagner, H., Ludwig, C., Crotjahn, L., Khan, M.S.Y., 1987. Biologically active saponins from *Dodonaea viscosa*. *Phytochemistry* 26, 697–701.

Wallace, D.J., 1996. The history of antimalarials. *Lupus* 5(1 Suppl.), S2–S3.

Wang, M.Y., West, B.J., Jensen, C.J., Nowicki, D., Su, C., Palu, A.K., Anderson, G., 2002. *Morinda citrifolia* (Noni): A literature review and recent advances in Noni research. *Acta Pharmacologica Sinica* 23, 1127–1141.

Webb, L.J., 1959. Some new records of medicinal plants used by the Aborigines of Tropical Queensland and New Guinea. *Proceedings of the Royal Society of Queensland* 71, 103–110.

Webb, L.J., 1969. The use of plant medicines and poisons by Australian Aborigines. *Mankind* 7, 137.

Weedon, D., Chick, J., 1976. Home treatment of basal cell carcinoma. *Medical Journal of Australia* 24, 928.

Wilkinson, J.M., Cavanagh, H.M.A., 2005. Antibacterial activity of essential oils from Australian native plants. *Phytotherapy Research* 19, 643–646.

Williams, C.J., 2013. *Medicinal Plants: An Antipodean Apothecary*, Vol. 4: Medicinal Plants in Australia. Rosenberg Publishing, Dural Delivery Centre, Dural, New South Wales, Australia.

Williams, L.A.D., Vasques, E., Reid, W., Porter, R., Kraus, W., 2003. Biological activities of an extract from *Cleome viscosa* L. (Capparaceae). *Naturwissenschaften* 90, 468–472.

Wink, M., 1999. Introduction: Biochemistry, role and biotechnology of secondary metabolites. In: Wink, M. (ed.), *Functions of Plant Secondary Metabolites and Their Exploitation in Biotechnology.* Sheffield Academic Press, Sheffield, U.K., pp. 1–16.

Wohlmuth, H., 1997. Bush medicines of western Sydney. *Australian Journal of Medical Herbalism* 9, 50–53.

Wood, P.B., 2008. Role of central dopamine in pain and analgesia. *Expert Review of Neurotherapeutics* 8, 781–797.

Wood, P.B., Schweinhardt, P., Jaeger, E., Dagher, A., Hakyemez, H., Rabiner, E.A., Bushnell, M.C., Chizh, B.A., 2007. Fibromyalgia patients show an abnormal dopamine response to pain. *European Journal of Neuroscience* 25, 3576–3582.

Woode, E., Ansah, C., Ainooson, G.K., Abotsi, W.M., Mensah, A.Y., Duweijua, M., 2007. Anti-inflammatory and antioxidant properties of the root extract of *Carissa edulis* (forsk.) Vahl (apocynaceae). *Journal of Science and Technology (Ghana)* 27, 6–15.

Wright, C.W., Allen, D., Phillipson, J.D., Kirby, G.C., Warhurst, D.C., Massiot, G., Men-Olivier, L.L., 1993. *Alstonia* species: Are they effective in malaria treatment? *Journal of Ethnopharmacology* 40, 41–45.

Young, L., 2007. *Lola Young: Medicine Woman and Teacher.* Fremantle Arts Centre Press, Fremantle, Western Australia, Australia.

Younos, C., Rolland, A., Fleurentin, J., Lanhers, M.C., Misslin, R., Mortier, F., 1990. Analgesic and behavioural effects of *Morinda citrifolia*. *Planta Medica* 56, 430–434.

Zandi, K., Taherzadeh, M., Tajbakhsh, S. et al., 2008. Antiviral activity of *Avicennia marina* leaf extract on HSV-1 and vaccine strain of polio virus in Vero cells. *International Journal of Infectious Diseases* 12(Suppl. 1), e298.

Zhong, H., Bowen, J.P., 2006. Antiangiogenesis drug design: Multiple pathways targeting tumor vasculature. *Current Medicinal Chemistry* 13, 849–862.

Zur, S.S., 2007. Letting go of data in Aboriginal Australia: Ethnography on 'rubber time'. *The Qualitative Report* 12, 583–593.

CHAPTER 17

Nutritional Characteristics and Bioactive Compounds in Australian Native Plants: A Review

David J. Williams and Mridusmita Chaliha

CONTENTS

INTRODUCTION

The market for functional foods and ingredients with enhanced nutritional and bioactive properties is growing rapidly (Park and Kim, 2014). To satisfy this demand, there has been a dramatic increase in the screening of so-called native foods that seek to identify plants with elevated levels of these health-promoting compounds (Shahidi, 2009). Many Australian native plants have a significant history as a food source and as traditional medicine by Australia's aboriginal people (Brand et al., 1983), and several of these are already known to have other significant functional properties (Zhao and Agboola, 2007).

The most researched health-promoting bioactives in plant foods are the antioxidants. They protect the body from oxidative damage by scavenging free radicals. Free radicals cause damage to the body's tissues, cells, nucleic acids, proteins and lipids, leading to the development of degenerative diseases such as cancer, cardio-vascular diseases, neural degeneration and diabetes (Day et al., 2009; Mohanty and

Cock, 2012). The major source of antioxidants in fact is plant-based food. Inclusion into the diet of foods rich in antioxidants or incorporation of extracts containing these into food could provide health benefits beyond those attributable to basic nutritional functions. Research over the last decade has shown (Konczak et al., 2009, 2010; Netzel et al., 2007) that many Australian native plant tissues represent rich sources of these bioactives whose antioxidant capacities often exceed many commonly consumed fruits including blueberry, the benchmark for antioxidant activity (Konczak et al., 2009).

Phenolic compounds together with the bioactive forms of vitamin C have been identified as the major contributors to antioxidant capacity (Netzel et al., 2007). The use of plants that have elevated levels of these antioxidant compounds has the added benefit of being seen as 'natural' and health-promoting to consumers increasingly sceptical of the use of synthetic antioxidants. Kakadu plum (*Terminalia ferdinandiana*), for example, is among the richest known sources of vitamin C and the phenolic compound ellagic acid and its derivatives (Konczak et al., 2014; Williams et al., 2014).

Similar to the development of 'natural' antioxidants there is an increasing desire to identify 'natural' substitutes for synthetic antimicrobials for use in the food industry and as possible alternatives to current medications. The tissues and extracted essential oils of many Australian native plants have a significant antimicrobial activity (Wilkinson and Cavanagh, 2005; Zhao and Agboola, 2007), several of which are particularly potent against a number of important food-borne pathogens, food spoilage bacteria and yeasts.

In this chapter, we provide a review of the scientific research that has evaluated the nutritional composition of 13 commercially significant Australian native plants. To assist in the comparison with modern cultivated plant foods, these native plants were divided into two categories based on the usage of their commercial products similar to that applied in Konczak et al. (2009): (1) *fruit*: Davidson's plum (*Davidsonia* spp.), desert limes (*Citrus glauca*), finger limes (*Citrus australasica*), Kakadu plum (*Terminalia ferdinandiana*), lemon aspen (*Acronychia* spp.), quandong (*Santalum acuminatum*), muntries (*Kunzea pomifera*) and riberry (*Syzygium luehmannii*); and (2) *herb/spice*: anise myrtle (*Syzygium anisatum*), bush tomato (*Solanum centrale*), lemon myrtle (*Backhousia citriodora*), Tasmanian pepper (*Tasmanian lanceolata*) and wattleseeds (*Acacia* spp.). Increasing interest in the compounds present in native plants led to the publication of tables of composition detailing the macro- and micro-nutrient contents of 440 native plant species (Brand-Miller et al., 1993). A thorough examination of the data presented in the tables suggested to the authors (Brand-Miller and Holt, 1998) that most native plants have similar nutrient composition to western foods in the same category with some notable exceptions, for example Kakadu plum. From their observations, it could be concluded that the most promising feature of these Australian native plants lies in their enhanced levels of bioactives, a necessary pre-requisite for any potential applications for both functional food and chemotherapeutic uses. Until 10 years ago, little attention was paid to the functional properties of Australian native plants, but since then, there have been an increasing number of studies. These have focused on identifying and quantifying these health-promoting compounds. This chapter also summarises these recent investigations with a view of providing up-to-date information that will assist in creating further commercial opportunities for these plants. We also identify issues that may be of concern and areas that require further research.

NUTRIENT COMPOSITION

Until 1993, only fragmentary information about the nutrient values of Australian native plants was available. In 1983, the nutrient contents of 28 plant tissues (both raw and cooked) that are commonly consumed by Australia's aboriginal people were evaluated (Brand et al., 1983). However, comparison of their data with supposedly similar but cultivated western fruit was made difficult by the limited number of samples tested. To overcome this replication shortfall and to provide up-to-date information on the nutrient intake of Australian aboriginal people living traditionally, three major scientific centres in Australia systematically analysed over 1200 samples of 500 plant foods. This substantial collaboration produced the Tables of Composition of Australian Aboriginal Foods (Brand-Miller et al., 1993).

A later review (Brand-Miller and Holt, 1998) completed the exercise of comparing the nutrient composition values presented in these tables with those reported for similar cultivated foods as eaten in Western countries (Paul and Southgate, 1978). Unfortunately for our purposes, desert lime, lemon aspen, muntries and riberry from the fruit group or the herbs/spices anise and lemon myrtle and Tasmanian pepper (both leaf and berry) did not feature in these tables. It was not until 2012 that information was published that provided a comprehensive set of nutritional values for these products (Read, 2012). Data retrieved from these two studies form the basis of our compilation of macro-nutrients (Table 17.1). The Read (2012) publication provided nutrient data on desert limes and lemon aspen as well as Kakadu plum which was excluded from the original comparison as presented by Brand-Miller and Holt (1998). By calculating an average nutrient composition of all the native fruit analysed (Table 17.1), it can be observed that the native fruits were slightly higher in protein and fat compared with the average for 17 types of cultivated western fruits. This may be attributed to the lower water content of the native fruit samples. The significantly higher carbohydrate and fibre contents of the native fruit cannot be explained away so readily.

As the focus of this book is directed at 13 commercially significant native plants whose major utilised parts are fruit, leaves and seeds, it is these tissues that will feature in this chapter. The nutrient data as provided by Read (2012) for anise myrtle and lemon myrtle as well as Tasmanian pepper augments these discussions (Table 17.1). Comparison with cultivated leaves such as spinach and lettuce showed that the cultivated leaves had higher moisture content, similar protein levels but much lower carbohydrate and fibre contents (Brand-Miller and Holt, 1998).

Acacia seeds are rich in nutrients and much higher in energy, protein and fat than any of the cereals such as wheat and rice (Brand-Miller and Holt, 1998). Their fat content was higher than most legumes (the same family as *Acacias*), but the values

Table 17.1   Macronutrient Composition (g/100 g FW) of Native Australian Fruit and Leaves Compared with Selected Cultivated Fruit and Vegetable Leaves, That Is, Spinach and Lettuce

Plant	Energy (kJ)	Moisture	Protein	Fat (g/100 g FW)	Total Carbohydrate	Dietary Fibre
Native fruit[a] (n = 319)	396	73	2.0	1.0	21	8.0
Cultivated fruit[b] (n = 21)	171	86	1.0	n.d.	8	3.0
Native herbs[c] (n = 31)	242	80	2.8	0.7	11	5.1
Cultivated leaves[d]	No data	93	2–3	n.d.	<1	3.0

Note: n.d., not detected.

[a] Values obtained from Brand-Miller and Holt (1998) with added data from Read (2012) for desert limes, Kakadu plum and lemon aspen whose data were modified to moisture of 73 g/100 g.
[b] Values from Paul and Southgate (1978).
[c] Values obtained from Brand-Miller and Holt (1998) with added data from Read (2012) for anise and lemon myrtle and Tasmanian pepper leaf.
[d] Values spinach and lettuce from Paul and Southgate (1978).

varied widely between samples. Brown et al. (1987) undertook an extensive investigation of the fatty acids in 19 edible *Acacias* and reported that the seed oils generally had a higher proportion of polyunsaturated fatty acids. In fact, six species contained such high proportions of oleic acid that it invited comparison to olive oil. The authors observed that this elevated unsaturated nature of the *Acacia* oils is desirable from a health point of view, but its propensity to be oxidised may present problems in food processing and storage.

A 2009 study systematically evaluated native Australian plants that included all of the fruit and herbs/spices selected in the current review except finger limes and muntries for minerals and antioxidant properties (Konczak et al., 2009). The results revealed that Australian native plants are richer sources of minerals such as Ca and trace elements such as Fe, Zn and Cu than similar cultivated plants. This conclusion confirmed a similar earlier view expressed by Brand-Miller and Holt (1998) following their inspection of the Tables of Composition of Australian Aboriginal Foods (1993). In the Konczak et al. (2009) study, lemon myrtle ranked highest for Ca content followed by Tasmanian pepper leaf and wattleseed. Slightly lower values were observed in the desert lime, riberry and Kakadu plum although they all contained significantly higher levels than the blueberry control. Anise myrtle was rich in Mg, while quandong fruit was abundant in Mg and Zn and lemon aspen high in Zn. The bush tomato, wattleseed and Tasmanian pepper leaf contained enhanced levels of Fe.

It may be timely to stress that the final nutrient composition of these plant products is determined by factors such as chemotype/cultivar, climate, soil type, preharvest practices, harvest maturity, postharvest treatments and storage conditions.

As most researchers seem to agree that Australian native plant foods have similar composition to western foods in the same category, with some notable exceptions, it is their elevated levels of physiologically active compounds that may offer the greatest opportunities for their development as commercial functional foods or chemotherapeutic agents.

BIOACTIVES WITH ANTIOXIDANT ACTIVITY

To date no single assay is universally accepted as a measure of the antioxidant capacity of plant food, because the antioxidant defence system and the involvement of the many types of free radicals generated in the body is complex (Huang et al., 2005). A variety of compounds within the plant will contribute to its overall antioxidant activity, so numerous assays have been developed, each focusing on different targets and mechanisms. This chapter will focus on two assay systems that have previously been used to screen Australian native plants for antioxidant properties. The oxygen radical absorbance capacity (ORAC) assay measures the ability of a food extract to reduce free radicals and, according to Konczak et al. (2009), is the most relevant system to human biology. Another assay system that evaluates the antiradical properties of plant extracts is the trolox equivalent antioxidant capacity (TEAC) which measures the reduction of the ABTS (2,2′-azino-bis-(3-ethylbenzthiazoline-6-sulphonic acid)) radical.

Netzel et al. (2007) indicated that the radical-scavenging activities of the evaluated native plant fruit were three- to fivefold higher using the TEAC assay than that of the benchmark blueberry control. Furthermore, Kakadu plum had by far the highest TEAC value followed by muntries and Tasmanian pepper leaf berries (Table 17.2). Of the relevant native plant products tested in the Zhao and Agboola (2007) investigations, quandong followed by anise and lemon myrtles had the greatest antiradical properties, also measured by the TEAC assay.

Table 17.2   Antioxidant Capacity (as Determined by the ORAC-H Assay; Expressed as μmol TEq/g FW and the TEAC Assay; Expressed as μmol TEq/g FW) of Selected Australian Native Fruit

Plant	ORAC-H (μmol TEq/g FW μmol TEq/g FW) values	References	TEAC (μmol TEq/g FW)	References
Native fruit[a]	44–501	Konczak et al. (2010)		
Native fruit[b]			13–123	Netzel et al. (2007)
Kakadu plum	315	Konczak et al. (2010)	204	Netzel et al. (2007)
Blueberry	77	Konczak et al. (2010)	39	

Note: ORAC-H + ORAC-L = ORAC-T values.
[a] Fruit tested: Davidson's plum, desert lime, finger lime, Kakadu plum, lemon aspen, quandong and riberry in Konczak et al. (2010).
[b] Fruit tested: Davidson's plum, finger lime, Kakadu plum, muntries, riberry and Tasmanian pepper leaf berries in Netzel et al. (2007).

A later more comprehensive assessment of the antioxidant properties was provided by Konczak et al. (2009) when they measured the contributions to the antioxidant properties of the water-soluble (hydrophilic) and lipid-soluble (lipophilic) fractions of native plant foods. The Kakadu plum and the quandong exhibited enhanced antioxidant capacities in comparison to blueberry as evaluated by the ORAC assay of the hydrophilic fraction (ORAC-H) (Table 17.2). Within the herb/spice category, this study reported that Tasmanian pepper leaf, anise and lemon myrtle exhibited high ORAC-H values, in fact higher than those determined for the Kakadu plum and quandong. An interesting feature of this study was the relatively high antioxidant activity attributed to the lipophilic fractions of some of the plants tested, for example lemon myrtle (43%), riberry (30%), lemon aspen (28%), Kakadu plum (26%), desert lime (20%) and Tasmanian pepper leaf (14%). These results are even more surprising when it is considered that for commonly consumed fruits including blueberries, most of the antioxidants were in the hydrophilic fraction (Wang et al., 1996; Table 17.2). The authors believe the elevated levels of lipophilic compounds such as carotenoids, vitamin E, lutein and tocopherols are the main reason for this activity, but the presence of essential oils with well-documented antioxidant properties may also contribute. Lemon myrtle, for example, has an essential oil composed mainly of the terpene citral (Ruberto and Baratta, 2000). In contrast, the measurements presented by Zhao and Agboola (2007) showed that the highest activity was found in the methanol extract of lemon myrtle with little activity in the hexane fraction. This suggested to these authors that the antioxidants were likely to be polar compounds.

Even with the possible significant contribution of these lipophilic compounds to the antioxidant properties of native plants, the hydrophilic compounds are probably the main source of their superior antioxidant capacities, in particular the bioactive forms of vitamin C and polyphenolics (Atkinson et al., 2005).

Using the values from the Tables of Composition of Aboriginal Foods and excluding the values of the exceptionally high Kakadu plum fruit, Brand-Miller and Holt (1998) concluded that the mean vitamin C (measured as ascorbic acid) content was lower than the average vitamin C in cultivated fruits (Paul and Southgate, 1978). However, Konczak et al. (2010) reported that finger limes and desert limes were good sources of ascorbic acid and were in fact two-fold higher than that measured in Californian oranges. A recent study measured both the active forms of vitamin C (ascorbic AA and dehydroascorbic acids DHAA) of Kakadu plum fruit and compared these levels to those present in strawberries and boysenberries (Williams et al., 2014). This comparison further underlined the exceptionally high levels present in the native fruit (Figure 17.1).

The water-soluble vitamin B9 (known also as folate) in combination with Zn, Mg, Ca and vitamin B12 has been reported to prevent genome damage caused by oxidative stress (Fenech, 2007). All of the evaluated native species in the Konczak et al. (2009) study contained folate with the desert lime being the richest source. Tasmanian pepper leaf, quandong, Kakadu plum, riberry and lemon aspen also contained significant amounts.

With the notable exception of Kakadu plum, phenolic compounds are the major source of antioxidant capacity with high positive correlations between antioxidant

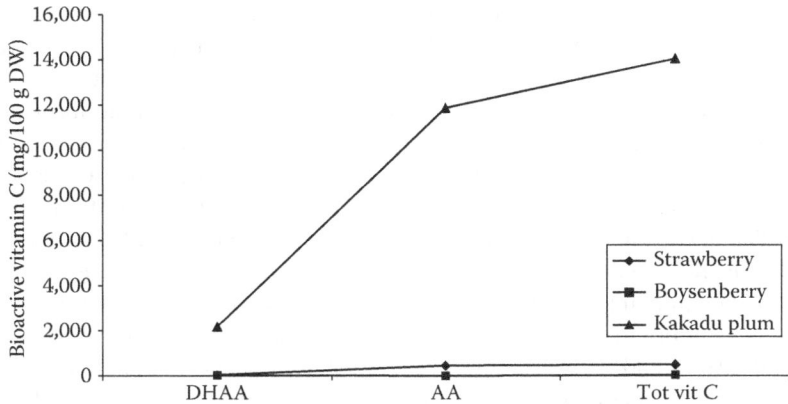

Figure 17.1 The proportion of dehydroascorbic acid (DHAA) to total vitamin C content in strawberries, boysenberries and whole Kakadu plum fruit. (Adapted from Williams, D.J. et al., *Food Res. Int.*, 66, 100, 2014.)

capacity as evaluated by the ORAC and TEAC assays and the level of total phenolics (Konczak et al., 2009; Netzel et al., 2007).

To fully understand the contribution of phenolic compounds to native plants' elevated antioxidant properties, Konczak et al. (2009) identified the major phenolic compounds present in 13 of these plants. Again the Kakadu plum fruit was a rich source of phenolics with over 50 compounds being identified. Another fruit that displayed high antioxidant properties was the quandong. In this fruit, together with the Davidson's plum, the major phenolics detected were the anthocyanins, plant pigments and potent antioxidants (Konczak et al., 2009). Anthocyanins were also found in the riberry extracts. Other phenolic compounds with antioxidant properties identified in the fruit category included quercetin, kaempferol, rutinosides and chlorogenic acid.

Tasmanian pepper leaf was the richest source of phenolic compounds within the herb/spice group followed closely by anise myrtle (Konczak et al., 2009). Unlike the Tasmanian pepper berry, the dried leaves contained only traces of anthocyanins, but the phenolic acid, chlorogenic acid, was measured at very high levels, up to 3% in the leaves. A later investigation identified the major phenolics in anise myrtle as being chlorogenic acid as well as the glycosides of quercetin or hesperitin. Myricetin was also detected but in lesser amounts (Konczak et al., 2010).

Only trace quantities of phenolic compounds were detected in the extracts of bush tomato with quercetin, kaempferol and luteolin and minor amounts of the phenolic acids such as chlorogenic, caffeic, ferulic, coumaric and hydrobenzoic acids being detected (Konczak et al., 2009).

Similarly, wattleseed extracts only had minute amounts with the major compounds being rutin, quercetin, kaempferol and luteolin as well as some chlorogenic acid (Konczak et al., 2009).

Recent testing in our laboratory (Williams et al., 2014) and by Guo et al. (2014) indicated that many of these native plants most notably anise myrtle and Kakadu plum contain significant amounts of the potent phenolic antioxidant, ellagic acid (EA) and its derivatives (Figure 17.2). The effectiveness of the three forms of EA, that is free EA, EA glycosides (EAGs) or the complex polymers known as ellagitanins (ETs), is due to their double bonds with an associated electron deficiency, making them highly reactive to molecular oxygen free radicals (Atkinson et al., 2005). The antioxidant efficiency of the EA forms is considered to be: ETs ≥ free EA > EAGs (reviewed in Williams et al., 2014). Investigations that focus on identifying EA forms in plant materials are made difficult by the fact that the forms differ widely in solubility: free EA is insoluble in water, while the ETs and EAGs are water soluble (Clifford and Scalbert, 2000). These solubility concerns could easily affect any extraction protocols if careful attention is not paid to solubilising the EA at all stages of the process.

To complete the assessment of bioactives that contribute to these plants' antioxidant properties, we will summarise the findings of those studies that have identified and quantified possible lipophilic candidates.

Vitamin E has potent antioxidant properties (Azzi, 2007) and of all the plant species evaluated in the Konczak et al. (2009) study, the highest levels of this vitamin were found in the anise myrtle extracts followed by lemon myrtle and Tasmanian pepper leaf. Significantly lower levels were detected in the fruit samples, with the Kakadu plum exhibiting the highest content closely followed by quandong.

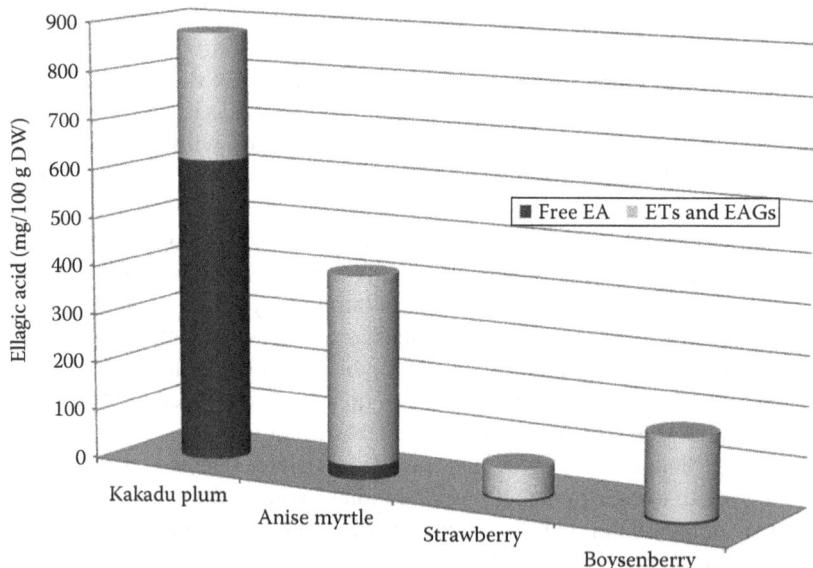

Figure 17.2 Concentration of EA forms (mg/100 g DW) in freeze–dried native fruit and herb powders. Values are means ± SD of triplicate analyses.

The carotenoid lutein has also well-recorded antioxidant properties as well as playing an important role in eye health (Richer et al., 2004). High levels of this bioactive were identified in anise myrtle, followed by lemon myrtle, Tasmanian pepper leaf, Kakadu plum, desert lime and Davidson's plum. In fact, all of these plant tissues contain more lutein than the Australian 'Hass' avocado, considered a primary source of this carotenoid (Zabaras et al., 2008).

It seems that many of these Australian native plants are rich sources of lipophilic and hydrophilic antioxidants, unlike the often used blueberry control and other traditionally consumed fruits where the bulk of the antioxidant bioactives is hydrophilic. Hydrophilic antioxidants may not accumulate in the body, while lipophilic antioxidants penetrate the lipoprotein cell membrane more easily and are therefore more readily available (Burton and Ingold, 1986). This led Konczak et al. (2009) to propose that native plants with both lipophilic and hydrophilic antioxidant compounds provide a more comprehensive protection from oxidative stress.

BIOACTIVES WITH ANTIMICROBIAL ACTIVITY

Until recently, most studies of Australian native plants have reported solely on the antioxidant capacities without examining other medicinally important bioactivities. This chapter will focus on recently ascertained antimicrobial properties and identification of the responsible active constituents.

In 2007, Zhao and Agboola assessed the antimicrobial properties of 19 Australian bushfoods that included Davidsons' plum, desert lime, lemon aspen, quandong and riberry as well as bush tomato and the fresh leaves of Tasmanian pepper, lemon and anise myrtle. Activity was measured against three classes of microorganisms: (1) selected food-borne human pathogenic bacteria, (2) food spoilage bacteria and (3) food spoilage yeasts. All the native plant products so tested (with one exception not relevant to this discussion) had measureable activity against all the microorganism classes, either inhibiting or slowing down their growth.

The strongest activity against common food-borne human pathogens was shown by the extracts from Tasmanian pepper leaves followed by the riberry extracts. Of the three types of extracting solvents used (methanol, hexane and water), the methanol extracts produced the strongest and broadest activities against this class of microbe. The strongest activity against common food spoilage bacteria was also exhibited by the methanol extract of Tasmanian pepper leaves. Extracts of anise and lemon myrtle, quandong, lemon aspen, bush tomato and Davidson's plum all showed significant activity. Cock and Mohanty (2011) found that the methanol extract of Kakadu plum fruit pulp had broad specificity, inhibiting the growth of 13 of the 14 bacteria tested. Unfortunately, individual pulp components responsible for this antimicrobial action were not identified. An interesting finding of the Zhao and Agboola (2007) study was that neither the methanol nor hexane solvent extracts of Tasmanian pepper leaves displayed any activity against the three lactic acid bacteria species tested. Even though these bacteria can cause spoilage in some products, they are beneficial organisms in foods like cheese and yoghurt as well as fermented meat and vegetable products such as salami and sauerkraut. Highest activity against the common food spoilage yeasts was found in the Tasmanian pepper leaves followed by lemon myrtle and bush tomatoes. Hexane was the most efficient in extracting the anti-yeast compounds, closely followed by methanol. The high antimicrobial potency of the Tasmanian pepper leaves comes as no surprise as one of its major constituents located primarily in the oil fraction is the methanol-soluble sesquiterpene, polygodial which has well-documented antimicrobial properties (Kubo and Taniguchi, 1988; McCallion et al., 1982).

Other native herbs and spices with commercially employed oil fractions are lemon and anise myrtle, and it was shown earlier that lemon myrtle oil has high anti-microbial potency (Lassak and McCarthy, 1983). Further investigations confirmed the effectiveness of different preparations of lemon myrtle against bacteria and fungi and concluded that the oil was an effective antibacterial and excellent antifungal agent (Wilkinson et al., 2003). Although the antibacterial activity of lemon myrtle oil was believed to be directly related to the high citral content of the oil (80%–98%), the results did not correlate with citral content, implying that other compounds within the oil may affect the susceptibility of the microorganisms. In fact, the antimicrobial activity of the oil was found to be greater than the citral alone, providing further support for this conclusion as well as indicating possible synergistic effects between the active ingredients.

While lemon myrtle oil has undergone extensive investigations into its antibacte-rial and antifungal activity (Hayes and Markovic, 2003; Wilkinson et al., 2003), only limited studies have measured this property in the essential oil of anise myrtle. Hood et al. (2003) established that this oil has activity against *Staphylococcus aureus* and *Escherichia coli*. A later study (Wilkinson and Cavanagh, 2005) examined the activity of 18 Australian essential oils including anise myrtle against five medically important bacteria and the yeast *Candida albicans*. However, this study reported the unusual finding that of the two commercial anise myrtle oil samples tested, only one was active against all the microbes, while the other was inactive towards *E. coli* and *C. albicans*. The authors suggested that this variability was due to the existence of two chemotypes of this plant, and it was possible that the samples tested were of differing chemical composition. Another interesting result was reported by Himejima and Kubo (1992) concerning the antifungal activity of polygodial which was very much enhanced by the addition of small amounts of *trans*-anethole, the main constituent of anise myrtle oil.

With the implementation of optimised antimicrobial testing methods (Sultanbawa et al., 2009) that addressed the problems of inconsistency associated with previous assays, the true value of these plants as antimicrobials with wide industrial applica-tions can now be comprehensively appraised.

BIOACTIVES WITH ANTI-INFLAMMATORY ACTIVITY

As the anti-inflammatory activities of ellagic (EA)- and chlorogenic acid–based extracts had been previously confirmed (Rosillo et al., 2012; Shan et al., 2009), indig-enous native herbs known to be rich sources of these polyphenolics were evaluated for their anti-inflammatory properties (Guo et al., 2014). Hydrophilic phenolic-rich extracts obtained from anise myrtle, lemon myrtle and Tasmanian pepper leaf were so tested. The findings that the EA-containing extracts of anise and lemon myrtle were more effective as anti-inflammatory agents than the chlorogenic acid containing Tasmanian pepper leaf suggested that enhanced anti-inflammatory activity in these herbs is based on EA and ellagitannin content. This conclusion could be tested by analysing extracts of Kakadu plum which possess far greater levels of EA (Williams et al., 2014).

BIOACTIVES WITH ANTI-NUTRIENT ACTIVITY

Anti-nutrients can reduce the digestibility of proteins and carbohydrates, interfere with mineral bioavailability and induce pathological changes in a number of enzymes, as well as binding nutrients, thereby rendering them unavailable (Speijers and Van Egmond, 1999). These compounds if present could negatively affect the overall value of consuming these native plants, thereby limiting opportunities as a functional food or as a source of health-promoting ingredients.

In 2001, Hegarty et al. examined chemical and toxicological information on major Australian native plant foods. Based on this assessment, further chemical analyses were conducted on selected bush food samples supplied by commercial growers and processors. Samples of the 13 selected plants included fresh or frozen fruit, fresh or dried leaves as well as fresh seeds. Anti-nutrients such as oxalates, saponins, cyanogens and alkaloids were measured.

The initial Hegarty et al. (2001) examination of existing chemical and toxicological information revealed that the data in the majority of cases were obtained on the non-edible parts; therefore, we will discuss potential anti-nutrient compounds in the edible portion of the 13 selected plants that were subsequently determined.

Oxalates were found in all the bush foods examined but at levels lower than those found in commercial leafy vegetables such as spinach and rhubarb with the exception of the leaves of anise and lemon myrtle in which the levels were two- to four-fold higher than those of spinach (Hegarty et al., 2001). However, since they are used as flavourings, they would not be consumed in similar quantities as leafy vegetables, thereby limiting the levels ingested.

None of samples had cyanogens above the limit of detection. Low levels of alkaloids were detected in several of the plants tested except lemon aspen. However, the levels in the whole lemon aspen fruit were much lower than those found in common foods such as soya beans (Fenwick and Oakenfull, 1983). Anise myrtle, lemon myrtle, finger limes, Tasmanian pepper as well as muntries also possessed some alkaloids. The authors decided that the alkaloids detected were not of great anti-nutritive concern because of the occurrence of similar alkaloids in commercial oranges and limes.

A possible constituent of the essential oils of some of the selected native plants is safrole, known to be carcinogenic in animals. Low levels of this compound were detected in the Tasmanian pepper berry.

Protease inhibitors (inhibitors of digestive enzymes) were located in the seeds of 38 *Acacia* species tested by Weder and Murray (1981) and a further 18 species tested by Kort (1985). Harwood (1994) therefore advised that foods incorporating *Acacia* seeds should not be eaten raw but steamed, boiled or baked to de-activate these compounds.

Hegarty et al. (2001) warned that non-toxicity of these plants should not be taken as a guarantee of safety, because all the constituents have not been fully catalogued and the degree of potential toxicity of a particular anti-nutrient may vary with other factors such as quantity consumed and personal tolerance.

CONCLUSIONS

Many of the selected 13 commercially significant Australian native plants have great potential as functional foods and as a source of health-promoting ingredients. Nearly all contain very high levels of antioxidants in both the hydrophilic and lipophilic fractions with contributing compounds such as ellagic acid and its derivatives, other phenolics, anthocyanins, carotenoids and the vitamins C and E. If this unique combination of bioactive compounds can be maintained during storage and processing, novel opportunities for developing effective 'natural' antioxidants may be realised.

Tasmanian pepper leaf, lemon and anise myrtle have great potential as antiseptics or surface disinfectants as well as a role in food preservation and prevention of microbial spoilage. Combinations of these native plant extracts could offer even greater antimicrobial benefits due to possible synergistic effects.

Since many of these plants are already being used as flavourings, their inclusion would have the added benefit of a 'natural' antimicrobial and food preservative as well as being a good source of other health-promoting bioactives and nutrients. The virtual absence of anti-nutrients should enhance these opportunities.

REFERENCES

Atkinson, C.J., Nestby, R., Ford, Y.Y., Dodds, P.A., 2005. Enhancing beneficial antioxidants in fruits: A plant physiological perspective. *BioFactors* 23, 229–234.

Azzi, A., 2007. Molecular mechanism of α-tocopherol action. *Free Radical Biology and Medicine* 43, 16–21.

Brand-Miller, J., James, K.W., Maggiore, P.M., 1993. *Tables of Composition of Australian Aboriginal Foods*. Aboriginal Studies Press, Canberra, Australian Capital Territory, Australia.

Brand-Miller, J.C., Holt, H.A., 1998. Australian aboriginal plant foods: A consideration of their nutritional composition and health implications. *Nutrition Research Reviews* 11, 5–23.

Brand-Miller, J.C., Mcdonnell, J., Lee, A., Cherikoff, V., Truswell, A.S., 1983. The nutritional composition of Australian aboriginal bushfoods. *Food Technology in Australia* 35, 293–298.

Brown, A., Cherikoff, V., Roberts, D., 1987. Fatty acid composition of seeds from the AustralianAcacia species. *Lipids* 22, 490–494.

Burton, G., Ingold, K.U., 1986. Vitamin E: Application of the principles of physical organic chemistry to the exploration of its structure and function. *Accounts of Chemical Research* 19, 194–201.

Clifford, M.N., Scalbert, A., 2000. Ellagitannins–nature, occurrence and dietary burden. *Journal of the Science of Food and Agriculture* 80, 1118–1125.

Cock, I.E., Mohanty, S., 2011. Evaluation of the antibacterial activity and toxicity of *Terminalia ferdinandia* fruit extracts. *Pharmacognosy Journal* 3, 72–79.

Day, L., Seymour, R.B., Pitts, K.F., Konczak, I., Lundin, L., 2009. Incorporation of functional ingredients into foods. *Trends in Food Science and Technology* 20, 388–395.

Fenech, M., 2007. Diet and genome stability: Effects of folate and alcohol. Presentation at *The Sebel Playford*, Adelaide, South Australia, Australia, August 21, 2007.

Fenwick, D.E., Oakenfull, D., 1983. Saponin content of food plants and some prepared foods. *Journal of the Science of Food and Agriculture* 34, 186–191.

Guo, Y, Sakulnarmrat, K., Konczak, I., 2014. Anti-inflammatory potential of native Australian herbs polyphenols. *Toxicology Reports* 1, 385–390.

Harwood, C., 1994. Human food potential of the seeds of some Australian dry-zone *Acacia* species. *Journal of Arid Environments* 27, 27–35.

Hayes, A., Markovic, B., 2003. Toxicity of Australian essential oil *Backhousia citriodora* (lemon myrtle). Part 2. Absorption and histopathology following application to human skin. *Food and Chemical Toxicology* 41, 1409–1416.

Hegarty, M.P., Hegarty, E.E., Wills, R.B.H., 2001. *Food Safety of Australian Plant Bushfoods*. Rural Industries Research and Development Corporation, RIRDC Publication No. 01/28, Canberra, Australian Capital Territory, Australia.

Himejima, M., Kubo, I., 1992. Antimicrobial agents from Licaria puchuri-major and their synergistic effect with polygodial. *Journal of Natural Products* 55, 620–625.

Hood, J.R., Wilkinson, J.M., Cavanagh, H.M., 2003. Evaluation of common antibacterial screening methods utilized in essential oil research. *Journal of Essential Oil Research* 15, 428–433.

Huang, D., Ou, B., Prior, R.L., 2005. The chemistry behind antioxidant capacity assays. *Journal of Agricultural and Food Chemistry* 53, 1841–1856.

Konczak, I., Maillot, F., Dalar, A., 2014. Phytochemical divergence in 45 accessions of *Terminalia ferdinandiana* (Kakadu plum). *Food Chemistry* 151, 248–256.

Konczak, I., Zabaras, D., Dunstan, M., Aguas, P., 2010. Antioxidant capacity and phenolic compounds in commercially grown native Australian herbs and spices. *Food Chemistry* 122, 260–266.

Konczak, I., Zabaras, D., Dunstan, M., Aguas, P., Roulfe, P., Pavan, A., 2009. *Health Benefits of Australian Native Foods: An Evaluation of Health-Enhancing Compounds*. Rural Industries Research and Development Corporation, RIRDC Publication No. 09/133, Canberra, Australian Capital Territory, Australia.

Kort, A., 1985. *Some Antinutritional Factors in Acacia Seeds*. Deakin University, Geelong, Victoria, Australia.

Kubo, I., Taniguchi, M., 1988. Polygodial, an antifungal potentiator. *Journal of Natural Products* 51, 22–29.

Lassak, E.V., Mccarthy, T., 1983. *Australian Medicinal Plants*. Methuen Australia, North Ryde, New South Wales, Australia.

Mccallion, R.F., Cole, A., Walker, J., Blunt, J., Munro, M., 1982. Antibiotic substances from New Zealand plants. *Planta Medica* 44, 134–138.

Mohanty, S., Cock, I.E., 2012. The chemotherapeutic potential of *Terminalia ferdinandiana*: Phytochemistry and bioactivity. *Pharmacognosy Reviews* 6, 29–36.

Netzel, M., Netzel, G., Tian, Q., Schwartz, S., Konczak, I., 2007. Native Australian fruits—A novel source of antioxidants for food. *Innovative Food Science and Emerging Technologies* 8, 339–346.

Park, H., Kim, H.S., 2014. Korean traditional natural herbs and plants as immune enhancing, antidiabetic, chemopreventive, and antioxidative agents: A narrative review and
perspective. *Journal of Medicinal Food* 17, 21–27.

Paul, A.A., Southgate, D.T., 1978. *McCance and Widdowson's The Composition of Foods.*
HMSO, London, U.K.

Read, C., 2012. *Nutritional Data for Australian Native Foods.* Rural Industries Research
and Development Corporation, RIRDC Publication No. 12/099, Canberra, Australian
Capital Territory, Australia.

Richer, S., Stiles, W., Statkute, L., Pulido, J., Frankowski, J., Rudy, D., Pei, K., Tsipursky, M.,
Nyland, J., 2004. Double-masked, placebo-controlled, randomized trial of lutein and
antioxidant supplementation in the intervention of atrophic age-related macular degeneration: The Veterans LAST study (Lutein Antioxidant Supplementation Trial).
Optometry 75, 216–230.

Rosillo, M.A., Sanchez-Hidalgo, M., Cardeno, A., Aparicio-Soto, M., Sanchez-Fidalgo,
S., Villegas, I., De La Lastra., C.A., 2012. Dietary supplementation of an ellagic
acid-enriched pomegranate extract attenuates chronic colonic inflammation in rats.
Pharmacological Research 66, 235–242.

Ruberto, G., Baratta, M.T., 2000. Antioxidant activity of selected essential oil components in
two lipid model systems. *Food Chemistry* 69, 167–174.

Shahidi, F., 2009. Nutraceuticals and functional foods: Whole *versus* processed foods. *Trends
in Food Science and Technology* 20, 376–387.

Shan, J., Fu, J., Zhao, Z., Kong, X., Huang, H., Luo, L., Yin, Z., 2009. Chlorogenic acid
inhibits lipopolysaccharide-induced cyclooxygenase-2 expression in RAW264.7
cells through suppressing NF-kappaB and JNK/AP-1 activation. *International
Immunopharmacology* 9, 1042–1048.

Speijers, G.J.A., Van Egmond, H.P., 1999. Natural toxins III. Inherent plant toxins. In: van
der Heijden, K. et al., (eds.), *International Food Safety Handbook.* Marcel Dekker,
New York, pp. 369–380.

Sultanbawa, Y., Cusack, A., Currie, M., Davis, C., 2009. An innovative microplate assay
to facilitate the detection of antimicrobial activity in plant extracts. *Journal of Rapid
Methods and Automation in Microbiology* 17, 519–534.

Wang, H., Cao, G., Prior, R.L., 1996. Total antioxidant capacity of fruits. *Journal of
Agricultural and Food Chemistry* 44, 701–705.

Weder, J.K.P., Murray, D.R., 1981. Distribution of proteinase inhibitors in seeds of Australian
acacias. *Zeitschrift für Pflanzenphysiologie* 103, 317–322.

Wilkinson, J.M., Cavanagh, H., 2005. Antibacterial activity of essential oils from Australian
native plants. *Phytotherapy Research* 19, 643–646.

Wilkinson, J.M., Hipwell, M., Ryan, T., Cavanagh, H.M., 2003. Bioactivity of *Backhousia
citriodora*: Antibacterial and antifungal activity. *Journal of Agricultural and Food
Chemistry* 51, 76–81.

Williams, D.J., Edwards, D., Pun, S., Chaliha, M., Sultanbawa, Y., 2014. Profiling ellagic acid
content: The importance of form and ascorbic acid levels. *Food Research International*
66, 100–106.

Zabaras, D., Konczak, I., Aguas, P., Giannikopoulos, G., 2008. Lipophilic bioactives in
Australian-grown "Hass" avocadoes. *Proceedings of the Fifth International Congress
of Pigments in Food*, Helsinki, Finland, August 14–16, 2008, pp. 192–194.

Zhao, J., Agboola, S., 2007. *Functional Properties of Australian Bushfoods.* Rural Industries
Research and Development Corporation, RIRDC Publication No. 07/030, Canberra,
Australian Capital Territory, Australia.

Figure 2.2 Photo of an anise myrtle (*Syzygium anisatum*) tree. (Photo by Gary Mazzorana. Copyright Australian Rainforest Products.)

Figure 2.3 Photo of a typical anise myrtle commercial plantation layout with lemon myrtle at the back. (Photo by Gary Mazzorana. Copyright Australian Rainforest Products.)

Figure 3.1 Photo of foliage, flowers and young fruit of bush tomato (*Solanum centrale*). (Photo by Slade Lee. Copyright Southern Cross University.)

Figure 3.3 Photo of hand harvesting of bush tomato (*Solanum centrale*) fruit. (Photo by Slade Lee. Copyright Southern Cross University.)

Figure 4.3 Photo of *Davidsonia jerseyana* fruit. (Photo by Margo Watkins. Copyright Rainforest Heart Orchard.)

Figure 5.2 Flowers of *Davidsonia jerseyana* showing the hirsute sepals and ovaries and the anthers before and after dehiscence.

Figure 6.2 Photo of a desert lime (*Citrus glauca*) tree. (Photo by Jock Douglas. Copyright Australian Desert Limes.)

Figure 6.4 Photo of ripe desert lime (*Citrus glauca*) fruits. (Photo by Yasmina Sultanbawa. Copyright University of Queensland.)

Figure 7.2 Photo of finger lime (*Citrus australasica*) tree with finger limes. (Photo by David Hancock. Copyright Skyscans.)

Figure 7.3 Photo of finger limes (*Citrus australasica*) in a range of skin colours. (Photo by David Hancock. Copyright Skyscans.)

Figure 8.2   Photo of Kakadu plum (*Terminalia ferdinandiana*) fruit. (Photo by Kim Courtenay. Copyright Kimberley Training Institute.)

Figure 8.3   Photo of Indigenous rangers sorting wild harvested Kakadu plum (*Terminalia ferdinandiana*) fruit. (Photo by Julian Gorman. Copyright Charles Darwin University.)

Figure 9.2 Photo of lemon aspen (*Acronychia acidula*) flowers. (Photo by Rus Glover. Copyright Woolgoolga Rainforest Products.)

Figure 9.4 Photo of Lemon aspen (*Acronychia acidula*) fruits. (Photo by Rus Glover. Copyright Woolgoolga Rainforest Products.)

Figure 10.3 Photo of lemon myrtle (*Backhousia citriodora*) leaf. (Photo by Gary Mazzorana. Copyright Australian Rainforest Products.)

Figure 10.4 Photo of a typical lemon myrtle commercial plantation layout and row plantings. (Photo by Gary Mazzorana. Copyright Australian Rainforest Products.)

Figure 11.2 Photo of a muntries (*Kunzea pomifera*) shrub with drip irrigation. (Photo by Yasmina Sultanbawa. Copyright University of Queensland.)

Figure 12.2 Photo of a Native Pepper (*Tasmannia lanceolata*) tree and its fruits. (Copyright Christopher Read/Diemen Pepper.)

Figure 13.2 Photo of a quandong (*Santalum acuminatum*) tree with fruits. (Copyright Ben Lethbridge.)

Figure 13.3 Photo of quandong (*Santalum acuminatum*) fruits. (Copyright Ben Lethbridge.)

Figure 14.2   Photo of a flowering riberry (*Syzygium luehmannii*) tree. (Photo by Rus Glover. Copyright Woolgoolga Rainforest Products.)

Figure 14.4   Photo of a riberry (*Syzygium luehmannii*) tree with fruits. (Photo by Rus Glover. Copyright Woolgoolga Rainforest Products.)

Figure 15.4 Photo of a wattle seed tree. (Photo by Yasmina Sultanbawa. Copyright University of Queensland.)

Figure 15.5 Photo of wattle seedpods and seeds. (Photo by Yasmina Sultanbawa. Copyright University of Queensland.)

CHAPTER 18

Australian Native Plants: Anti-Obesity and Anti-Inflammatory Properties

David J. Williams, Mridusmita Chaliha and Yasmina Sultanbawa

CONTENTS

INTRODUCTION

The prevalence of obesity is increasing at an alarming rate worldwide and places a major economic burden on health-care systems (Popkin, 2009; World Health Organisation, 1998). Obesity occurs when energy uptake is greater than energy expenditure, resulting in an accumulation of excess body fat with subsequent damaging physiological changes; this co-occurrence is often referred to as metabolic syndrome (Marinou et al., 2010; Piper, 2011; Singla et al., 2010). These alterations in conjunction with a linked chronic inflammatory state result in an increased risk of insulin resistance and the manifestation of type 2 diabetes as well as cardiovascular disease (Grundy, 2012).

At the cellular level, obesity is distinguished by an increase in the number and size of adipocytes (fat storage cells) that have differentiated from pre-adipocytes in the adipose tissue (Furuyashiki et al., 2004). This transition from undifferentiated pre-adipocytes into mature adipocytes constitutes the adipocyte life cycle (Figure 18.1), and treatments that regulate both the size and number of adipocytes may provide a valuable addition to reduced dietary energy in combating obesity.

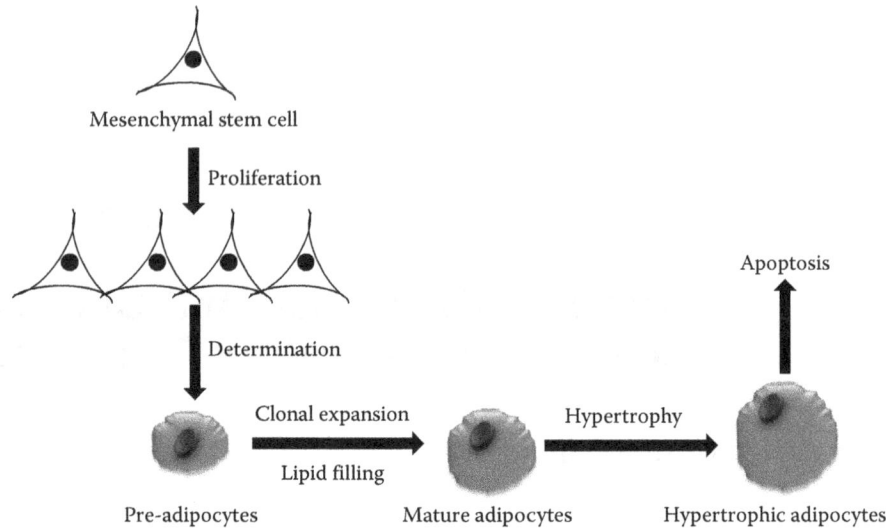

Figure 18.1 Summary of adipocyte life cycle. (Adapted from Rayalam, S. et al., *J. Nutr. Biochem.*, 19, 717, 2008.)

    While the strategy of reducing dietary fat content combined with increased physi-
cal activity has been shown to be effective in preventing obesity (Astrup, 2001; World
Health Organisation, 2007), numerous studies have shown that this simple message is
being ignored and alternative strategies are being sought (Kruger et al., 2004; Stern
et al., 1995). Evidence gained primarily from cell culture studies is supporting the view
that certain plant-based phytochemicals can modify the life cycle of adipocytes and
hence have the potential to reduce adipose tissue and obesity (reviewed in Rayalam
et al., 2008). Furthermore, these authors proposed that by targeting several points in
the adipocyte life cycle with different phytochemicals, a very effective obesity treatment could be devised. The mechanisms by which these compounds achieve this
adipose tissue reduction are diverse, ranging from suppression of growth, inhibition
of pre-adipocyte differentiation, stimulation of lipolysis to the apoptosis of existing
adipocytes (Gonzalez-Castejon and Rodriguez-Casado, 2011). Also due to the close
association between obesity and chronic inflammation, phytochemicals with strong
anti-inflammatory activity may also possess efficacy in counteracting the negative
aspects of obesity. Several classes of polyphenols appear to be the most likely candidates for anti-obesity agents through studies suggesting a profound impact on altering the adipocyte life cycle (Mulvihill and Huff, 2010; Yun, 2010). Even though the
polyphenols have received most attention, other classes of compounds, namely carotenoids and certain terpenes, have been implicated in the prevention of inflammation-
associated diseases such as obesity (Guri et al., 2007; Krinsky et al., 2004).
    Fruit and vegetables provide the major dietary source for these compounds,
with the types and levels varying markedly between species and even cultivars

(Nuutila et al., 2003; Singh et al., 2007). An increasing body of literature is high-lighting that tissue extracts from many Australian native plants are rich in phenolic compounds, especially phenolic acids, flavonoids and anthocyanins in addition to possessing potent anti-inflammatory activities (Cock, 2013; Konczak et al., 2010; Williams et al., 2014). Apart from a study that monitored the development of obe-sity in mice while fed a diet of Illawarra plum fruit (*Podocarpus elatus*), surprising little attention has been paid to utilising Australian native plants' enhanced phenolic load as likely anti-obesity agents (Symmonds and Fenech, 2009). Providing further impetus to this approach are the findings of Tan et al. (2011) when they reported that polyphenol-rich extracts from Illawarra plum exhibited significant anti-proliferative activity against various cancer cell lines.

With this backdrop, the chapter will review the proposed mechanisms of action of phytochemicals on obesity-related pathways and identify commercial Australian native plant sources that possess elevated levels of phytochemicals that are capable of modulating the adipocyte life cycle and/or reduce inflammatory responses.

Synergistic interactions with combinations of phytochemicals have previously been investigated for the treatment of some cancers (Chan et al., 2003; Hermalswarya and Doble, 2006). However, such synergistic interactions among dietary phytochem-icals acting on adipocytes have received only limited attention (Adams and Cory, 1998; Yang et al., 2007). So far, these studies have been encouraging, with results indicating an enhanced induction of apoptosis and suppression of adipogenesis by phytochemicals used in combination. Results from such studies and their applicabil-ity to Australian native plants will also be examined.

PHYTOCHEMICALS WITH ANTI-OBESITY AND ANTI-INFLAMMATORY PROPERTIES

The development and maintenance of obesity involves many complex molecu-lar mechanisms and connected cell signalling pathways; so, to provide an in-depth description of the possible phytochemical interactions is well beyond the scope of this chapter. Therefore, this chapter is limited to a summary of proposed mecha-nisms of action of the major plant-based phytochemicals.

Polyphenols are a class of phytochemicals widespread in plants that have dem-onstrated one or more potential anti-obesity effects (Badimon et al., 2010; Mulvihill and Huff, 2010). The strongest current evidence of phenolics possessing these prop-erties are phenolic acid derivatives including chlorogenic, ferulic, *p*-coumaric, gallic and caffeic acids and related compounds (Camire et al., 2009, Pan et al., 2010); the flavonoid sub-classes: flavonols – quercetin, kaempferol, myricetin, and isorham-netin (Yun, 2010); anthocyanins – cyanidin which has also been shown to possess strong anti-inflammatory activity and flavones, most notably luteolin and apigenin (Gonzalez-Castejon and Rodriguez-Casado, 2011).

A recent study (Son et al., 2010) evaluated the effects on the lipid metabolism of mice after feeding ferulic acid. This dietary phenolic acid suppressed the weight gain due to the high fat diet and inhibited fatty acid biosynthesis. Earlier, Hsu and

Yen (2006) assessed the inhibitory effect of dietary phenolic acids on the formation of mouse pre-adipocytes. Chlorogenic and coumaric acids caused significant inhibition of cell growth as well as enhancing apoptosis. Gallic acid, while not affecting the adipocyte cell cycle, did increase the number of apoptotic cells. In a complementary study, these authors reported a significant direct correlation between the inhibition of adipocyte proliferation and the antioxidant activity of these phenolic acids (Hsu et al., 2006), confirming a previous observation that plant-based extracts with high antioxidant activity possessed preventative effects against obesity (Tsuda et al., 2003).

Another phenolic acid receiving considerable attention mainly for its potent antioxidant activity is ellagic acid (EA) and its derivatives that includes glycosides and the complex ellagitannins (Atkinson et al., 2005). While this phytochemical does not appear to have undergone a detailed examination of potential anti-obesity properties, its enhanced anti-inflammatory capacity has already been recognised (Rosillo et al., 2012). Guo et al. (2014) evaluated the anti-inflammatory properties of native Australian herbs and reported those rich in EA were significantly more active than those with high levels of chlorogenic acid.

Quercetin and its glycoside, rutin, are dietary flavonols found in many plant-based foods with well-established effects on the adipocyte life cycle. They have been shown to inhibit adipogenesis (Strobel et al., 2005) and to induce apoptosis in mouse pre-adipocytes (Fang et al., 2008). In this latter study, kaempferol also exhibited anti-obesity properties but to a lesser extent (Fang et al., 2008). Ahn et al. (2008) provided useful insights into the molecular mechanisms by which quercetin influences the regulation of fat cell differentiation and apoptosis. Other researchers have revealed that this flavonol also provides protective effects against obesity-related inflammation (Al-Fayez et al., 2006; Chuang et al., 2010). In these, quercetin instigated attenuation of inflammation markers in human adipocytes and reduction of circulating markers in animal models.

Park et al. (2008) exposed human adipocytes to quercetin in combination with the isoflavone, genistein and the stilbene, reservatrol. The combined treatments caused enhanced inhibition of lipid accumulation in maturing adipocytes, far greater than the responses to individual compounds.

Park et al. (2009) established the anti-adipogenic properties of the flavone, luteolin on application to murine 3T3-L1 pre-adipocytes. This treatment caused decreased lipid accumulation and inhibited differentiation. An earlier study (Kuppusamy and Das, 1992) had shown that the addition of a similar flavone, apigenin, induced lipolysis in rat adipocytes.

Flavanones, a class of flavonoids found in high concentrations in citrus fruit, have shown promising evidence in preventing weight gain and other components of metabolic syndrome. The main dietary representatives of this sub-class are naringenin and hesperetin, which are known cholesterol-lowering agents as well as compounds that decrease pre-adipocyte proliferation (Borradaile et al., 1999; Hsu and Yen, 2006).

Another flavonoid sub-class with potential for harnessing anti-obesity effects is the anthocyanins, responsible for the red, blue and purple colours in fruit and vegetables (Clifford, 2000). In a 2008 study, Tsuda reported that anthocyanins exhibited significant anti-inflammatory properties when applied to enlarged adipose tissues. A further possible anti-obesity mechanism associated with anthocyanins was described by Sasaki et al. (2007). They observed that cyanidin, a common anthocyanin in foods, reduced blood glucose levels and down-regulated inflammatory protein cytokines in the adipose tissue of mice. Recent studies have established that an increase in expression of these inflammatory molecules in adipose tissue contributes to the development of insulin resistance (Kamei et al., 2006; Sartipy and Loskutoff, 2003).

Carotenoids, though not polyphenols, but rather a sub-class of terpenoids, have been reported to possess both anti-obesity and anti-inflammatory abilities (Gonzalez-Castejon and Rodriguez-Casado, 2011). They are responsible for the yellow, orange and red colour of many vegetables. α-Carotene is one of the most abundant carotenoids in the diet and can be converted in the body to an active form of vitamin A. β-Carotene inhibits inflammatory gene expression in lipopolysaccharide-stimulated macrophages. Possible anti-obesity roles for both these carotenes have been proposed based on the finding that the plasma of overweight and obese children had significantly lower levels of α-carotene and β-carotene when compared to healthy weight children (Burrows et al., 2009).

Terpenes from plants comprise a group of compounds with wide-ranging biological effects that include anti-viral and anti-inflammatory properties (Zhang and Demain, 2005). The anti-inflammatory activities of several medicinal plants are believed to arise from the presence of one or more terpene lactones (Zhang and Demain, 2005). Abscisic acid, a natural sesquiterpene, has displayed efficacy in the treatment of diabetes and obesity-related inflammation (Guri et al., 2007).

With the exception of this observational study, most of the evidence supporting the effects of dietary phytochemicals on obesity comes from mechanistic studies using cell lines or animal models (Hsu and Yen, 2006; Morikawa et al., 2007). This type of research assists in generating hypotheses for studies in humans and adds weight to any move towards a more plant-based diet for the prevention of chronic lifestyle-related disease. Indeed, the targeting of several points in the adipocyte life cycle by dietary phytochemicals has been proposed as a potentially effective obesity treatment approach (Badimon et al., 2010). Direct evidence of effects from human clinical trials is urgently required to confirm the anti-obesogenic effects of diets high in phytochemicals.

MULTIPLE PHYTOCHEMICAL COMBINATIONS
IN AUSTRALIAN NATIVE PLANTS

Complex mechanisms are involved in regulating adipose tissue development by dietary phytochemicals, it follows that exposure of adipocytes to multiple phytochemicals found in a high plant-based diet could result in enhanced or even synergistic effects. In fact, according to Rayalam et al. (2008), the earlier monotherapy approach has resulted in a distinct lack of success. Research over the last decade has shown that many Australian native plant tissues represent rich sources of these phytochemicals in often unique combinations mainly due to the plants' adaptations to diverse weather conditions experienced in Australia (Konczak et al., 2009; Netzel et al., 2007; Tan et al., 2010; Williams et al., 2014).

In this section, we summarise the scientific research that has evaluated the phytochemical composition of 13 commercially significant Australian native plants with a view of highlighting prospective candidates as anti-obesity agents. To assist in this assessment, they were divided into two categories based on the usage of their commercial products similar to that applied in Konczak et al. (2009): (1) *fruit*: Davidson's plum (*Davidsonia* spp.), desert limes (*Citrus glauca*), finger limes (*Citrus australasica*), Kakadu plum (*Terminalia ferdinandiana*), lemon aspen (*Acronychia* spp.), quandong (*Santalum acuminatum*), muntries (*Kunzea pomifera*) and riberry (*Syzygium luehmannii*); and (2) *herb/spice*: anise myrtle (*Syzygium anisatum*), bush tomato (*Solanum centrale*), lemon myrtle (*Backhousia citriodora*), Tasmanian pepper (*Tasmanian lanceolata*) and wattleseeds (*Acacia* spp.).

AUSTRALIAN NATIVE FRUIT SOURCES OF ANTI-OBESOGENIC PHYTOCHEMICALS

The total phenolic content as well as their composition profiles as documented in the scientific literature for the selected Australian native fruit is presented in Table 18.1. Even though most dietary phenolic compounds are present in the glycoside form, for the sake of simplicity, only the aglycone moiety is reported.

Inspection of this table's total phenolic contents shows that Kakadu plum is by far the richest source; in fact, Konczak et al. (2009) reported that its level was 4.7 times higher than that found in blueberries (often used as a benchmark for phenolic-induced anti-oxidant activity). The level of phenolic compounds in Davidson plum extracts was also higher than blueberry (1.5-fold), while the quandong exhibited similar values. Although the total phenolic content measurements in Konczak et al. (2009) were higher than that reported in the earlier Netzel et al. (2007) study for the first two fruits, both confirmed that these native tissues were rich sources of phenolic compounds. Furthermore, both studies established that Kakadu plum possesses an exceptionally rich mixture of phenolic compounds including the potential anti-obesity phytochemicals; quercetin, hesperitin, kaemferol and luteolin (Konczak et al., 2009) as well as exhibiting high levels of the potent anti-inflammatory bioactives, EA and ellagitannins (Williams et al., 2014). Kakadu plum was also seen to possess elevated quantities of the carotenoid, lutein, in fact at higher levels than the Australian 'Hass' avocado, seen as a primary source of lutein (Konczak et al., 2009). With only slightly lower total phenolic contents, both quandong and Davidson's plums also displayed a diverse array of potential anti-obesity phytochemicals, with the major constituents being anthocyanins (Table 18.1). Davidson's plum extracts also contained appreciable levels of lutein (Konczak et al., 2009). According to

Table 18.1 Total Phenolic Content (mg/Gallic Acid Equivalents/g DW) and Identified Polyphenolic Phytochemicals with Documented Anti-Obesity and Anti-Inflammatory Properties Present in Selected Commercial Australian Native Fruit

Australian Native Fruit	Total Phenolic Content[a] and Identified Phytochemicals	References
Davidsons' plum (*Davidsonia* spp.)	50.25 mg/GA eq./g DW Myricetin Quercetin and rutin Cyanidin Delphinidin Peonidin	Konczak et al. (2009) Konczak et al. (2009) Netzel et al. (2007)
Desert limes (*Citrus glauca*)	9.36 mg/GA eq./g DW Lutein	Vuong et al. (2014) Vuong et al. (2014)
Finger limes (*Citrus australasica*)	9.43 mg/GA eq./g DW Cyanidin Pelargodin	Netzel et al. (2007) Netzel et al. (2007)
Kakadu plum (*Terminalia ferdinandiana*)	158.57 mg/GA eq./g DW EA and Egallitannins Quercetin Hesperitin Kaempferol Luteolin	Konczak et al. (2009) Konczak et al. (2009) Tan et al. (2011) Vuong et al. (2014) Williams et al. (2014)
Lemon aspen (*Acronychia* spp.)	10.49 mg/GA eq./g DW Chlorogenic, caffeic, coumaric and ferulic acids Kaempferol Quercetin and rutin Luteolin	Konczak et al. (2009) Konczak et al. (2009)
Quandong (*Santalum acuminatum*)	32.87 mg/GA eq./g DW Quercetin and rutin Kaempferol Cyanidin Pelargonidin	Konczak et al. (2009) Konczak et al. (2009)
Muntries (*Kunzea pomifera*)	76.06 mg/GA eq./g DW Cyanidin Delphinidin	Netzel et al. (2007) Netzel et al. (2007) Tan et al. (2011)
Riberry (*Syzgium* spp.)	23.62 mg/GA eq./g DW Gallic and chlorogenic acids Quercetin and rutin Cyanidin	Konczak et al. (2009) Konczak et al. (2009) Vuong et al. (2014)

[a] Total phenolic content using Folin–Ciocalteu assay (Singelton and Rossi, 1965) and all values corrected for ascorbic acid concentration and moisture.

Netzel et al. (2007), the level of total phenolics in muntries was 2.6-fold of that of the blueberry control. The main contributor to this elevated value was not, according to these authors, anthocyanins but other unidentified phenolics. Until there is a more comprehensive assessment of this fruits' phenolic compounds, only limited statements can be provided regarding its eligibility as an anti-obesity ingredient, but it does show promise.

AUSTRALIAN NATIVE HERB AND SPICE SOURCES
OF ANTI-OBESOGENIC PHYTOCHEMICALS

The phenolic compound profiles for the selected Australian native herbs and spices are presented in Table 18.2. Again the phenolic compounds are presented in the aglycone form.

Within this category, Tasmanian pepper leaf contained the highest levels of phenolics followed by anise and lemon myrtle, again all with contents greater or on par with phenolic-rich blueberry control (Table 18.2; Konczak et al., 2009).

The high chlorogenic acid content in conjunction with appreciable levels of quercetin/rutin and the carotenoid indicates that the leaves of the Tasmanian pepper would be an ideal candidate for further investigation as an anti-obesity agent. A similar rich-mix phenolic profile plus the addition of EA derivatives and elevated levels of lutein makes the two myrtles worthy of further study. The presence of terpenes in the lipophilic fraction of these three herbs could be of additional benefit in

Table 18.2   Total Phenolic Content (mg/Gallic Acid Equivalents/g DW) and Identified Polyphenolic Phytochemicals with Documented Anti-Obesity and Anti-Inflammatory Properties Present in Selected Commercial Australian Native Herbs and Spices

Australian Native Herbs and Spices	Total Phenolic Content[a] and Identified Phytochemicals	References
Anise myrtle (*Syzgium anisatum*)	55.93 mg/GA eq./g DW Chlorogenic acid EA and ellagitannins, quercetin Myricetin Hesperetin	Konczak et al. (2009) Konczak et al. (2009) Sakulnarmrat and Konczak (2012)
Bush tomato (*Solanum central*)	12.40 mg/GA eq./g DW Chlorogenic, ferulic and p-coumaric acids Quercetin and rutin	Konczak et al. (2009) Konczak et al. (2009) Konczak et al. (2010)
Lemon myrtle (*Backhousia citriodora*)	31.44 mg/GA eq./g DW EA and ellagitannins Myricetin Hesperetin Naringen	Konczak et al. (2009) Konczak et al. (2009) Sakulnarmrat and Konczak (2012)
Tasmanian pepper leaf (*Tasmanian lanceolata*)	102.06 mg/GA eq./g DW Chlorogenic acid Quercetin and rutin Lutein Cyanidin	Konczak et al. (2009) Cock (2013) Konczak et al. (2009) Sakulnarmrat and Konczak (2012)
Wattleseeds (*Acacia* spp.)	0.76 mg/GA eq./g DW Chlorogenic acid Quercetin and rutin Kaempferol Leutolin	Konczak et al. (2009) Konczak et al. (2009) Konczak et al. (2010) Tan et al. (2010)

[a] Total phenolic content using Folin–Ciocalteu assay (Singelton and Rossi, 1965) and all values corrected for ascorbic acid concentration and moisture.

terms of utilisation of the plant tissues' adipocyte-modulating properties. Although the credentials of these plants' major terpenes (anise myrtle – sesquiterpene, anethole; lemon myrtle – monoterpene, citral; Tasmanian pepper leaf – sesquiterpene, polygodial) are yet to be assessed in terms of anti-obesity properties, other similar natural monoterpenes have shown efficacy in the treatment of diabetes and metabolic syndrome-related inflammation (Moteki et al., 2002; Zhang and Demain, 2005), while Salminen et al. (2008) observed considerable anti-inflammatory activities for several plant-based sesquiterpenes.

CONCLUSIONS AND FUTURE RESEARCH DIRECTIONS

Scientific investigations using cell culture and animal model studies demonstrate that phytochemicals derived from plant tissues can induce lipolysis, decrease lipid accumulation and induce apoptosis in adipose tissue. These mechanisms indicate potential anti-obesity properties that lend themselves to testing in human clinical studies. In addition to possible effects on adipocytes themselves, the anti-inflammatory properties reported for some of these phytochemicals suggest a powerful adjunct to dietary energy restriction in obesity-related chronic disease management.

This chapter does not provide an exhaustive list of phytochemicals with possible anti-obesity properties so far recorded in edible plants. As research identifies other compounds that modify the adipocyte life cycle or possess high anti-inflammatory activity, and knowledge of the human adipocyte life cycle expands, additional plants containing different combinations of phytochemicals with even greater anti-obesity potential may emerge. Recent investigations into the edible plants used as food and medicines by the Australian Aboriginal people indicated many are rich sources of these physiologically active compounds. Many of these plants possess a wide variety of anti-obesity phytochemicals providing the opportunity of synergistic and enhanced effects. The fruit of the Kakadu plum followed by quandong and Davidson's plum appears to offer the best opportunities for the future development of anti-obesity agents from Australian native fruit, while Tasmanian pepper leaf, anise and lemon myrtle from the classification of herbs/spices also show promise.

However, before any conclusive naming of native Australian plants as anti-obesity agents, there is a vital requirement for future studies to verify whether these anti-obesity properties due to these plant-based phytochemicals translate from their relatively proven *in vitro* actions into *in vivo* performance especially in humans.

REFERENCES

Adams, J.M., Cory, S., 1998. The Bcl-2 protein family: Arbiters of cell survival. *Science* 281, 1322–1326.

Ahn, J., Lee, H., Kim, S., Park, J., Ha, T., 2008. The anti-obesity effect of quercetin is mediated by the AMPK and MAPK signalling pathways. *Biochemical Biophysical Research Communication* 373, 545–549.

Al-Fayez, M., Cai, H., Tunstall, R., Steward, W.P., Gescher, A.J., 2006. Differential modulation of cyclooxygenase-mediated prostaglandin production by the putative cancer chemopreventive flavonoids tricin, apigenin and quercetin. *Cancer Chemotherapy and Pharmacology* 58, 816–825.

Astrup, A., 2001. Healthy lifestyles in Europe: Prevention of obesity and type II diabetes by diet and physical activity. *Public Health Nutrition* 4, 499–515.

Atkinson, C.J., Nestby, R., Ford, Y.Y., Dodds, P.A., 2005, Enhancing beneficial antioxidants in fruits: A plant physiological perspective. *BioFactors* 23, 229–234.

Badimon, L., Vilahur, G., Padro, T., 2010. Nutraceuticals and atherosclerosis: Human trials. *Cardiovascular Therapeutics* 28, 202–215.

Borradaile, N.M., Carroll, K.K., Kurowska, E.M., 1999. Regulation of HepG2 cell apolipoprotein B metabolism by the citrus fruit flavanones hesperetin and naringenin. *Lipids*, 34, 591–598.

Burrows, T.L., Warren, J.M., Colyvas, K., Garg, M.L., Collins, C.E., 2009. Validation of overweight children's fruit and vegetable intake using plasma carotenoids. *Obesity* 17, 162–168.

Camire, M.E., Kubow, S., Donnelly, D.J., 2009. Potatoes and human health. *Critical Reviews in Food Science and Nutrition* 49, 823–840.

Chan, M.M., Fong, D., Soprano, K.J., Holmes, W.F., Heverling, H., 2003. Inhibition of growth and sensitization to cisplatin-mediated killing of ovarian cancer cells by polyphenolic chemopreventive agents. *Journal of Cellular Physiology* 194, 63–70.

Chuang, C.C., Martinez, K., Xie, G., Kennedy, A., Bumrungpert, A., Overman, A., Jia, W., McIntosh, M.K., 2010. Quercetin is equally or more effective than resveratrol in attenuating tumor necrosis factor-{alpha}-mediated inflammation and insulin resistance in primary human adipocytes. *American Journal of Clinical Nutrition* 92, 1511–1521.

Clifford, M.N., 2000. Anthocyanins-nature, occurrence and dietary burden. *Journal of the Science of Food and Agriculture* 80, 1063–1072.

Cock, I.E., 2013. The phytochemistry and chemotherapeutic potential of *Tasmanian lanceolata* (Tasmanian pepper): A review. *Pharmacognosy Communications* 3, 1–13.

Fang, X.K., Gao, J., Zhu, D.N., 2008. Kaempferol and quercetin isolated from *Euonymus alatus* improve glucose uptake of 3T3-L1 cells without adipogenesis activity. *Life Science* 82, 615–622.

Furuyashiki, T., Nagayasu, H., AokI, Y., Bessho, H., Hashimato, T., Kanazawa, K., Ashida, H., 2004. Tea catechin suppresses adipocyte differentiation accompanied by down-regulation of PPARgamma2 and C/EBPalpha in 3T3-L1 cells. *Biotechnology Biochemistry* 68, 2353–2359.

Gonzalez-Castejon, M., Rodriguez-Casado, A., 2011. Dietary phytochemicals and their potential effects on obesity: A review. *Journal of Pharmacological Research* 64, 438–455.

Grundy, S., 2012. Pre-diabetes, metabolic syndrome and cardiovascular risk. *Journal of the American College Cardiology* 59, 635–643.

Guo, Y., Sakulnarmrat, K., Konczak, I., 2014. Anti-inflammatory potential of native Australian herbs polyphenols. *Toxicology Reports* 1, 385–390.

Guri, A.J., Hontecillas, R., Si, H., Liu, D., Bassaganya-Riera., 2007. Dietary abscisic acid ameliorates glucose tolerance and obesity-related inflammation in db/db mice fed high fat diets. *Journal of Clinical Nutrition* 26, 107–116.

Hermlswarya, S., Doble, M., 2006. Potential synergism of natural products in the treatment of cancer. *Phytotherapy Research* 20, 239–249.

Hsu, C.L., Huang, S.L., Yen, G.C., 2006. Inhibitory effect of phenolic acids on the prolif-
eration of 3T3-L1 preadipocytes in relation to their antioxidant activity. *Journal of Agricultural and Food Chemistry* 54, 4191–4197.

Hsu, C.L., Yen, G.C., 2006. Introduction of cell apoptosis in 3T3-L1 pre-adipocytes by flavonoids is associated with antioxidant activity. *Molecular Nutrition and Food Research* 50, 1072–1079.

Kamei, N., Tobe, K., Suzuki, R. et al., 2006. Overexpression of monocyte chemoattractant
protein-1 in adipose tissues causes macrophage recruitment and insulin resistance. *Journal of Biological Chemistry* 281, 26602–26614.

Konczak, I., Zabaras, D., Dunstan, M., Aguas, P., 2010. Antioxidant capacity and phenolic
compounds in commercially grown native Australian herbs and spices. *Food Chemistry* 122, 260–266.

Konczak, I., Zabaras, D., Dunstan, M., Aguas, P., Roulfe, P., Pavan, A., 2009. *Health Benefits
of Australian Native Foods: An Evaluation of Health-Enhancing Compounds*. Rural Industries Research and Development Corporation, RIRDC Publication No. 09/133, Canberra, Australian Capital Territory, Australia.

Krinsky, N., Mayne, S.T., Sies, H., 2004. *Carotenoids in Health and Disease*. Marcel Dekker,
New York.

Kruger, J., Galuska, D.A., Serdula, M.K., Jones, D.A., 2004. Attempting to lose weight:
Specific practices among U.S. adults. *American Journal of Preventive Medicine* 26, 402–406.

Kuppusamy, U.R., Das, N.P., 1992. Effects of flavonoids on cyclic AMP phosphodiesterase
and lipid mobilization in rat adipocytes. *Biochemical Pharmacology* 44, 1307–1315.

Marinou, K., Tousoulis, D., Antonopoulos, A.S., Stefanadi, E., Stefanadis, C., 2010. Obesity
and cardiovascular disease: From pathophysiology to risk stratification. *International Journal of Cardiology* 138, 3–8.

Morikawa, K., Ikeda, C., Nonaka, M., Suzuki, I. 2007. Growth arrest and apoptosis induced by
quercetin is not linked to adipogenic conversion of human preadipocytes. *Metabolism: Clinical and Experimental* 56, 1656–1665.

Moteki, H., Hibasami, H., Yamada, Y., Katsuzaki, H., Imai, K., Komiya, T., 2002. Induction
of apoptosis by 1,8-cineol in two human leukemia cell lines but not in human cancer cell line. *Oncology Reports* 9, 757–760.

Mulvihill, E.E., Huff, M.W., 2010. Antiatherogenic properties of flavonoids: Implications for
cardiovascular health. *Canadian Journal of Cardiology* 26(Suppl A), 17A–21A.

Netzel, M., Netzel, G., Tian, Q., Schwartz, S., Konczak, I., 2007. Native Australian fruits—
A novel source of antioxidants for food. *Innovative Food Science and Emerging Technologies* 8, 339–346.

Nuutila, A.M., Puupponen-Pimia, R., Aarni, M., Oksman-Caldentey, K.M., 2003. Comparison
of antioxidant activities of onion and garlic extracts by inhibition of lipid peroxidation and radical scavenging activity. *Food Chemistry* 81, 485–493.

Pan, M-H., Lai, C-S., Ho, C-T., 2010. Anti-inflammatory activity of natural dietary
flavonoids. *Food and Function* 1, 15–31.

Park, H.J., Kim, S.H., Kim, Y.S., 2009. Luteolin inhibits adipogenic differentiation by regulating PPAR gamma activation. *Biofactors* 35, 373–379.

Park, H.J., Yang, J.Y., Amabati, S., 2008. Combined effects of genistein, quercetin and resveratrol in human and 3T3-L1 adipocytes. *Journal of Medicinal Food* 11, 773–783.

Piper, A.J., 2011. Obesity hypoventilation syndrome—The big and the breathless. *Sleep
Medicinal Review* 15, 79–89.

Popkin, B.M., 2009. *The World Is Fat: The Fads, Trends, Policies, and Products that Are Fattening the Human Race.* Avery Trade/Penguin Group, New York.

Rayalam, S., Della-Fera, M.A., Baile, C.A., 2008. Phytochemicals and regulation of the adipocyte life cycle. *Journal of Nutritional Biochemistry* 19, 717–726.

Rosillo, M.A., Sanchez-Hidalgo, M., Cardeno, A., Aparicio-Soto, M., Sanchez-Fidalgo, S., Villegas, I., De La Lastra, C.A., 2012. Dietary supplementation of an ellagic acid-enriched pomegranate extract attenuates chronic colonic inflammation in rats. *Pharmacological Research* 66, 235–242.

Sakulnarmrat, K., Konczak, I., 2012. Composition of native Australian herbs polyphenolic-rich fractions and in vitro inhibitory activities against key enzymes relevant to metabolic syndrome. *Food Chemistry* 134, 1011–1019.

Salminen, A., Lehtonen, Suuronen, T., Kaamiranta, K., Huuskonen, J., 2008. Natural inhibitors NF-κB signalling with anti-inflammatory and anti-cancer potential. *Cell Molecular Life Science* 65, 2979–2999.

Sartipy, P., Loskutoff, D.J., 2003. Monocyte chemoattractant protein 1 in obesity and insulin resistance. *Proceedings of the National Academy of Sciences of the United States of America* 100, 7265–7270.

Sasaki, R., Nishimura, N., Hoshino, H. et al., 2007. Cyanidin 3-glucoside ameliorates hyperglycemia and insulin sensitivity due to down regulation of retinol binding protein 4 expression in diabetic mice. *Biochemical Pharmacology* 74, 1619–1627.

Singelton, V.L., Rossi, J.A., 1965. Colorimetry of total phenolics with phosphomolybdic-phosphotungstic acid reagents. *American Journal of Enology and Viticulture* 16, 144–158.

Singh, J., Upadhyay, A.K., Prasad, K., Bahadur, A., Rai, M., 2007. Variability of carotenes, vitamin C, E and phenolics in Brassica vegetables. *Journal of Food Composition and Analysis* 20, 106–112.

Singla, P., Bardoloi, A., Parkash, A.A., 2010. Metabolic effects of obesity: A review. *World Journal Diabetes* 1, 76–88.

Son, M.J., Rico, C.W., Nam, S.H., Kang, M.Y., 2010. Effect of oryzanol and ferulic acid on the glucose metabolism of mice fed with a high-fat diet. *Journal of Food Science* 76, H7–H10.

Stern, J.S., Hirsch, J., Blair, S.N., Foreyt, J.P., Frank, A., Kumanyika, S.K., 1995. Weighing the options: Criteria for evaluating weight-management programs. The Committee to Develop Criteria for Evaluating the Outcomes of Approaches to Prevent and Treat Obesity. *Obesity Research* 3, 591–604.

Strobel, P., Allard, C., Perez-Acle, T., Calderon, R., Aldunate, R., Leighton, F., 2005. Myricetin, quercetin and catechin-gallate inhibit glucose uptake in isolated rat adipocytes. *Biochemical Journal* 386, 471–478.

Symmonds, E., Fenech, M., 2009. Prevention of obesity in mice by the Australian fruit Illawarra plum. Nutrition Society of Australia, Annual Scientific Meeting, Newcastle, New South Wales, Australia.

Tan, A.C., Konczak, I., Ramzan, I., Sze, D.M.Y., 2011. Native Australian fruit polyphenols inhibit cell viability and induce apoptosis in human cancer cell lines. *Nutrition and Cancer* 63, 444–455.

Tan, A.C., Konczak, I., Sze, D.M.Y., Ramzan, I., 2010. Towards the discovery of novel phytochemicals for disease prevention from native Australian plants: An ethnobotanical approach. *Asia Pacific Journal of Clinical Nutrition* 19, 330–334.

Tsuda, T., 2008. Regulation of adipocyte function by anthocyanins; possibility of preventing the metabolic syndrome. *Journal of Agricultural and Food Chemistry* 56, 642–646.

Tsuda, T., Horio, F., Uchida, K., Aoki, H., Oskawa, T., 2003. Dietary cyanidin 3-*o*-β-d-glucoside-rich purple corn colour prevents obesity and amelorates hyperglycemia in mice. *Journal of Nutrition* 133, 2125–2130.

Vuong, Q.V., Hirun, S., Phillips, P.A., Chuen, T.L.K., Bowyer, M.C., Goldsmith, C.D., Scarlett, C.J., 2014. Fruit-derived phenolic compounds and pancreatic cancer: Perspectives from Australian native fruits. *Journal of Ethnopharmacology* 152, 227–242.

Williams, D.J., Edwards, D., Pun, S., Chaliha, M., Sultanbawa, Y., 2014. Profiling ellagic acid content: The importance of form and ascorbic acid levels. *Food Research International* 66, 100–106.

World Health Organization, 1998. *Obesity: Preventing and Managing the Global Epidemic—Report of a WHO Consultation on Obesity.* World Health Organization, Geneva, Switzerland.

World Health Organization, 2007. *Global Strategy on Diet, Physical Activity and Health.* World Health Organization, Geneva, Switzerland.

Yang, J.Y., Della-Fera, M.A., Hausman, D.B., Baile, C.A., 2007. Enhancement of ajoene-induced apoptosis by conjugated linoleic acid in 3T3-L1 adipocytes. *Apoptosis* 14, 388–397.

Yun, J.W., 2010. Possible anti-obesity therapeutics from nature—A review. *Phytochemistry* 71, 1625–1641.

Zhang, L., Demain, A.L., 2005. *Natural Products: Drug Discovery and Therapeutic Medicine.* Humana Press, New York.

CHAPTER 19

Food Preservation and the Antimicrobial Activity of Australian Native Plants

Yasmina Sultanbawa

CONTENTS

INTRODUCTION

Modern consumers are wary of the use of chemical preservatives in food products. Food technologists and processors therefore face the challenge of finding other sources of safe and effective preservatives. Plant-derived extracts that have antimicrobial properties have the potential to fulfil this need for 'natural' and 'safe' preservatives (Burt, 2004).

Another challenge faced by the food industry is the rise of a group of food-related microorganisms that are not only antibiotic resistant but are able to survive the traditional food-processing and preservation methods (Gyawali and Ibrahim, 2014), which have been attributed to the misuse of antibiotics in animal production as a major factor. In the medical sector too, there is a need for new antimicrobials, in view of today's pathogenic organisms becoming increasingly resistant to available antibiotics, such as the multi-drug resistant *Staphylococcus aureus*.

Antimicrobials are chemical substances that inhibit microbial growth or cause microbial cell death after exposure to the chemical. Antimicrobials can be described as traditional (chemical preservatives) and novel substances labelled 'natural'. These natural antimicrobials can be obtained from plants, animals, microbial and mineral sources. Antimicrobials that are sourced from plants which include vegetables, fruits and herbs/spices are termed as plant extracts or plant antimicrobials (Sultanbawa, 2011). Plant extracts contain phytochemicals, which in addition to being important for the functioning of the plant, also serve as a defence against microorganisms and other predators. Phytochemicals in plants are broadly classified as phenolic compounds, terpenoids and essential oils, alkaloids, lectins and polypeptides (Balasundram et al., 2006; Cowan, 1999). Phenolic compounds are known for their bioactive properties; in particular, those occurring naturally have been reported to be very effective as food preservatives due to their antimicrobial properties (Tajkarimi et al., 2010). Plants or plant by-products with known or suspected antimicrobial activity can be assessed for this activity as well as their potential as natural preservatives in food and other applications, using standard microbiological methods (Sultanbawa, 2014). In this search for new antimicrobials, many Australian native food plants have shown promise as plant antimicrobials for food applications.

This chapter is a review of the antimicrobial properties of the 13 commercially grown Australian native food plants and their application in different foods systems as natural preservatives to extend storage life of the product. These 13 species have been categorised into two groups, native herbs/spices and native fruits, a division that was previously adopted by Konczak et al. (2009). Native herbs/spices include anise myrtle (*Syzygium anisatum*), bush tomato (*Solanum centrale*), lemon myrtle (*Backhousia citriodora*), Tasmanian pepper (*Tasmannia lanceolata*) and wattle-seeds (*Acacia* spp.). Native fruits include Davidsons' plum (*Davidsonia* spp.), desert limes (*Citrus glauca*), finger limes (*Citrus australasica*), Kakadu plum (*Terminalia ferdinandiana*), lemon aspen (*Acronychia* spp.), quandong (*Santalum acuminatum*), muntries (*Kunzea pomifera*) and riberry (*Syzygium luehmannii*).

ANTIMICROBIAL PROPERTIES OF AUSTRALIAN NATIVE HERBS AND SPICES

Lemon myrtle, Anise myrtle and Tasmanian pepper have shown the most effective and promising antimicrobial activity for application in food, feed and cosmetic industries. Tables 19.1 through 19.3 refer to the in vitro antimicrobial activity reported for these three native herbs.

The antimicrobial activity of *B. citriodora* is related to the high citral content (Lis-Balchin et al., 1998); however, the importance of other minor components, such

Table 19.1   In Vitro Testing of Anise Myrtle Leaf Extracts against Microorganisms

Anise Myrtle	Inhibited Microorganism	References
Ethanol, water and acetone extracts	*Staphylococcus aureus*, *Escherichia coli*	Sultanbawa et al. (2015)
Polyphenol-rich extracts of ethanol, water and acetone	*Staphylococcus aureus*, *Escherichia coli*, *Listeria monocytogenes*, *Shewanella putrefaciens*, *Acinetobacter baumannii*, *Enterobacter aerogenes*, *Pseudomonas aeruginosa*, *Proteus vulgaris*, *Geotrichum candidum*	Sultanbawa et al. (2015)
Essential oil	*E. coli*, *S. aureus*, *Salmonella typhimurium*, *Candida albicans*, *Alcaligenes faecalis*	Wilkinson and Cavanagh (2005)
Essential oil	*E. coli*, Methicillin-resistant *Staphylococcus aureus* (*MRSA*), *Staphylococcus epidermis*, *Enterobacter aerogenes*, *Proteus vulgaris*, *Samonella typhimurium*, *Salmonella enteritidis*, *Enterococcus faecalis*	Hood et al. (2003)
Ethanol extracts	*Enterococcus faecalis*, *E. coli*, *Listeria monocytogenes*, *Pseudomonas aeruginosa*, *S. enteritidis*, *S. typhimurium*, *S. aureus*	Dupont et al. (2006)
Methanol extracts	*Bacillus subtilis*, *Vibrio cholera*	Zhao and Agboola (2007)

as linalool, β-myrcene in combination with citral for enhanced antimicrobial activity, has also been reported. The complete inhibition of *S. aureus* (0.156%) and *E. coli* (0.313%) by lemon myrtle essential oil at a much lower concentration than the pure citral compound for *S. aureus* (0.625%) and *E. coli* (2.5%), indicates the synergistic effects of citral with other compounds contributing to the antibacterial activities (Sultanbawa et al., 2009). Another study on the antibacterial and antifungal activity of lemon myrtle essential oils against 21 microorganisms concluded that the essential oils were more effective as an antifungal agent (Wilkinson et al., 2003).

Anethole is the major volatile compound in anise myrtle essential oil (Zafeiropoulou et al., 2010). Anethole has shown enhanced antifungal activity in combination with polygodial (a major volatile compound in Tasmanian pepper leaf) and dodecanol, against *Saccharomyces cerevisiae* and the pathogenic yeast *Candida albicans* (Fujita et al., 2007).

Polygodial is the principal antimicrobial compound in Tasmanian pepper leaf extracts and is very effective against fungi (Table 19.3). The fungicidal activity of polygodial was significantly increased when combined with anethole and sorbic acid, indicating a synergistic effect of these compounds (Fujita and Kubo, 2005a,b).

Table 19.2   In Vitro Testing of Lemon Myrtle Leaf Extracts against Microorganisms

Lemon Myrtle (*Backhousia citriodora*)	Inhibited Microorganism	References
Ethanol, water and acetone extracts	*Staphylococcus aureus, Escherichia coli*	Sultanbawa et al. (2015)
Polyphenol-rich extracts of ethanol, water and acetone	*Staphylococcus aureus, Escherichia coli, Listeria monocytogenes, Shewanella putrefaciens, Acinetobacter baumannii, Enterobacter aerogenes, Pseudomonas aeruginosa, Proteus vulgaris, Geotrichum candidum*	Sultanbawa et al. (2015)
Hexane extracts	*Aeromonas hydrophila, Bacillus cereus, Listeria monocytogenes, S. aureus, Salmonella enteritidis, Vibrio cholera, Yersinia enterolica, Acinetobacter baumannii, Bacillus subtilis, Pseudomonas aeruginosa, Psychrobacter phenylpyruvica, Candida albicans, C. colliculosa, C. lipolytica, C. stellata, Hanseniaspora uvarum, Pichia anómala, P. membranifaciens, Rhodotorula mucilaginosa, Schizosaccharomyces octosporus*	Zhao and Agboola (2007)
Essential oil	*Staphylococcus aureus, Escherichia coli*	Sultanbawa et al. (2009)
Essential oil	*Monilinia fructicola*	Lazar-Baker et al. (2011)
Essential oil	*Clostridium perfringens*	Zrustova et al. (2006)

The major compounds in the polyphenol-rich fractions of the following native herbs were

- *Anise myrtle*: ellagitannin, ellagic acid (EA) glycoside, EA, catechin, myricetin and quercetin
- *Lemon myrtle*: ellagitannin, EA glycoside, EA, hesperetin, myricetin and quercetin
- *Tasmanian pepper leaf*: chlorogenic acid, quercetin 3-rutinoside, quercetin glucosides

The percent growth inhibition of the tested organisms using the polyphenol-rich fraction of anise myrtle and lemon myrtle at a concentration of 0.75 µg/mL ranged from 26% to 100%, while for Tasmanian pepper leaf, inhibition ranged from 5% to 52% (Sultanbawa et al., 2015; Tables 19.1 through 19.3). These results indicate the potential of using different extracts of the aforementioned native herbs as natural antimicrobials in different foods, according to the relevant food spoilage or disease-causing organisms relevant to that commodity.

Table 19.3   In Vitro Testing of Tasmanian Pepper Leaf Extracts against Microorganisms

Tasmanian Pepper Leaf (*Tasmannia lanceolata*)	Inhibited Microorganism	References
Ethanol, water and acetone extracts	*Staphylococcus aureus*, *Escherichia coli*	Sultanbawa et al. (2015)
Polyphenol-rich extracts of ethanol, water and acetone	*Staphylococcus aureus*, *Escherichia coli*, *Listeria monocytogenes*, *Shewanella putrefaciens*, *Acinetobacter baumannii*, *Enterobacter aerogenes*, *Pseudomonas aeruginosa*, *Proteus vulgaris*, *Geotrichum candidum*	Sultanbawa et al. (2015)
Water, ethanol, hexane extracts	*Escherichia coli*, *Salmonella typhimurium*, *Listeria monocytogenes*, *Staphylococcus aureus*	Weerakkody et al. (2010)
Water, ethanol, hexane extracts	*Aeromonas hydrophila*, *Bacillus cereus*, *Clostridium perfringens*, *E. coli* 0157:H7, *Listeria monocytogenes*, *Shigella sonnei*, *S. aureus*, *Vibrio cholera*, *Yersinia enterolica*, *Acinetobacter baumannii*, *Bacillus subtilis*, *Pseudomonas aeruginosa*, *Psychrobacter phenylpyruvica*, *Candida albicans*, *C. colliculosa*, *C. lipolytica*, *C. stellata*, *Hanseniaspora uvarum*, *Pichia anómala*, *P. membranifaciens*, *Rhodotorula mucilaginosa*, *Schizosaccharomyces octosporus*	Zhao and Agboola (2007)
Polygodial	*Sclerotinia libertiana*, *Mucor mucedo*, *Rhizopus chinensis*, *Aspergillus niger*, *Penicillium crustosum*, *Zygosaccharomyces bailii*, *Saccharomyces cerevisiae*, *Salmonella choleraesuis*	Fujita and Kubo (2005a,b); Kubo et al. (2001); Kubo and Taniguchi (1988)

ANTIMICROBIAL PROPERTIES OF AUSTRALIAN NATIVE FRUITS

Kakadu plum, Davidson plum, muntries, riberries, lemon aspen, quandong, finger and desert lime are the native fruits that are commercially produced for value addition. Of these native fruits, extracts from Kakadu plum and Davidson plum have shown promising antimicrobial activity (Cusack et al., 2012a; Sultanbawa et al., 2015).

Kakadu plum fruit and leaf extracts have shown broad-spectrum antimicrobial activity against food-related organisms as given in Table 19.4. Two studies (Brossier et al., 2012; Cock and Mohanty, 2011) have reported on the antibacterial activity

Table 19.4   In Vitro Testing of Kakadu Plum Leaf and Fruit Extracts against Food-Related Microorganisms

Kakadu Plum	Inhibited Microorganisms	Minimum Inhibitory Concentrations (MIC)	References
Methanol extracts (fruit)	Gram-negative bacteria: *Alcaligenes feacalis, Aeromonas hydrophilia, Citrobacter freundii, Klebsiella pneumoniae, Proteus mirabilis, Pseudomonas fluroscens, Escherichia coli, Salmonella newport, Shigella sonnei* Gram-positive bacteria: *Bacillus cereus, Staphylococcus aureus, Staphylococcus epidermidis, Streptococcus pyogenes*	35–925 µg/mL using the disc diffusion assay	Cock and Mohanty (2011)
Ethanol extracts (leaves)	Gram-positive bacteria: *Staphylococcus aureus, Bacillus cereus*	Mixture of lemon myrtle 35% (w/w), Kakadu plum 30% (w/w), finger lime 20% (w/w) leaves, wheat grass sprouts 15% (w/w). 31–125 mg/mL using the broth dilution assay	Shami et al. (2013)
Edible film (fruit)	Gram-negative bacteria: *Escherichia coli, Pseudomonas aeruginosa, Acinetobacter baumannii* Gram positive bacteria: *Staphylococcus aureus,* methicillin-resistant *Staphylococcus aureus (MRSA), Listeria monocytogenes, Bacillus cereus, Bacillus subtilis*	Film composition – Kakadu plum powder 3% (w/w), carboxymethylcellulose (CMC) 1.5% (w/w) and glycerol 2% (w/w)	Brossier et al. (2012)
Ethanol, water and acetone extracts (fruit)	*Staphylococcus aureus, Escherichia coli*	8.75% (v/v) using the microplate assay	Sultanbawa, et al. (2015)

of Kakadu plum fruit, and indicated that extracts obtained using water, ethyl acetate, chloroform and hexane were more effective against Gram-positive bacteria as opposed to Gram-negative bacteria. An antimicrobial film formed from Kakadu plum extract exhibited a similar trend, being more effective against Gram-positive bacteria (Brossier et al., 2012). In contrast, the methanol extracts of the fruit were equally effective against both the Gram-negative and Gram-positive bacteria. A mix of Australian native plant leaves comprising of lemon myrtle, Kakadu plum, finger lime and Australian wheat grass sprout is available as a commercial food supplement. The ethanolic extracts of these natural food blend exhibited good antibacterial activity against *Staphylococcus aureus, Escherichia coli* and *Bacillus cereus* but had no effect on *Psuedomonas aeruginosa* (Shami et al., 2013). As mentioned earlier, the Kakadu plum fruit is a very good source of EA and gallic acid (GA) (Konczak et al., 2010; Williams et al., 2014). EA and GA as pure compounds have shown activity as antibacterial (Fogliani et al., 2005) and antifungal (Rangkadilok et al., 2012) agents. The high EA and GA contents of Kakadu plum would have contributed to the elevated antibacterial activity. Kakadu plum is also a rich source of ascorbic acid, which has been shown to enhance the antibacterial activity of other polyphenolic compounds by reducing the oxidative inactivation of these polyphenols (Cock and Mohanty, 2011). This suggests a pronounced synergistic effect of ascorbic acid and polyphenolic compounds present in Kakadu plum. The addition of plant phytochemicals that promote these synergistic interactions could be an exciting area for the future development of functional food ingredients.

In summary, the rich nutritional and phytochemical profiles of Kakadu plum demonstrate the potential of promoting this traditional Australian fruit as a functional food and ingredient. The significantly high antioxidant and broad-spectrum antimicrobial activities clearly indicate its potential in other food systems to enhance quality and improve safety. Synergistic combinations of polyphenolic compounds and ascorbic acid within the Kakadu plum matrix open new research opportunities in developing more effective antioxidant and antimicrobial formulations.

Davidson plum species have also shown broad-spectrum antimicrobial activity (Sultanbawa et al., 2015; Zhao and Agboola, 2007). The ethanol and acetone extracts of Davidson's plum at a concentration of 8.75% (v/v) showed complete inhibition of both *S. aureus* and *E. coli* and in excess of 90% inhibition was observed with the water extract (Table 19.5). The polyphenol-rich extracts of ethanol, methanol, water and acetone at 8.75% (v/v) inhibited the growth of the following organisms: *Staphylococcus aureus, Escherichia coli, Listeria monocytogenes, Shewanella putrefaciens, Acinetobacter baumannii, Enterobacter aerogenes, Psuedomonas aeruginosa, Proteus vulgaris* (Table 19.5). The major phenolic compounds identified

Table 19.5   In Vitro Testing of Davidson Plum Extracts against Microorganisms

Davidson Plum (*Davidsonia pruriens*)	Inhibited Microorganism	References
Ethanol, water and acetone extracts	*Staphylococcus aureus, Escherichia coli*	Sultanbawa et al. (2015)
Polyphenol-rich extracts of ethanol, methanol, water and acetone	*Staphylococcus aureus, Escherichia coli, Listeria monocytogenes, Shewanella putrefaciens, Acinetobacter baumannii, Enterobacter aerogenes, Pseudomonas aeruginosa, Proteus vulgaris*	Sultanbawa et al. (2015)
Methanol extracts	*Aeromonas hydrophila, Bacillus cereus, Clostridium perfringens, E. coli 0157:H7, Shigella sonnei, S. aureus, Vibrio cholera, Yersinia enterolica, Acinetobacter baumannii, Bacillus subtilis, Pseudomonas aeruginosa, Psychrobacter phenylpyruvica octosporus*	Zhao and Agboola (2007)

in the polyphenol-rich fraction included EA, EA derivatives, quercetin, rutin, myricetin and flavonoids including delphinidin sambubioside, and cyanidin sambubioside. The high level of EA could have contributed to the antimicrobial activity in Davidson plum (Sultanbawa et al., 2015).

CASE STUDIES: INCORPORATION OF AUSTRALIAN NATIVE PLANT EXTRACTS IN FOOD SYSTEMS

Terminalia ferdinandiana (Kakadu Plum) as a Natural Preservative for Prawns

To avoid the use of chemical additives in their products, the farmed prawn industries in Australia were seeking a natural preservative to extend the storage life of cooked chilled prawns. Extensive trials at the Health and Food Sciences Precinct pilot plant in Brisbane demonstrated the efficacy of using Kakadu plum as a natural antimicrobial agent. These studies have resulted in the formulation of a proprietary natural preservative using Kakadu plum to extend the shelf life of cooked chilled prawns. On-farm trials with this plant extract formulation have been successfully evaluated and commercial dipping of prawns has been done for the past 2 years (2013, 2014). In 2014, 600 t of prawns having a value of AUS$10 million were dipped in this solution and successfully marketed in Queensland (Sultanbawa, personal communication).

Davidsonia pruriens and *Terminalia ferdinandiana* as a Natural Preservative for Kangaroo Meat

At present, the kangaroo meat industry uses sulphites as a chemical preservative to extend the shelf life of comminuted or minced meat for pet food. Due to the health hazards associated with the use of sulphites in foods, there is a need to replace sulphites with alternatives such as natural antimicrobials from plant extracts. Different mixes of Kakadu and Davidson plum powder were incorporated into minced kangaroo meat using vacuum and modified atmosphere packaging conditions and shelf life was extended from 14 to 29 days under chilled storage conditions. These two native plum extracts also absorbed the off-odour notes produced in the minced meat during chilled storage and acted as a natural antioxidant (Cusack et al., 2012b).

Backhousia citriodora, *Terminalia ferdinandiana* and *Tasmannia lanceolata* as a Natural Preservative in Different Food Systems

Sauces and Marinades for Fish Fillets

This study evaluated the effect of oregano essential oil as a natural preservative in combination with sauces made from Australian native plants with antimicrobial properties, such as *Backhousia citriodora* (lemon myrtle), *Tasmannia lanceolata* (Tasmanian pepper leaf) and *Terminalia ferdinandiana* (Kakadu plum). Treated Crimson Snapper (*Lutjanus erythropterus*) fish fillets were stored in modified atmosphere packaging while shelf life studies were conducted. In a second study, combinations of Oregano essential oil, water extracts of lemon myrtle, Tasmanian pepper leaf and Kakadu plum were screened for antimicrobial activity against *S. aureus* and *E. coli* using a microtitre assay. Crimson Snapper (*Lutjanus erythropterus*) fish fillets were dipped in an aqueous solution of 0.1% v/v oregano essential oil (Treatment A). Two sauces were prepared using the following combinations of Australian native plants and other herbs and spices: Sauce A – lemon myrtle flakes (0.15% w/w) and Kakadu plum powder (0.5% w/w) (Treatment B) and Sauce B – ground lemon myrtle (0.2% w/w), ground Tasmanian pepper leaf powder (0.25% w/w) and Kakadu plum powder (0.5% w/w) (Treatment C).

Two other sauces were prepared with traditional chemical preservatives in place of the native plant extracts. Potassium sorbate (0.665% w/w) and sodium benzoate (0.059% w/w) were used at permitted levels as given in Standard 1.3.1 of the Australia New Zealand Food Standard Code. The base mix for the sauces consisted of xanthan gum, starch and water and was taken as the control. The sauces were challenged with the following microorganisms: *S. aureus* and *E. coli* (10^5 cfu/g), *Saccharomyces cerevisiae* (10^5 cfu/g) and *Aspergillus niger* (10^5 cfu/g) and stored at 4°C and 25°C for 30 days. The fish fillets were dipped in oregano essential oil and then coated with sauces A or B, placed in modified atmosphere packaging and stored at 4°C for 12 days.

Results for the antimicrobial screening of the plants extracts revealed complete inhibition of *S. aureus* and *E. coli* at 0.078% of oregano essential oil. Lemon myrtle, Tasmanian pepper leaf and Kakadu plum ethanol extracts were completely inhibitory at 2.1, 3.2 and 3.9 mg/gallic acid equivalents (GAEs)/g respectively. Sauces incorporating native plant ingredients or chemical preservatives had a total bacterial count < 10 cfu/mL and yeast and mould counts < 100 cfu/mL at the end of 30 days when stored at 4°C and 25°C. No growth of *S. aureus* and *E. coli* was observed in any of the sauces after being challenged with the aforementioned bacteria. Growth of *Aspergillus niger* and *Saccharomyces cerevisiae* was detected in the control after 7 days, while there was no growth observed in the sauces containing native plant extracts and traditional chemical preservatives. There was a significant drop > log 1 cfu/g in total viable count and anaerobe count after the oregano essential oil dip. This method of oregano dip in combination with sauces made with plant antimicrobial extracts and MAP resulted in a shelf life extension of 6 days more than that reported in previous studies, before the total counts went above log 7 cfu/g (Sultanbawa et al., 2010).

Fish Fillets

There is an increasing need by the fish industry to prolong shelf life of this highly perishable commodity to enable long distance transportation and to ensure the quality of the product on arrival at retail markets. To address this issue, Sultanbawa et al. (2012)

used oregano and lemon myrtle essential oils, singly and in combination to assess the extension in storage life of chilled Barramundi (*Lates calcarifer*, also known as Asian sea bass) fillets. Barramundi fillets were sprayed or dipped in 0.1% (v/w) oregano and a mix of 0.05% (v/w) oregano and 0.025% (v/w) lemon myrtle essential oils, vacuum packed and stored for 14–16 days at 4°C. Addition of oregano and a mixture of oregano and lemon myrtle extended the shelf life of the dipped fish fillets by 4 days and sprayed samples by up to 3 days.

Fresh Cut Fruits

Papaya (*Carica papaya* L.), a tropical fruit, is subject to a variety of physical and chemical changes after harvest. The use of fresh-cut papaya as a ready-to-eat product in the food service sector such as in aged care homes is very limited owing to the challenges in maintaining its quality and microbiological safety during storage. Liu et al. (2014) reported that lemon myrtle essential oil at 0.01%, 0.5% and 0.1% levels did not significantly change the physicochemical characteristics of the fresh-cut papaya but delayed the microbial growth at chilled temperatures of 4°C. The shelf life of lemon myrtle-treated papaya was extended by a further 4 days before total counts exceeded 10^5 cfu/g. Food Standards Australia New Zealand has set the microbiological quality of ready-to-eat foods at SPC < 10^6 cfu/g for the 'satisfactory' category, and <10^7 cfu/g for the 'marginal' category (Food Standards Australia New Zealand, 2001). In addition, gas chromatographic analysis revealed higher major volatiles such as linalool in the lemon myrtle-treated samples which could enhance fresh papaya flavour in comparison to the sample without lemon myrtle.

CONCLUSIONS AND FUTURE DIRECTIONS

The combination of native plant extracts and different packaging formats extended the shelf life of different food systems without the use of chemical sanitisers or preservatives. This clearly demonstrates the potential of natural antimicrobials, such as native plant products, as natural preservatives in multifactorial preservation models incorporating packaging and low temperature storage. The Australian native food industry, with its access to the diverse and rich Australian flora, has an enormous potential to contribute to this market for natural ingredients. An extensive range of Australian native plants is available for further research and development, having significant health-promoting and other functional properties in addition to their unique flavour profiles.

REFERENCES

Balasundram, N., Sundram, K., Samman, S., 2006. Phenolic compounds in plants and agri-industrial by-products: Antioxidant activity, occurrence, and potential uses. *Food Chemistry* 99, 191–203.

Brossier, M., Cusack, A., Edwards, D., Kelly, M., Sultanbawa, Y., 2012. Antimicrobial films from Kakadu plum. *Proceedings of the 12th Government Food Analysts' Conference*, Brisbane, Australia, pp. 194–195.

Burt, S., 2004. Essential oils: Their antibacterial properties and potential applications in foods – A review. *International Journal of Food Microbiology* 94, 223–253.

Cock, I.E., Mohanty, S., 2011. Evaluation of the antibacterial activity and toxicity of Terminalia ferdinandia fruit extracts. *Pharmacognosy Journal* 3, 72–79.

Cowan, M.M., 1999. Plant products as antimicrobial agents. *Clinical Microbiology Reviews* 12, 564–582.

Cusack, A., Chaliha, M., Currie, M., Burt, P., Matkinyidze, K., Sultanbawa, Y., 2012a. Shelf-life extension of communited meat using natural antimicrobials. *Seventh Dubai International Food Safety Conference and IAFP's First Middle East Food Safety Symposium*, Convention and Exhibition Centre, Dubai.

Cusack, A., Chaliha, M., Currie, M., Sultanbawa, Y., 2012b. Antibacterial activity of selected Australian native fruit extracts. *Seventh Dubai International Food Safety Conference and IAFP's First Middle East Food Safety Symposium*, Convention and Exhibition Centre, Dubai.

Dupont, S., Caffin, N., Bhandari, B., Dykes, G.A., 2006. In vitro antibacterial activity of Australian native herb extracts against food-related bacteria. *Food Control* 17, 929–932.

Fogliani, B., Raharivelomanana, P., Bianchini, J.-P., Bouraïma-Madjèbi, S., Hnawia, E., 2005. Bioactive ellagitannins from *Cunonia macrophylla*, an endemic Cunoniaceae from New Caledonia. *Phytochemistry* 66, 241–247.

Fujita, K., Fujita, T., Kubo, I., 2007. Anethole, a potential antimicrobial synergist, converts a fungistatic dodecanol to a fungicidal agent. *Phytotherapy Research* 21, 47–51.

Fujita, K., Kubo, I., 2005a. Multifunctional action of antifungal polygodial against Saccharomyces cerevisiae: Involvement of pyrrole formation on cell surface in antifungal action. *Bioorganic and Medicinal Chemistry* 13, 6742–6747.

Fujita, K., Kubo, I., 2005b. Naturally occurring antifungal agents against *Zygosaccharomyces bailii* and their synergism. *Journal of Agricultural and Food Chemistry* 53, 5187–5191.

Gyawali, R., Ibrahim, S.A., 2014. Natural products as antimicrobial agents. *Food Control* 46, 412–429.

Hood, J.R., Wilkinson, J.M., Cavanagh, H.M.A., 2003. Evaluation of common antibacterial screening methods utilized in essential oil research. *Journal of Essential Oil Research* 15, 428–433.

Konczak, I., Zabaras, D., Dunstan, M., Aguas, P., 2010. Antioxidant capacity and hydrophilic phytochemicals in commercially grown native Australian fruits. *Food Chemistry* 123, 1048–1054.

Konczak, I., Zabaras, D., Dunstan, M., Aguas, P., Roulfe, P., Pavan, A., 2009. *Health Benefits of Australian Native Foods: An Evaluation of Health-Enhancing Compounds*. Rural Industries Research and Development Corporation, RIRDC Publication No. 09/133, Canberra, Australian Capital Territory, Australia.

Kubo, I., Fujita, K., Lee, S.H., 2001. Antifungal mechanism of polygodial. *Journal of Agricultural and Food Chemistry* 49, 1607–1611.

Kubo, I., Taniguchi, M., 1988. Polygodial, an antifungal potentiator. *Journal of Natural Products* 51, 22–29.

Lazar-Baker, E.E., Hetherington, S.D., Ku, V.V., Newman, S.M., 2011. Evaluation of commercial essential oil samples on the growth of postharvest pathogen Monilinia fructicola (G. Winter) Honey. *Letters in Applied Microbiology* 52, 227–232.

Lis-Balchin, M., Deans, S.G., Eaglesham, E., 1998. Relationship between bioactivity and chemical composition of commercial essential oils. *Flavour and Fragrance Journal* 13, 98–104.

Liu, D., Cusack, A., Chaliha, M., Pun, S., Stanley, R., Sultanbawa, Y., 2014. Fresh-cut papaya with lemon myrtle as a value added product. *29th International Hoticultural Congress*, Brisbane, Queensland, Australia.

Rangkadilok, N., Tongchusak, S., Boonhok, R., Chaiyaroj, S.C., Junyaprasert, V.B., Buajeeb, W., Akanimanee, J., Raksasuk, T., Suddhasthira, T., Satayavivad, J., 2012. In vitro antifungal activities of longan (Dimocarpus longan Lour.) seed extract. *Fitoterapia* 83, 545–553.

Shami, A.M., Philip, K., Muniandy, S., 2013. Synergy of antibacterial and antioxidant activities from crude extracts and peptides of selected plant mixture. *BMC Complementary and Alternative Medicine* 13, 1–11.

Sultanbawa, Y., 2011. Plant antimicrobials in food applications: Minireview. In: Méndez-Vilas, A. (ed.), *Science Against Microbial Pathogens: Communicating Current Research and Technological Advances*. Formatex Research Center, Badajoz, Spain, pp. 1084–1093.

Sultanbawa, Y., 2014. Plant extracts as natural antimicrobials in food preservation. In: Ravishankar Rai, V., Jamuna Bai, A. (eds.), *Microbial Food Safety and Preservation Techniques*. CRC Press/Taylor & Francis Group, Boca Raton, FL, pp. 373–382.

Sultanbawa, Y., Currie, M., Cusack, A., Mayze, J., Slattery, S., 2012. Potential of using oregano and lemon myrtle essential oils to extend the shelf-life of fish fillets. *Seventh Dubai International Food Safety Conference and IAFP's First Middle East Food Safety Symposium*, Convention and Exhibition Centre, Dubai.

Sultanbawa, Y., Cusack, A., Currie, M., Davis, C., 2009. An innovative microplate assay to facilitate the detection of antimicrobial activity in plant extracts. *Journal of Rapid Methods and Automation in Microbiology* 17, 519–534.

Sultanbawa, Y., Sanderson, J., Chaliha, M., Cusack, A., Currie, M., Reeves, R., Exley, P., Paulo, C., Mayze, J., Forrest, A., 2010. Effects of plant extract and MAP on fresh crimson snapper (*Lutjanus erythropterus*) fillets coated with sauce during chilled storage. *Proceedings of the International Conference on Antimicrobial Research (ICAR2010)*, Valladolid, Spain, pp. 127–128.

Sultanbawa, Y., Williams, D., Chaliha, M., Konczak, I., Smyth, H., 2015. *Changes in Quality and Bioactivity of Native Foods During Storage*. Rural Industries Research and Development Corporation, RIRDC Publication No. 15/010, Canberra, Australian Capital Territory, Australia.

Tajkarimi, M.M., Ibrahim, S.A., Cliver, D.O., 2010. Antimicrobial herb and spice compounds in food. *Food Control* 21, 1199–1218.

Weerakkody, N.S., Caffin, N., Turner, M.S., Dykes, G.A., 2010. In vitro antimicrobial activity of less-utilized spice and herb extracts against selected food-borne bacteria. *Food Control* 21, 1408–1414.

Wilkinson, J.M., Cavanagh, H.M.A., 2005. Antibacterial activity of essential oils from Australian native plants. *Phytotherapy Research* 19, 643–646.

Wilkinson, J.M., Hipwell, M., Ryan, T., Cavanagh, H.M.A., 2003. Bioactivity of *backhousia citriodora*: Antibacterial and antifungal activity. *Journal of Agricultural and Food Chemistry* 51, 76–81.

Williams, D.J., Edwards, D., Pun, S., Chaliha, M., Sultanbawa, Y., 2014. Profiling ellagic acid content: The importance of form and ascorbic acid levels. *Food Research International* 66, 100–106.

Zafeiropoulou, T., Evageliou, V., Gardeli, C., Yanniotis, S., Komaitis, M., 2010. Retention of trans-anethole by gelatine and starch matrices. *Food Chemistry* 123, 364–368.
Zhao, J., Agboola, S., 2007. *Functional Properties of Australian Bushfoods.* Rural Industries Research and Development Corporation, RIRDC Publication No. 07/030, Canberra, Australian Capital Territory, Australia.
Zrustova, J., Mares, P., Brooker, J.D., 2006. Secondary plant metabolites to control growth of *Clostridium perfringens* from chickens. *British Poultry Abstracts* 2, 26–27.

CHAPTER 20

Unique Flavours from Australian Native Plants

Heather Smyth and Yasmina Sultanbawa

CONTENTS

INTRODUCTION

Australian chefs have been experimenting and exploring this country's native food flavours for decades, while the Aborigines of Australia have been aware of these unique flavours and their properties for over 40,000 years. These native foods can be combined to give a range of amazing flavour combinations, as an example: from Desert and Finger Lime cool drinks to a main meal of Emu with sweet chilli sauce made from Davidson's plum and a dessert of lemon myrtle and macadamia nut cheesecake with a roasted wattle seed coffee to complete the meal. This simple menu gives you an idea of the potential of these novel and new flavours and the enormous opportunity it presents to promote Australian cuisine globally.

These unique flavours of Australian plant foods need to become mainstream flavours that would be able to penetrate diverse markets. Many of the new flavours that are in the market have originated from restaurant menus and by people travelling abroad and experiencing flavours of different countries. A good example of a new flavour becoming a mainstream flavour is the Thai chilli sauce call 'Sriracha'. Although this has been a staple seasoning in Thailand for years, the sauce began appearing on U.S. high-end restaurant menus in 2012, and gradually made its way onto retail shelves as condiment sauces. By the end of 2013, it was available in lower-end restaurants and fast food chains (Michail, 2015).

The global flavour market is predicted to reach $15.2 billion by 2020 according to a new report by Allied Market Research. The natural flavours segment held most of the shares in terms of value, with the organic processed food and beverages driving growth in this segment (IFT, 2015). Changing consumer preference for natural food additives has also resulted in the increased use of natural flavourings by the food and beverage industry. This trend has been reflected in the changes to legislation in the European Union to include new definitions for natural flavours and processes that can be used in the preparation of natural flavouring substances (Marriott, 2012). According to Regulation (EC) No. 1334/2008, a 'natural flavouring substance' shall mean a flavouring substance obtained by appropriate physical, enzymatic or microbiological processes from material of vegetable, animal or microbiological origin either in the raw state or after processing for human consumption by one or more of the traditional food preparation processes such as cooking, baking, drying, chopping, evaporation, fermentation, grinding, infusion, roasting, steeping, emulsification, distillation, freezing and refrigeration.

Australian native food flavours would fit very well in the 'natural flavour substance' category as these native plant food ingredients are mostly processed using traditional food processing methods. Furthermore, this segment of the flavour market has the highest value, and this is favourable to the native food industry as they can target niche markets with the current quantities of production. This clearly indicates the potential that native food flavours have to contribute to this natural flavour market.

This chapter will review the current 13 commercial native plant foods that are used as flavour ingredients either as the whole plant food or as extracts. These 13 species have been categorised into two groups: native herbs/spices and native fruits.

Native herbs/spices include anise myrtle (*Syzygium anisatum*), bush tomato (*Solanum centrale*), lemon myrtle (*Backhousia citriodora*), Tasmanian pepper (*Tasmannia lanceolata*) and wattleseeds (*Acacia* spp.). Native fruits include Davidsons' plum (*Davidsonia* spp.), desert limes (*Citrus glauca*), finger limes (*Citrus australasica*), Kakadu plum (*Terminalia ferdinandiana*), lemon aspen (*Acronychia* spp.), quandong (*Santalum acuminatum*), muntries (*Kunzea pomifera*) and riberry (*Syzygium luehmannii*).

NATIVE PLANT FOODS AND INGREDIENTS: SENSORY DESCRIPTION

Typically, these native fruits and herbs/spices have strong flavours and aromas that are very intense and as such cannot be consumed as a whole food, and they are more suited as an ingredient to flavour a food (Smyth et al., 2012). The challenge for the Australian native food industry was to develop a sensory terminology that could be used across the industry and such a terminology would be imperative if these native flavours were to be promoted internationally in global markets. A world first 'Australian native food flavour wheel' was developed by a group of scientists from the University of Queensland and the Department of Agriculture and Fisheries, Queensland, led by Dr Heather Smyth. This research was identified as a research priority in the research and development plan of the peak industry body for Australian native foods, the Australian Native Food Industry Ltd (ANFIL) and the Rural Research and Development Corporation (RIRDC), an organisation that funded this research study. The 13 commercial native plant foods reviewed in this chapter were ascribed sensory descriptions after a thorough sensory study conducted in 2009. Details of the sensory terms are given in Tables 20.1 and 20.2 for both native

Table 20.1   Sensory Description of Australian Native Fruit

Native Fruit	Product Format	Sensory Description of Aroma and Flavour
Davidson plum (*Davidsonia jerseyana*)	Frozen fruit	Aroma is earthy like fresh beetroot with slight pickled and chemical notes. Flavour is intensely tart and astringent.
Davidson plum (*Davidsonia pruriens*)	Frozen fruit	Aroma of rosella and stewed rhubarb, some musk and candy notes. Flavour is intensely tart and astringent.
Desert lime (*Citrus glauca*)	Frozen fruit	Aroma of brown lime citrus, fermented notes, some pickled, stewed fruit and cut grass. Flavour is tart with some astringency and bitterness.
Finger lime (*Citrus australasica*) variety with green skin and colourless pulp	Frozen fruit	Aroma of fresh citrus with some cooked notes. Taste is citrus, tart with some astringency and bitterness.
Finger lime (*Citrus australasica*) variety with red skin and red-coloured pulp	Frozen fruit	Aroma of fresh and cooked citrus with slight fermented notes. Taste is citrus, tart with some astringency and bitterness.
Kakadu plum (*Terminalia ferdinandiana*)	Frozen fruit	Aroma of stewed apples and pears, some cooked citrus, pickled and fermented notes. Taste is tart and bitter with a strong stewed fruit flavour intensity.
Lemon aspen (*Acronychia acidula*)	Frozen fruit	Aroma of fresh citrus, conifer leaf, some chemical notes. Flavour of citrus, tart, some astringency and bitterness
Muntries (*Kunzia pomifera*)	Frozen fruit	Aroma of moist fruit mince, spice, bush honey and buttery. Taste is sweet.
Quandong (*Santaluum acuminatum*)	Frozen fruit	Aroma of dry lentils or beans with some earthy and fermented notes. A strong flavour intensity, tart with a salty taste
Riberry (*Syzygium luehmannii*)	Frozen fruit	Aroma of sweet spiced tea, with some musk notes, bush honey and resinous. Flavour is tart and astringent with some sweetness.

Source: Modified from Smyth, H.E. et al., *J. Sens. Stud.*, 27, 471, 2012.

Table 20.2   Sensory Description of Australian Native Herbs/Spices

Native Herb/Spice	Product Format	Sensory Description of Aroma and Flavour
Anise myrtle (*Syzygium anisatum*)	Dried milled leaf	Aroma of aniseed, menthol and herbal. Flavour of aniseed, some sweetness and slightly cooling on the palate.
Lemon myrtle (*Backhousia citriodora*)	Dried milled leaf	Aroma of lemon candy, perfumed with some menthol notes. Flavour is strong lemon with some sweetness and cooling on the palate.
Bush tomato (*Solanum centrale*)	Dried ground fruit	A savoury caramelised aroma of carob, some cereal notes. A savoury flavour with some sweetness.
Tasmanian pepper berry (*Tasmannia lanceolata*)	Dried ground fruit	Aroma of bush scrub with perfumed, fruity candy notes. Lingering heat on palate.
Tasmanian pepper leaf (*T. lanceolata*)	Dried milled leaf	Aroma of Australian bushland, dry paperbark and herbal. Intense heat, which slowly develops on palate.
Wattle seed (*Acacia victoriae*)	Dried ground fruit	Aroma of crushed nuts, cereal-like and slightly rancid. A savoury wheat biscuit flavour on the palate.
Wattle seed (roasted) (*A. victoriae*)	Dried, roasted ground seed	Aroma of toasted coffee grounds, sweet spice, raisin and chocolate. A savoury flavour, black coffee and some bitterness.

Source: Modified from Smyth, H.E. et al., *J. Sens. Stud.*, 27, 471, 2012.

fruit and herbs/spices (Smyth, 2010; Smyth et al., 2012). The sensory language developed in this study used scientific descriptive sensory methods and a trained panel of eleven experienced assessors. The panellists were asked to smell, taste, describe and ultimately rate the intensity of each flavour for the commercial native plant species. The sensory data were interpreted to develop discrete, accurate and precise sensory descriptions of each product, and the sensory language used by the panel was compiled into a lexicon, better known as a 'flavour wheel'. The benefits include a useful means of describing the flavours of native food products. Marketers of native foods can use this information to promote these commercial products using a sensory language that is accurate and user-friendly. Mainstream food industry can understand these native flavours which would lend to more effective design and development of food products that incorporate native flavours. Successful promoting of native foods has the potential to grow the industry by stimulating interest both in Australia and abroad.

NATIVE PLANT EXTRACTS AS FLAVOUR INGREDIENTS

A wide range of natural flavour preparations are available for some of the commercial native plant species. The flavours from these plant materials are extracted using green extraction technologies such as distillation and supercritical carbon dioxide extraction to produce essential oils (EOs), herb and spice extracts. The native fruits, herbs and spices are processed using traditional methods, such as drying, freezing, grinding, milling, roasting, pickling and cooking. This again fits well in the natural flavour substance space as defined by the European Union. Commercial products using native fruit, herb/spice ingredients in food, cosmetic and therapeutics are given in Table 20.3.

Table 20.3   Commercial Products with Native Plant Ingredients

	Commercial Products
Native herb/spice	
Lemon myrtle	Dried milled herb, EO, herbal teas, spice, yoghurt, ice cream, confectionary, biscuits, breads, curries, pastas, cheeses, sauces, liqueurs, beverages
	Personal care products such as shampoos, soaps, creams, toothpaste, conditioners, hand wash, food preservative
Anise myrtle	Dried milled herb, EO, herbal teas, spice, candies, health (cough mixtures), cosmetic and personal care products.
Tasmanian pepper	Fresh, frozen and dried whole berries, dried milled leaf, EO, food preservative or therapeutic ingredient. Spicy ingredient in flour mixes, relishes, sauces, soups and stews, in meats, flavoured pastas, pates and cheese
Bush tomato	Dried milled spice, herbal blends, pasta, relishes, chutneys, dressings, sauces, sprinkles (dukkahs)
Wattle seed	Roasted milled flour, cakes, breads, casseroles, curries, ice-cream, sauces, marinades and caffeine-free coffee
Native fruit	
Davidson plum	Fresh and frozen whole fruits, frozen puree, jams, chutneys, sauces, cordials, ice-cream, wines and liqueurs, whole fruit compote.
Desert lime	Fresh and frozen whole fruit, EO, canned product, preserves, brined peel, candied products, pickles, pulp as a base for soft drinks, preserves, confections, cakes, sauces, dressings, beverage, concentrate, syrup, dehydrated juice.
Finger lime	Fresh and frozen whole fruit, garnish in seafood and desserts, sprinkled on soups or added to salads. The pulp is used in dressings, jams, and sauces, rind used in cakes and muffins, range of personal care products.
Kakadu plum	Fresh and frozen whole fruits and puree, jams, chutneys, sauces, sports drinks, personal care products such as cleansers, body lotions, hand cream and lip balm, natural preservative for prawns.
Lemon aspen	Fresh and frozen whole fruit, beverage, sauces, dressings, jellies, chutneys, relishes, dried and ground spice, flavouring for mineral water, fruit wine.
Riberry	Fresh and frozen whole fruits, jams, conserves, chutneys, relishes, ice-cream, yoghurt, chocolates, cakes and sauces.
Muntries	Fresh, frozen and dried whole fruits, pies, jams, sauces, fruit straps, and ice cream.
Quandong	Fresh, frozen and dried whole fruits, pies, jams, chutneys, sauces, fruit cordial, ice cream, liqueur, dried fruit straps or leathers.

Source: Clarke, M., *Australian Native Food Industry Stocktake*, Rural Industries Research and Development Corporation, RIRDC Publication No. 12/066, Canberra, Australian Capital Territory, Australia, 2012.

Lemon Myrtle Essential Oil

Lemon myrtle EO in Australia and overseas is used as a source of lemon flavouring in the food and beverage industry. The main contributing compounds to the lemon and sweet aromas of lemon myrtle EO are neral (citral a) and geranial (citral b), citronellal and linalool (Forbes-Smith and Paton, 2002). Lemon myrtle EO is used as a lemon flavour replacement in milk-based products, such as cheesecake or ice cream to prevent curdling associated with the acidity of citrus fruits such as lemon (Horn et al., 2012). The dried and milled leaves of lemon myrtle are used in specialty cheeses in Australia; these cheeses are semi-hard, vacuum packed and are matured for less than 3 months and there is a growing market for cheese with novel flavours (Agboola and Radovanovic-Tesic, 2002). Lemon myrtle EO has a high content of the aldehyde citral > 90% (Pengelly, 2003). Citral, with a Flavour and Extract Manufacturers Association (FEMA) No. 2045, is used as a flavour ingredient in foods up to an average level of 200 ppm and it is generally regarded as safe (GRAS) (Smith et al., 2001).

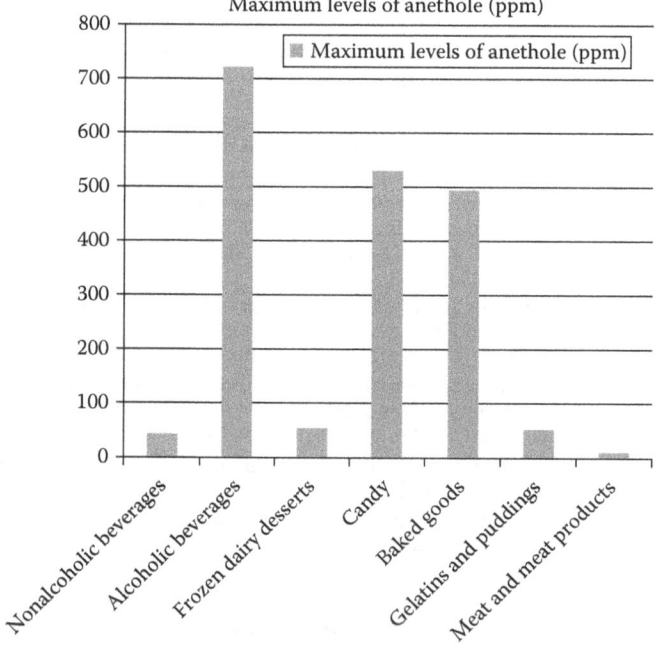

Figure 20.1 Average maximum usage levels for anethole as a flavouring substance in different food categories. (Modified from Taylor, S.L. and Dormedy, E.S., The role of flavoring substances in food allergy and intolerance, in Steve, L.T., ed., *Advances in Food and Nutrition Research*, Academic Press, 1998, pp. 1–44.)

Anise Myrtle Essential Oil

Anise myrtle EO has over 90% anethole (Blewitt and Southwell, 2000) and is comparable to other sources of anethole-rich EO such as fennel, anise seed and star anise. The EO is used in the food and beverage industry, in mouth care products, in pharmaceutical preparation and in personal care products like soaps. Anethole is used as a flavouring agent in the food and beverage industry as given in Figure 20.1 (Taylor and Dormedy, 1998). The aniseed flavour can be used in both sweet and savoury dishes at the levels allowed by regulations.

TASMANIAN PEPPER LEAF EXTRACT

The EO from the Tasmanian pepper leaf is extracted by a solvent extraction process unlike anise and lemon myrtle EOs which uses a steam distillation process. After the plant material has been extracted with a solvent like hexane, it produces a waxy aromatic compound referred to as the concrete. This concrete is soluble in ethanol, and it is this extract at a concrete yield of 6% or more that is used in commercial applications (Menary et al., 1999).

Polygodial is a sesquiterpene that was reported as a compound contributing to the pungent taste in *Tasmannia lanceolata* leaves (Loder, 1962; Read and Menary, 2000).

Tasmannia lanceolata extract has been approved as a GRAS flavouring ingredient FEMA No. 4755, under the conditions of their intended use in food flavourings in accordance with the 1958 Food Additives Amendment to the Federal Food, Drug and Cosmetic Act. The accepted levels as a flavouring substance in food are given in Figure 20.2 (Marnett et al., 2013). *Tasmannia lanceolata* extract is used in both sweet and savoury products in the food industry. These extracts have also been used as a flavour enhancer in chewing gum to enhance the mint, peppermint and spearmint (Menary et al., 2003). Polygodial has been used in wasabi to enhance the pungent effect of allyl isothiocyanate and interestingly to improve the sensory properties of artificial sweeteners. These effects could be attributed to the pungency and trigeminal stimulatory effect of polygodial (Starkenmann et al., 2011). Polygodial is a good example of the potential uses of pungent molecules in different sensory applications.

Australian Finger Lime

Three cultivars of fresh finger lime fruits, Alstonville, Judy's Ever bearing and Durham's Emerald, were assessed for their volatile components. This study reported that Australian native finger limes known to be unique for caviar-like pulp, shape, peel, pulp colour and flavour showed a similar distinctive characteristic at a molecular level. The data indicated unique chemical compositions with unusual ratios of major volatile metabolites: limonene/sabinene in cultivar Alstonville, limonene/citronellal/isomenthone in cultivar Judy's Ever bearing and limonene/citronellal/citronellol in

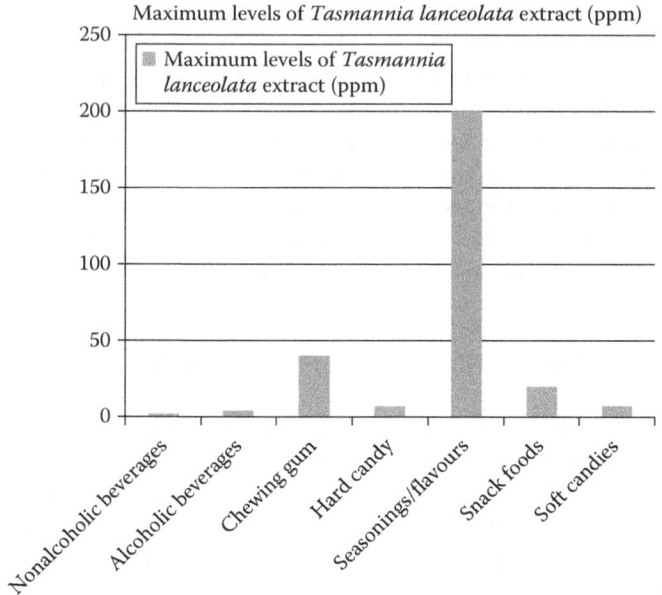

Maximum levels of *Tasmannia lanceolata* extract (ppm)

Figure 20.2   Average maximum use levels (ppm) for *Tasmannia lanceolata* extract as a flavouring substance in different food categories. (Modified from Marnett, L.J. et al., *Food Technol.*, 67, 38, 2013.)

cultivar Durham's Emerald. These volatiles were identified for the first time in these cultivars of finger limes (Delort et al., 2015).

Wattleseed

While unroasted wattleseed has a nutty and cereal-like flavour, roasted wattleseed when ground to a powder looks like coffee powder and has an intense coffee–chocolate–hazelnut flavour. Some of the major compounds identified after the roasting process were pyrazines, aldehydes/ketones, furans, pyridines, terpenes, pyrroles and amines. Pyrazines formed by reactions such as the Maillard reaction and Strecker degradation during the heating process are described as having a sweet, bitter and corn-like aromas, and the alkyl substitution of pyrazines can lead to aromas of burnt, grassy, roasted, pungent and nutty characters (Forbes-Smith and Paton, 2002). Wattleseed in its roasted form has the potential to be promoted as a decaffeinated coffee. It can also be used in baked products to give a chocolate or hazelnut flavour.

Lemon Aspen

The major compounds contributing to the flavour of lemon aspen juice were the monoterpenes and sesquiterpenes. The predominant monoterpene was limonene followed by 3-carene, terpinolene, β-myrcene, α-pinene and γ-terpinene. The main sesquiterpenes were santalene, α-bergomotene. Lemon aspen fruit contains volatile compounds that contribute to the citrus flavour notes such as limonene and α-pinene; it also had other volatiles contributing to herb, grass, pineapple and pine. Some of the products developed from lemon aspen are sorbets and beverages. An interesting effect was observed when a beverage was developed using lemon aspen juice and canned pineapple juice. Sensory tests indicated that addition of lemon aspen juice decreased the sweetness and increased the bitterness of the beverage (Forbes-Smith and Paton, 2002).

CONCLUSIONS AND FUTURE RESEARCH DIRECTIONS

It is clear that one of the main commercial values of Australian native plant foods is their unique and intense flavour. These ingredients could be minor components in food formulations but contribute to distinctly different food products both as sweet and savoury items.

Further research is needed to characterise and identify unique molecules in the Australian native food plants with novel flavour properties that can be used by the food and beverage industry, similar to the success of the pungent molecule polygodial in Tasmanian pepper which has very different applications.

Growth of the native food industry would be a major economic benefit to growers, processors and the indigenous communities that participate in the wild harvest. The use of these novel native foods in mainstream food and beverage industries would be key to the commercial success of these products. Towards this end, research should be done to support commercialisation and create new markets for the thirteen commercial species discussed in this chapter and the many more native food plants that need to be studied for their novel properties. We are at the beginning of a very exciting journey of biodiscovery which will hugely benefit the stakeholders involved in the native food industry.

REFERENCES

Agboola, S.O., Radovanovic-Tesic, M., 2002. Influence of Australian native herbs on the maturation of vacuum-packed cheese. *LWT-Food Science and Technology* 35, 575–583.

Anonymous, 2015. Global flavours market to reach $15.2 billion by 2020. Institute of Food Technologists. www.ift.org/food-technology/daily-news/2015 (accessed May 13 2015).

Blewitt, M., Southwell, I.A., 2000. *Backhousia anisata* Vickery, an alternative source of (E)-anethole. *Journal of Essential Oil Research* 12, 445–454.

Clarke, M., 2012. *Australian Native Food Industry Stocktake*. Rural Industries Research and Development Corporation, RIRDC Publication No. 12/066, Canberra, Australian Capital Territory, Australia.

Delort, E., Jaquier, A., Decorzant, E., Chapuis, C., Casilli, A., Frérot, E., 2015. Comparative analysis of three Australian finger lime (*Citrus australasica*) cultivars: Identification of unique citrus chemotypes and new volatile molecules. *Phytochemistry* 109, 111–124.

Forbes-Smith, M., Paton, J.E., 2002. *Innovative Products from Australian Native Foods.* Rural Industries Research and Development Corporation, RIRDC Publication No. 02/109, Canberra, Australian Capital Territory, Australia.

Horn, T., Barth, A., Ruehle, M., Haeser, A., Juerges, G., Nick, P., 2012. Molecular diagnostics of Lemon Myrtle (*Backhousia citriodora* versus *Leptospermum citratum*). *European Food Research and Technology* 234, 853–861.

Loder, J.W., 1962. Occurrence of sesquiterpenes polygodial and guaiol in leaves of *Drimys lanceolata* (Poir.) Baill. *Australian Journal of Chemistry* 15, 389–390.

Marnett, L.J., Cohen, S.M., Fukushima, N.J., Gooderham, S.S., Hecht, S.S., Rietjens, I.M.C.M., Smith, R.L. et al., 2013. GRAS flavoring substances 26. *Food Technology* 67, 38–56.

Marriott, R., 2012. Natural flavourings from green chemistry for foods and beverages. In: Baines, D., Seal, R. (eds.), *Natural Food Additives, Ingredients and Flavourings.* Woodhead Publishing, Cambridge, U.K., pp. 260–278.

Menary, R.C., Dragar, V.A., Garland, S.M., 1999. *Tasmannia Lanceolata – Developing a New Commercial Flavour Product.* Rural Industries Research and Development Corporation, RIRDC Publication No. 99/124, Canberra, Australian Capital Territory, Australia.

Menary, R.C., Dragar, V.A., Thomas, S., Read, C.D., 2003. *Mountain Pepper Extract: Tasmannia lanceolata Quality Stabilisation and Registration.* Rural Industries Research and Development Corporation, RIRDC Publication No. 02/148, Canberra, Australian Capital Territory, Australia.

Michail, N., 2015. Flavour migration: From offbeat idea to mainstream favourite. food navigator.com (accessed 19 February 2015).

Pengelly, A., 2003. Antimicrobial activity of lemon myrtle and tea tree oils. *Australian Journal of Medical Herbalism* 15, 9–11.

Read, C., Menary, R., 2000. Analysis of the contents of oil cells in *Tasmannia lanceolata* (Poir.) A. C. Smith (Winteraceae). *Annals of Botany* 86, 1193–1197.

Smith, R.L., Doull, J., Feron, V.J., Goodman, J.I., Munro, I.C., Newberne, P.M., Portoghese, P.S. et al., 2001. GRAS flavoring substance 20. *Food Technology* 55, 34–55.

Smyth, H., 2010. *Defining the Unique Flavours of Australian Native Foods.* Rural Industries Research and Development Corporation, RIRDC Publication No. 10/062, Canberra, Australian Capital Territory, Australia.

Smyth, H.E., Sanderson, J.E., Sultanbawa, Y., 2012. Lexicon for the sensory description of Australian native plant foods and ingredients. *Journal of Sensory Studies* 27, 471–481.

Starkenmann, C., Cayeux, I., Birkbeck, A.A., 2011. Exploring natural products for new taste sensations. *CHIMIA International Journal for Chemistry* 65, 407–410.

Taylor, S.L., Dormedy, E.S., 1998. The role of flavoring substances in food allergy and intolerance. In: Steve, L.T. (ed.), *Advances in Food and Nutrition Research.* Academic Press, New York, pp. 1–44.

PART III

Commercial Applications

CHAPTER **21**

Native Australian Plant Extracts: Cosmetic Applications

Hazel MacTavish-West

CONTENTS

INTRODUCTION

Australian flora is unique due to its geographical isolation and therefore offers the prospect of discovery of new compounds, with potential value as food or medicine. This chapter is an overview to highlight some of the directions that have to be taken and the issues that have to be addressed to enable Australian natural products to become ingredients in products used by the global cosmetics industry.

The cosmetics industry is large, global and growing, especially in Latin America, India and China. Among current trends, one which shows no sign of waning includes interest in plant-derived natural ingredients. Novel, new ingredients are much in demand, with many recent ones arising from the Amazon, following discoveries of new 'superfoods'. Ethnobotanical evidence is another source of information providing direction for the utilisation of natural ingredients. Australian native plants are on the radar of many companies involved in this sector; however, the research required to prove their worth, as well as supply issues are currently limiting the opportunity.

THE COSMETICS INDUSTRY

The definition of 'cosmetics' in the United States is 'articles intended to be rubbed, poured, sprinkled, or sprayed on, introduced into, or otherwise applied to the human body for cleansing, beautifying, promoting attractiveness, or altering the appearance'. In the EU, it is 'any substance or preparation intended for placing in contact with the various external parts of the human body (epidermis, hair system, nails, lips and external genital organs) or with the teeth and the mucous membranes of the oral cavity with a view exclusively or principally to cleaning them, perfuming them or protecting them in order to keep them in good condition, change their appearance or correct body odours'.

The cosmetics industry incorporates all the products found in a typical vanity cupboard: skincare moisturisers, deodorants, sun protection cream, self-tanning lotion, shampoo, make-up, toothpaste, hair dye, hair gel and so on. The five main categories in the cosmetics industry and their contribution to the overall industry are facial skincare (27%), personal care (23%) make-up and hair care (20% each) and fragrance (10%). Companies operating in this industry generally fall into one of two types: (1) manufacture and supply of finished products, with their own trademarked brand; or (2) production of ingredients and raw materials, sometimes with their own patent-protected ingredients. Marketing of finished products is all about promoting the specific formulation, format, packaging or specific claims. Marketing of raw materials and ingredients, which may include oils, surfactants, detergents, stabilisers, preservatives, colours, fragrances, minerals, plant extracts, amino acids, etc., is more about provenance, purity/quality, price, bioactivity and/or efficacy. Extracts from Australian native plants would predominantly be sold to ingredients companies, whilst ensuring all supporting evidence and information of relevance to finished product manufacturers was made available.

The cosmetics industry is expanding rapidly. Although the heart of the industry was historically in Europe, the Asian marketplace is now one of the largest emerging sectors. Events such as In-Cosmetics (www.in-cosmetics.com), journals/websites such as *Personal Care Magazine* (www.personalcaremagazine.com) and Household and Personal Products Industry (www.happi.com) and the website www.cosmeticsbusiness.com are good starting points for the new entrant to this industry. Exact figures for the size of the industry vary depending on the sectors included. For example, at US$100 billion sales globally (2013), skincare, which is the sector that most Australian plant extracts could market into, accounts for a quarter of the total beauty industry, and is expected to grow to US$117 billion by 2017. The Asia Pacific region generates half of total revenues, while Latin America is currently the fastest growing region (Euromonitor International, 2014 data). Anti-aging products are the fastest growing category in skincare, worth US$23 billion in 2012 and set to increase by over US$6 billion by 2017. The United States, China, South Korea and Japan are the main drivers. The predominant U.S. companies in facial skincare, which includes skin moisturisers, cleansers, facial products, anti-acne and anti-aging products, are Procter and Gamble (PandG) and Unilever. The world's largest beauty brands are, in reducing order of size: Olay, Avon, L'Oreal, Neutrogena, Nivea, Lancôme, Dove and Estée Lauder. Other personal care products in which Australian native plant extracts may also find applications include toothpaste and deodorants. The dominant companies for these products in the United States are PandG, L'Oreal and Unilever.

Natural ingredients such as plant extracts have been used in cosmetic products for a very long time, and for many there is good safety and toxicology information. Nature provides substances which are highly active, and which can therefore be harmful and require safety testing. In the excitement to get new plant extracts into cosmetic products, sometimes the requirements to prove safety have been overlooked, in the past. New products are critical to the success of the cosmetics industry. Whole new trends may develop on the back of a new source of ingredients, such as 'superfruits' from the Amazon with high anti-oxidant activity, skin-lightening bioactives, new novel ingredients such as retinoic acid, new targeted consumer groups such as men's grooming or new delivery methods, that is a packaging change. Longer-term trends include interest in plant-derived natural ingredients, clean labelling and organics. European legislation restricts use of new ingredients (EU Regulation 1223/2009), with a ban on animal testing throughout the development of ingredients and in finished products, and a positive list of approved ingredients (http://ec.europa. eu/consumers/sectors/cosmetics/cosing/ingredients/index_en.htm). Despite this, new novel ingredients are much in demand, although the hurdles to overcome prior to volume sales in Europe are high.

COSMETIC CLAIMS: SAFETY AND EFFICACY

Cosmetic products need to meet two criteria: safety and efficacy. Safety of cosmetic ingredients and finished products is mandatory, globally. Safety claims are generally made by using the terms 'dermatologically tested' and/or 'hypoallergenic'. The assessment of safety-in-use of cosmetic ingredients and products based on natural ingredients is complex and requires information related to toxicity of both ingredients used, and the extent and route(s) of exposure. The latter involves assessing the types of products in which the ingredient may be used, the loading level, how much consumers would use, how often they would use it, the area of the body to which the product will be applied and potential misuse which may increase exposure. This enables a calculation to be made regarding the amount likely to enter the body. Other sources of exposure, from other products, also need to be considered. Toxicity is evaluated based on knowledge of topical effects (skin, mucous membranes, eye irritation, phototoxicity, skin sensitisation and photosensitisation), and if significant absorption occurs, on systemic effects data: mutagenicity, carcinogenicity, embryotoxicity, teratogenicity, reproductive toxicity as well as general toxicity and effects on specific organs. Thus, adequate absorption studies via the skin and/or other relevant exposure routes need to be undertaken. There are internationally accepted guidelines such as the OECD guidelines for testing chemicals and complying with principles of good laboratory practice. Efficacy claims made on cosmetic products generally relate to either (1) functionality, which means the benefit ascribed to the product (i.e. 'anti-wrinkle', or 'moisturising') or (2) the presence or absence of a specific ingredient (i.e. 'contains a specific natural ingredient, mineral or plant extract', or 'paraben-free').

Perceptions of aging skin include dullness and dryness, wrinkling and sagging, altered pore size and texture, skin thinning, uneven pigmentation and development of dark spots. Many characteristics of aging skin are due to exposure to light, both visible and UV wavelengths (Watson et al., 2007). Marketing terms used to describe the efficacy of specific products include: dark spot correction, sun protection, pore minimising, skin tone evening, skin brightening or illuminating, skin smoothing, moisturising, skin protection, concealment and a reduction in the appearance of lines. Cosmetic ingredients need to fit within the definition of a cosmetic: no physiological changes can be claimed; the effect must be seen on epidermal layers only, and comply with any specific regulations, such as SPF claim legislation. 'Anti-ageing' claims such as a 'reduction in the appearance of wrinkles and fine lines', or a 'visible skin lightening effect', may come from the inclusion of bioactives including anti-oxidants and moieties that inhibit melanin synthesis, or inhibition of activity of collagenase or elastase enzymes, for example.

Functionality claims relating to efficacy need to be strongly substantiated by (1) clinical data showing an actual improvement in skin parameters objectively measured by dermatologists; (2) by ex-vivo or in-vivo studies where the impact is measured instrumentally and (3) by consumer data based on the perceived benefits from a sample population. Laboratory studies on the functionality of cosmetic ingredients are increasingly complex, and blur the lines between cosmetics and medicine. For example, tests may relate to upregulating specific genes, stimulating protein production and cell regeneration and the structural impact on collagen fibre bundles. Despite this, consumer testing is often the most easily understandable from a marketing perspective, as demonstrated by the UK sell-out of Boots No. 7 Protect and Perfect Intense Beauty Serum in response to independent consumer testing by a BBC team (http://news.bbc.co.uk/2/hi/health/8022644.stm). The active ingredients are claimed to be retinyl palmitate, anti-oxidants, peptides and alfalfa extract.

As the largest organ of the body, skin protects the internal environment, and skin diseases account for 34% of disease worldwide; such diseases can be persistent and difficult to treat, although they are rarely life threatening. Many plants used in medicine for centuries have become popular ingredients in cosmetic formulations, particularly those relating to traditional treatment of skin disorders. Ethnopharmacological evidence may drive discovery of new extracts and bioactives, and may also support marketing claims made on finished products. There is a general lack of information on plants used in ethnomedicine for skincare, limited research on the activity of crude extracts and even less information on active constituents (Lall and Kishore, 2014). Plant extracts already used in topical preparations may have anti-microbial, anti-inflammatory or anti-oxidative activity.

For example, carotenoids may provide a source of retinoic acid, flavonoids and polyphenols may provide UV protection, anti-inflammatory activity and metal-chelating properties. Medical claims relating to a reduction in inflammation, scarring, for treatment of burns or skin diseases, such as acne, may also be relevant to some cosmetic products. However, if pharmaceutical ingredients are added to cosmetics, even if they are on the GRAS list (GRAS = Generally Regarded As Safe), no claim on the efficacy related to the drug's activity can be made on the cosmetic product. There are a group of new products which have combined cosmetic claims (cleaning, beautification), with health care claims related to curing and healing problems: so-called 'cosmeceuticals'. How these straddle the various legislative issues is product specific and may be country specific.

Complete information on the chemical composition of complex mixtures in typical plant extracts is not normally available or expected for cosmetic ingredients; attention is generally placed on specific components, for which quantitative data is required. There are some undesirable components in some plant extracts, known to give rise to hypersensitivity reactions. Allergenic constituents in some common plants include menthol in *Mentha piperita*, eucalyptol in *Eucalyptus globulus* and thymol in *Thymus vulgaris*. For these and other potential allergenic components, appropriate labelling of the potential risk is required, and maximum usage levels may be set: for example menthol is avoided in baby products and should not exceed 2% in other products. Other potential undesirable components, some of which may have carcinogenic bioactivity, include oestrogenic substances (seen in carrot, ivy, hop and sage extracts); safrole (detected in over 50 species) which must not exceed 50 mg/kg in products for adults; pyrrolizidine alkaloids which may be hepatotoxic (found in borage); quercetin and its glucosides, for which there is contradictory data and many rich dietary sources; coumarins which can produce liver damage; escin, which is a triterpenic saponin and has a maximum use in finished products of 1%; quinine and its salts which should not exceed 0.2% in hair lotions and 0.5% in shampoos and hair rinses; methyl salicylate; estragole; sanguinarine; capsaicin and hypericine. It is advised to screen plant extracts for the quantitative presence of these and to evaluate current legislation. In addition, essential oils may have an irritant action when applied at full strength to the skin, and some oils cause hypersensitivity and photodermatitis reactions, in addition to hazards from excessive inhalation. Cosmetic tars may also have considerable allergenic potential, particularly from their potential content of benzene, toluene, xylene and styrene; usage is generally restricted to anti-dandruff shampoos.

EXTRACTION TECHNOLOGIES

Plant preparations may have very different properties and effects depending on the part used and the extraction process. Apart from use of the plant part in its pure state, that is dried and milled, the types of plant preparations used in cosmetic products may include essential oils, extracts, resins and tars, tinctures and waters (aromatic distilled waters). Generally, industrially produced, standardised preparations are used.

1. Essential oils or essences are made by distillation of fresh or dried plant materials in a steam current. The insoluble fraction in water is then collected after condensation. The resulting products is often purified in various ways, for example by dissolving it in highly concentrated hydroalcoholic solutions, by fractional distillation or by crystallisation. So-called concrete essences are extracted by light organic solvents (e.g. hexane), followed by purification with absolute ethanol, resulting in an 'absolute essence'.

2. Extracts are made by extracting the plant material by a suitable solvent and subsequently concentrating under particular temperature and vacuum conditions. For cosmetics: water, hydroalcoholic solutions or propylene glycol are the most commonly used solvents. Such extracts are often preferred over dry (powdered) extracts or soft (solid) ones. The concentration of the extract is expressed via the extract/plant material ratio in weight; the functional principles of the preparation may also be referred to, that is anti-oxidant activity per gram.

3. Resins are viscous exudates collected from plants directly, sometimes after wounding of the plant. Tars are dark brown or nearly black liquids obtained from wood of various trees by distillation, particularly *Pinus* spp.

4. Tinctures are liquid preparations made by maceration or digestion of plant material in a suitable solvent. Unlike extracts, there is no concentration step. In addition to water, hydroalcoholic solutions and propylene glycol, lipids (i.e. vegetable oils), and fluid esters are the solvents most commonly used. The concentration is also expressed as tincture/plant ratio in weight, and the functionality of the tincture may be highlighted.

5. Waters (aromatic distilled waters) are aqueous solutions containing essential oils, obtained by condensation of the steam used when obtaining the essential oils by steam distillation.

The type of information required on plant preparations for cosmetics includes name of the plant, plant part used, type of extract and extraction solvent, purification/concentration, plus organoleptic characteristics, solubility, pH, loss on drying, ash, total solids, specific gravity, refraction index, specific rotation power and alcohol (or other solvent) content. Potential impurities, particularly in complex and concentrated plant extracts, may include pesticides, solvent residues, heavy metals and microbes, especially pathogenic bacteria.

It is the water-soluble fraction, and either the steam-distilled or solvent-extracted fat-soluble fractions of Australian native plants that are potentially the most interesting for the cosmetics industry, in the author's opinion. The water-soluble fraction may contain diverse and comparatively high content of polyphenolics (anthocyanidins, catechins, flavonoids, tannins and procyanidins), with their inherent UV-absorbing and high anti-oxidant bioactivity. The fat-soluble fractions may have natural fragrance, plus anti-microbial or anti-fungal and other relevant bioactivity, related to terpenes and lactones, among other moieties. Most phytochemicals, including polyphenolics and terpenoids, are produced in response to plant stress. Plant stress goes hand in hand with growth in an Australian environment, with its inherent drought, heat, soils with low mineral levels and, of course, high visible and UV radiation. In addition, the flora of Australia is diverse, often unique and, in some cases, comparatively ancient. The essential oils of many Australian and other plants have considerable anti-bacterial and anti-fungal activities. Those examined in particular include essential oils from *Eucalyptus* spp. and *Melaleuca alternifolia*. Solvent extracted 'concrete' from *Boronia megastigma* Nees. has found use as a fragrance component in hair conditioner, and newly GRAS-registered extracts from *Tasmannia lanceolata* leaves are being examined for a range of bioactivities of relevance to cosmetics. There are a plethora of papers comparing the anti-oxidant activity of all types of extracts from a wide range of plant parts from Australian endemic plants. As research progresses, specific bioactivity of extracts becomes more the focus. For example, inhibition of growth in specific cancer cell lines, or activity of specific enzymes.

There are already a number of Australian companies operating in this space, with product ranges that usually include essential oils and other extracts: some with in-house consumer brands and some with global customers. In addition, there is some excellent research occurring in a number of institutes in Australia on the diverse and emerging activities of extracts from native flora, for food and non-food applications.

FOOD AS SKINCARE

There has historically been a good correlation between trends in foods and trends in cosmetic ingredients, as evidenced by the sequential rise in interest in natural ingredients, plant-derived ingredients, marine plants, superfruits and, more recently, trends incorporating vegetable-derived ingredients as part of skin-applied cosmetics. The term 'nutricosmetics' was coined relatively recently to describe products such as tablets, capsules, pills, liquids, granules and foods which are taken orally to improve health and beauty. Nutricosmetics represent the intersection of nutrition and personal care. An example is Imedeen®, a patented biomarine complex that promises to optimise skin health and prevent the early signs of aging. This is a rapidly growing category within the cosmetics industry, particularly in countries such as Japan, where there are speciality stores and departments dedicated to nutricosmetics. Such products are typically sold through pharmacies in Western Europe. This sector of the industry was valued at US$1.5 billion in 2007 and is estimated to be US$4 billion by 2015 (Taeymans et al., 2014).

Other areas of bioactivity claimed by nutricosmetic products include anti-aging, repair and prevention, sun protection, pigmentation, skin whitening and slimming. Natural ingredients of interest to the nutricosmetics sector include carotenoids, polyphenols, vitamins, soy extracts (isoflavones), micronutrients, glycopolyglycans, amino acids, other plant-based elements and polyunsaturated fatty acids from fish oils. Specific examples include green tea, coenzyme-Q10, superfruits such as Acai, grape seed extract, lutein, lycopene, vitamins A, C and E and zinc and selenium. New food and beverage products have emerged with a combination of nutritional and health claims related to prevention or treatment of a disease: so-called 'nutriceuticals' or functional foods. In Australia, FSANZ (Food Standards Australia, New Zealand) has legislation which applies to these products.

Research illustrating the potential for Australian native plants as new ingredients in products in these sectors of the cosmetics industry is emerging. There are now several robust studies of the polyphenolic and vitamin content and anti-oxidant activity of water and solvent-based extracts from leaves and berries of several species (Konczak et al., 2014; Netzel et al., 2006, 2007). Species showing the most promise include fruits of *Tasmannia lanceolata* (Native, or Mountain Pepper), *Eugenia carissoides* (Cedar Bay Cherries), *Syzygium luehmannii* (Riberry) and especially Kakadu plum (*Terminalia ferdinandiana*), with its exceedingly high content of tannins, ellagic acid, ascorbic acid and total phenolics. Many of the fruits have at least three times more total phenolics than blueberry, with concomitantly greater anti-oxidant activity (Netzel et al., 2007). Kakadu plum shows significant anti-oxidative, anti-inflammatory and propoptotic anti-cancer activities (Tan et al., 2011). Generally, nutricosmetics incorporate dried powders of the whole plant material or specific tissues, not plant extracts per se. This reduces the cost of production and can make this area of the industry more accessible to growers, although safety and toxicity testing is crucial.

When Australian native fruit high in phenolics and anti-oxidant activity are extracted, often the extracts are imbued with a deep purple colour from the anthocyanins present. Lack of the presence of anthocyanins should in no way be linked with lack of potential cosmetic activity however, as extracts from *Eugenia carissoides* and *Pleiogynium timorense* (Burdekin Plum) had low anthocyanin content but were still rich in total phenolics and anti-oxidant activity (Netzel et al., 2007). In addition, we have found in our own research that aqueous extracts from leaves of *Tasmannia lanceolata* had stronger anti-tyrosinase, anti-collagenase and anti-elastase bioactivities in *in vitro* studies, compared with extracts from the berries, which were rich in anthocyanins (MacTavish-West, 2014).

OTHER ASPECTS

There is considerable cost incurred in developing, testing and producing a new plant extract to bring into the global cosmetic arena. Not least is managing control over the potential applications for the extract, because in many cases, exclusivity to specific final product manufacturers may be part of the licensing agreement. It is not possible to patent plants, or any extract from them. However, formulations containing plant extracts or pure compounds may be patented if they have a specific application. For example, a root extract from *Pothomorphe umbellate* (pariparoba), a shrub native to the Atlantic biomes, has been patented for pharmaceutical and cosmetic applications, due to its anti-aging and photo-protective activities (Biavatti et al., 2007).

Other issues relating to development of products from countries including Australia are international treaties regulating movement of materials, including the Convention on International Trade in Endangered Species of Wild Fauna and Flora (CITES, www.cites.org) which Australia has been part of since 1976; the Convention on Biological Diversity (CBD, www.cbd.int) and the Nagoya Protocol. Many cosmetic manufacturing companies require proof or certification regarding sustainable and ethical sourcing of ingredients. Another pertinent regulatory issue relates to organic certification, which is particularly challenging for wild-crafted product. For some plant products from Australia, the sheer distance between harvest locations and processing plants is a matter for serious consideration, in terms of managing crops throughout the season, availability of harvest labour, appropriate storage to maintain quality and costs for all of the above, plus transport.

RELEVANT RESEARCH

Australian native flora has not been extensively researched for potential cosmetic applications, largely due to its almost overwhelming diversity, although there are some indicative papers and reviews. The ethnomedicinal use of various Australian plants by Aboriginal communities for relief from diverse skin disorders is excellently described by Lassak and McCarthy (1983), and has been extensively reviewed by Ian E. Cock in many publications and will not be repeated herein, largely. In addition, the reader is referred to the considerable work of Palombo (2011) reviewing plant extracts with activity against oral bacteria, and Palombo and Semple (2001) screening many native Australian plants for anti-bacterial activity, and related papers.

A non-exhaustive summary of relevant recent research undertaken on the cosmetic potential of Australian native plants is presented in Table 21.1. For the purposes of this summary, activity related to anti-microbial, anti-fungal, anti-inflammatory, anti-oxidant, anti-septic or specific activity relevant to potential anti-aging claims in reasonably recent, peer-reviewed work is included. Activity relating to treatment of skin disorders including warts, cancers, leprosy etc., undisclosed 'skin diseases', rheumatism or pain, is not, except where the identified active constituents may be indicative of potential bioactivity of cosmetic (i.e. epidermal) relevance. Plants not categorically native to Australia have not been extensively reviewed.

OPPORTUNITIES AND CHALLENGES

The opportunities within the global cosmetics industry are wide ranging, continually evolving and potentially large, both in terms of volume required and value returned. Whether for topical treatments for skin cancer, to mediate the visual effects of aging or reduce skin pigmentation or as orally ingested capsules to procure beauty from within, there are abundant opportunities for plants rich in diverse, complex and largely uncharacterised natural chemistry. All of the factors which make Australian native plants resilient, unique and a constant source of discovery, especially their geographic isolation and harsh growing conditions, imbibe them

Table 21.1   A Non-Exhaustive Summary of Relevant Recent Research Undertaken on the Cosmetic Potential of Australian Native Plants

Species	Family	Plant Part/Extract Type	Bioactivity Claimed	Potential Active Ingredient[a]	References
Acacia falcata	Mimosaceae	Bark, embrocation	Skin diseases	Tannins.	Lassak and McCarthy (1983)
Acacia implexa	Mimosaceae	Bark, embrocation	Skin diseases	Tannins.	Lassak and McCarthy (1983)
Acacia tetragonophylla	Mimosaceae	Ashes from bark-free	Anti-septic	Unknown.	Lassak and McCarthy (1983)
Acacia victoriae (Wattleseed)	Mimosaceae	Fruit/seed	Anti-oxidant	Phenolics.	Konczak et al. (2009)
Acronychia acidula (Lemon Aspen)	Rutaceae	Leaves	Anti-oxidant	Phenolics.	Konczak et al. (2009)
Amyema quandang	Loranthaceae	Leaves, ethanol extract	Anti-bacterial (including vs. MRSA and VRE isolates)	Unknown.	Palombo and Semple (2002)
Araucaria bidwillii (Bunya nuts)	Araucariaceae	Fruit	Anti-oxidant	Phenolics.	Zhao and Agboola (2007)
Backhousia citriodora (Lemon myrtle)	Myrtaceae	Leaves Water, methanol and hexane extracts	Anti-oxidant Anti-bacterial and anti-fungal Some toxicity issues	Phenolics. Citral (geranial, neral).	Konczak et al. (2009) Zhao and Agboola (2007) Hayes and Markovic (2001, 2003)
Barringtonia racemosa	Lecythidaceae	Root, pulverised	Anti-septic	Unknown.	Lassak and McCarthy (1983)
Callistris glaucophylla (Aus. White Cypress)	Cupressaceae	Leaf essential oil	Anti-bacterial	Unknown.	Wilkinson and Cavanagh (2005)
Carica papaya	Caricaceae	Seed. Various solvent extracts. Water extract	Anti-oxidant	Phenolics, flavonoids.	Zhou et al. (2011), Panzarini et al. (2014)
Centipeda cunninghamii	Asteraceae	Plant decoction	Eye inflammation, skin infections	Myriogenin and cis-chrysanthenyl acetate are present.	Lassak and McCarthy (1983)

(Continued)

Table 21.1 (Continued) A Non-Exhaustive Summary of Relevant Recent Research Undertaken on the Cosmetic Potential of Australian Native Plants

Species	Family	Plant Part/Extract Type	Bioactivity Claimed	Potential Active Ingredient[a]	References
Citrus glauca (Desert, Wild Lime)	Rutaceae	Fruit Water, methanol and hexane extracts	Anti-oxidant Anti-bacterial	Phenolics. Unknown.	Konczak et al. (2009) Zhao and Agboola (2007)
Davidsonia pruriens	Cunoniaceae	Fruit Water, methanol and hexane extracts	Anti-oxidant Anti-bacterial	Phenolics. Unknown.	Konczak et al. (2009) Zhao and Agboola (2007)
Davidsonia jerseyana	Cunoniaceae	Fruit	Anti-oxidant	Phenolics.	Konczak et al. (2009)
Entada phaseoloides	Mimosaceae	Spongy trunk fibres, infusion	Skin diseases	Unknown.	Lassak and McCarthy (1983)
Eremophila alternifolia	Myoporaceae	Leaves, ethanol extract	Anti-bacterial (including vs. MRSA and VRE isolates)	Unknown. Terpenes or sterols.	Palombo and Semple (2002) Shah et al. (2004)
Eremophila duttonii	Myoporaceae	Leaves, ethanol extract	Anti-bacterial (including vs. MRSA and VRE isolates)	Unknown. A sterol, terpene or sugar.	Palombo and Semple (2002) Shah et al. (2004)
Eucalyptus globulus	Myrtaceae	Leaf essential oil	Anti-bacterial Anti-fungal	1,8-cineole.	Bachir Raho and Benali (2012) Many other references
Eucalyptus gummifera	Myrtaceae	Gum	Astringent, ring-worm	Phenolics.	Lassak and McCarthy (1983)
Eucalyptus olida (Forest berry)	Myrtaceae	Leaves Water, methanol and hexane extracts	Anti-oxidant, anti-bacterial, anti-fungal	Phenolics.	Zhao and Agboola (2007)
Eucalyptus stragiana (Lemon Ironbark)	Myrtaceae	Leaves Water, methanol and hexane extracts	Anti-oxidant activity, anti-bacterial, anti-fungal	Phenolics.	Zhao and Agboola (2007)

(Continued)

Table 21.1 (*Continued*) A Non-Exhaustive Summary of Relevant Recent Research Undertaken on the Cosmetic Potential of Australian Native Plants

Species	Family	Plant Part/Extract Type	Bioactivity Claimed	Potential Active Ingredient[a]	References
Eucalyptus spp.	Myrtaceae	Leaves Water extracts	Anti-bacterial and anti-fungal	Unknown.	Zhao and Agboola (2007)
Ficus opposita	Moraceae	Latex	Ring-worm	Unknown.	Lassak and McCarthy (1983)
Grevillea robusta (Aus silver oak)	Proteaceae	Leaf methanol extract	Skin lightening, anti-chloasma agents	Arbutin derivatives, grevillosides.	Yamashita-Higuchi et al. (2014)
Kunzea ericoides (Kanuka)	Myrtaceae	Leaf essential oil	Anti-bacterial, anti-fungal Anti-inflammatory	Alpha-pinene, viridifiorol, viridiflorene.	Chen et al. (2014) and references cited therein
Lepidosperma viscidum	Cyperaceae	Stem base, ethanol extract	Anti-bacterial (including vs. MRSA and VRE isolates)	Unknown.	Palombo and Semple (2002)
Leptospermum petersonii	Myrtaceae	Leaf essential oil	Anti-fungal	Volatiles. Neral, geranial.	Hood et al. (2010) Kim and Park (2012)
Leptospermum scoparium (Manuka)	Myrtaceae	Leaf essential oil	Anti-bacterial, anti-fungal	Sesquiterpene hydrocarbons, triketones.	Chen et al. (2014) and references cited therein. Porter and Wilkins (1999)
		Honey	Anti-inflammatory	Methylglyoxal.	Hammond and Donkor (2013)
Litsea glutinosa	Lauraceae	Leaf decoction or juice	Eye and skin infections	Alkaloids possibly.	Lassak and McCarthy (1983)
Melaleuca alternifolia (Tea tree)	Myrtaceae	Leaf essential oil	Anti-microbial Anti-viral Anti-inflammatory, anti-fungal Anti-psoriasis. Head lice	Terpinen-4-ol.	Gnatta et al. (2013) Li et al. (2013) Ninomiya et al. (2013) Many other references
Mentha astralis	Lamiaceae	Leaves Water, methanol and hexane extracts	Anti-oxidant activity, anti-bacterial	Phenolics.	Zhao and Agboola (2007)

(Continued)

Table 21.1 (Continued)   A Non-Exhaustive Summary of Relevant Recent Research Undertaken on the Cosmetic Potential of Australian Native Plants

Species	Family	Plant Part/Extract Type	Bioactivity Claimed	Potential Active Ingredient[a]	References
Myristica insipida	Myristicaceae	Bark resin	Ring-worm	Unknown.	Lassak and McCarthy (1983)
Pittosporum phillyraeoides	Pittosporaceae	Fruits, decoction	Eczema	Triterpenoid saponins.	Lassak and McCarthy (1983)
Planchonia careya	Lecythidaceae	Leaves. Water and methanol extracts	Anti-bacterial. Wound healing	Triterpenes and triterpene saponins. Specific compounds identified.	McRae et al. (2008)
Podocarpus elatus (Illawarra plum)	Podocarpaceae	Fruit Water, methanol and hexane extracts	Anti-oxidant, anti-bacterial	Phenolics.	Zhao and Agboola (2007)
Pongamia pinnata	Fabaceae	Seed oil	Skin diseases	Pongamol and flavonoids.	Lassak and McCarthy (1983)
Prostanthera incisa (Cut leaf mint bush)	Lamiaceae	Leaves Water, methanol and hexane extracts	Anti-oxidant, anti-bacterial, anti-fungal	Phenolics.	Zhao and Agboola (2007)
Santalum acuminatum (Quandong)	Santalaceae	Fruit Water, methanol and hexane extracts	Anti-oxidant Anti-bacterial	Phenolics. Unknown.	Konczak et al. (2009) Zhao and Agboola (2007)
Santalum lanceolatum	Santalaceae	Leaf	Anti-bacterial, anti-inflammatory	Sesquiterpene, alcochol, lanceol.	Lassak and McCarthy (1983)
Sarcostemma austral	Asclepiadaceae	Sap	Skin rashes, anti-bacterial	Proteolytic enzymes, alkaloids, steroidal saponins, triterpenes.	Lassak and McCarthy (1983)
Scaevola spinescens	Goodeniaceae	Stem decoction, leaf smoke	Skin diseases	Furocoumarin.	Lassak and McCarthy (1983)

(Continued)

Table 21.1 (Continued) A Non-Exhaustive Summary of Relevant Recent Research Undertaken on the Cosmetic Potential of Australian Native Plants

Species	Family	Plant Part/Extract Type	Bioactivity Claimed	Potential Active Ingredient[a]	References
Securinega melanthesoides	Euphorbiaceae	Leaf (young) infusion	Skin rashes, anti-bacterial	Alkaloids.	Lassak and McCarthy (1983)
Solanum centrale (Bush tomato)	Solanaceae	Fruit Water, methanol and hexane extracts	Anti-oxidant Anti-bacterial	Phenolics. Unknown.	Konczak et al. (2009) Zhao and Agboola (2007)
Striga curviflora	Scrophulariaceae	Plant, water extract	Skin diseases	Unknown.	Lassak and McCarthy (1983)
Syzygium anisatum (Anise/Aniseed myrtle, Anisata, *Anethola anisata*)	Myrtaceae	Leaves Water, methanol and hexane extracts Essential oil	Anti-oxidant Anti-bacterial	Phenolics. Unknown.	Konczak et al. (2009) Zhao and Agboola (2007) Wilkinson and Cavanagh (2005)
Syzygium paniculatum (lilly pilly)	Myrtaceae	Fruit. Crude extract	Anti-oxidant Others	Phenolics. Flavonoids.	Vuong et al. (2014)
Syzygium luehmannii (Riberry)	Myrtaceae	Fruit Water, methanol and hexane extracts	Anti-oxidant Anti-bacterial	Phenolics. Unknown.	Konczak et al. (2009) Zhao and Agboola (2007)
Tasmannia lanceolata (Native Pepperberry)	Winteraceae	Fruit Water, methanol and hexane extracts	Anti-oxidant Anti-bacterial and anti-fungal	Phenolics. Unknown.	Konczak et al. (2009) Zhao and Agboola (2007)

(Continued)

Table 21.1 (Continued) A Non-Exhaustive Summary of Relevant Recent Research Undertaken on the Cosmetic Potential of Australian Native Plants

Species	Family	Plant Part/Extract Type	Bioactivity Claimed	Potential Active Ingredient[a]	References
Tasmannia lanceolata (Native Pepperberry, Mountain Pepper)	Winteraceae	Leaves Water, methanol and hexane extracts	Anti-oxidant Anti-bacterial, anti-fungal	Phenolics, carotenoids. Polygodial. Unknown.	Konczak et al. (2009) Zhao and Agboola (2007)
Tetragonia tetragoniodes (Warrigal greens)	Aizoaceae	Leaves	Anti-oxidant	Phenolics.	Zhao and Agboola (2007)
Terminalia ferdinandiana (Kakadu Plum)	Combretaceae	Fruit	Anti-oxidant	Phenolics, carotenoids.	Konczak et al. (2009, 2014)
Thespia populnea	Malvaceae	Tree bark decoction, unripe fruit juice	Skin diseases	Thespesin.	Lassak and McCarthy (1983)
Thryptomine calycina	Myrtaceae	Leaf essential oil	Anti-bacterial	Unknown.	Wilkinson and Cavanagh (2005)

[a] The potential active ingredient may only reflect the chemistry as yet defined, rather than proven specific bioactivity.

with extreme potential as rich, unique sources of bioactive molecules with activity of interest to this industry. We know that use of Australian native plants in cosmetics is being considered globally, not least due to their anti-microbial activity and potential relevance for use as natural preservatives in cosmetic formulations (Primi Putri, 2010).

The challenges ahead include better characterisation of the plants and their extracts, as lack of knowledge will always limit their potential due to both safety and efficacy issues. Scientific research could extend the range of relevant tests undertaken in screening studies, beyond simply 'anti-oxidative activity'. The producer-based industry needs to better understand the nature of usage legislation such as REACH in Europe and perhaps could take a more co-operative approach to overcome these and other hurdles related to marketing, for the more promising species. The other major issue is shoring up supply and ensuring that a robust, sustainable, secure and cost-effective harvesting, production and transport system is in place for the crops, migrating over time from wild harvested to more managed plantations where feasible. Eventually, for some species, propagation of specific genotypes may be required to reduce the inherent variation that exists in natural stands and even some plantations. For some opportunities, the future is definitely rosy.

DISCLAIMER

The author is not under any contractual arrangements with any manufacturers of cosmetic ingredients or finished cosmetic products, at the time of writing.

REFERENCES

Bachir Raho, B., Benali, M., 2012. Antibacterial activity of the essential oils from the leaves of *Eucalyptus globulus* against *Escherichia coli* and *Staphylococcus aureus*. *Asian Pacific Journal of Tropical Biomedicine* 2, 739–742.

Biavatti, M.W., Marensi, V., Leite, S.N., Reis, A., 2007. Ethnopharmacognostic survey on botanical compendia for potential cosmeceutic species from Atlantic Forest. *Revista Brasileira de Farmacognosia* 17, 650–653.

Chen, C.C., Yan, S.H., Yen, M.Y., Wu, P.F., Liao, W.T., Huang, T.S., Wen, Z.H., Wang, H.M.D., 2014. Investigations of kanuka and manuka essential oils for *in vitro* treatment of disease and cellular inflammation caused by infectious microorganisms. *Journal of Microbiology, Immunology and Infection*. doi:http://dx.doi.org/10.1016/j.jmii.2013.12.009.

Euromonitor International. 2014. www.euromonitor.com.

Gnatta, J.R., Pinto, F.M., Bruna, C.Q., Souza, R.Q., Graziano, K.U., Silva, M.J., 2013. Comparison of hand hygiene antimicrobial efficacy: *Melaleuca alternifolia* essential oil *versus* triclosan. *Revista Latino-Americana de Enfermagem* 21, 1212–1219.

Hammond, E.N., Donkor, E.S., 2013. Antibacterial effect of Manuka honey on *Clostridium difficile*. *BMC Research Notes* 6, 188.

Hayes, A.J., Markovic B., 2001. Toxicity of Australian essential oil *Backhousia citriodora* (Lemon myrtle). Part 1. Antimicrobial activity and in vitro cytotoxicity. *Food and Chemical Toxicology* 40, 535–543.

Hayes, A.J., Markovic, B., 2003. Toxicity of Australian essential oil *Backhousia citriodora* (Lemon myrtle). Part 2. Absorption and histopathology following application to human skin. *Food and Chemical Toxicology* 41, 1409–1416.

Hood, J.R., Burton, D.M., Wilkinson, J.M., Cavanagh, H.M.A., 2010. The effect of *Leptospermum petersonii* essential oil on *Candida albicans* and *Aspergillus fumigatus*. *Medical Mycology* 48, 922–931.

Kim, E., Park I., 2012. Fumigant antifungal activity of Myrtaceae essential oils and consituents from *Leptospermum petersonii* against three *Aspergillus* species. *Molecules* 17, 10459–10469.

Konczak, I., Maillot, F., Dalar, A. 2014. Phytochemical divergence in 45 accessions of *Terminalia ferdinandiana* (Kakadu plum). *Food Chemistry* 151, 248–256.

Konczak, I., Zabaras, D., Dunstan, M., Aguas, P., Roulfe, P., Pavan, A., 2009. *Health Benefits of Australian Native Foods: An Evaluation of Health-Enhancing Compounds*. Rural Industries Research and Development Corporation, RIRDC Publication No. 09/133, Canberra, Australian Capital Territory, Australia.

Lall, N., Kishore, N., 2014. Are plants used for skin care in South Africa fully explored? *Journal of Ethnopharmacology* 153, 61–84.

Lassak E.V., McCarthy, T., 1983. *Australian Medicinal Plants*. Reed New Holland, Sydney, New South Wales, Australia.

Li, X., Duan, S., Chu, C., Xu, J., Zeng, G., Lam, A.K.Y., Zhou, J. et al., 2013. *Melaleuca alternifolia* concentrate inhibits *in vitro* entry of influenza virus into host cells. *Molecules* 18, 9550–9566.

MacTavish-West, H. 2014. Bioactive extracts for the personal care industry. Final report, Rural Industries Research and Development Corporation, Canberra, Australian Capital Territory, Australia.

McRae, J.M., Yang, Q., Crawford, R.J., Palombo, E.A. 2008. Antibacterial compounds from *Planchonia careya* leaf extracts. *Journal of Ethnopharmacology* 116, 554–560.

Netzel, M., Netzel, G., Tian, Q., Schwartz, S., Konczak, I., 2006. Sources of antioxidant activity in Australian native fruits. Identification and quantification of anthocyanins. *Journal of Agricultural and Food Chemistry* 54, 9820–9826.

Netzel, M., Netzel, G., Tian, Q., Schwartz, S., Konczak, I., 2007. Native Australian fruits—A novel source of antioxidants for food. *Innovative Food Science and Emerging Technologies* 8, 339–346.

Ninomiya K., Hayama, K., Ishijima, S.A., Maruyama, N., Irie, H., Kurihara, J., Abe, S., 2013. Suppression of inflammatory reactions by terpinen-4-ol, a main constituent of tea tree oil, in a murine model of oral candidiasis and its suppressive activity to cytokine production of macrophages in vitro. *Biological and Pharmaceutical Bulletin* 36, 838–844.

Palombo, E.A., 2011. Traditional medicinal plant extracts and natural products with activity against oral bacteria: Potential application in the prevention and treatment of oral diseases. *Evidence Based Complementary and Alternative Medicine*. Article ID: 680354. http://dx.doi.org/10.1093/ecam/nep067.

Palombo, E.A., Semple, S.J., 2001. Antibacterial activity of traditional Australian medicinal plants. *Journal of Ethnopharmacology* 77, 151–157.

Palombo, E.A., Semple, S.J., 2002. Antibacterial activity of Australian plant extracts against methicillin-resistance *Staphylococcus aureus* (MRSA) and vancomycin-resistant enterococci (VRE). *Journal of Basic Microbiology* 42, 444–448.

Panzarini, E., Dwikat, M., Mariano, S., Vergallo, C., Dini, L., 2014. Administration dependent antioxidant effect of *Carica poapaya* seeds water extract. *Evidence-Based Complementary and Alternative Medicine*. Article ID: 281508. http://dx.doi.org/10.1155/2014/281508.

Porter, N.G., Wilkins, A.L., 1999. Chemical, physical and antimicrobial properties of essential oils of *Leptospermum scoparium* and *Kunzea ericoides*. *Phytochemistry* 50, 407–415.

Primi Putri, T. 2010. *Potential Use of Edible Fruits and Australian Native Plants*. LAP Lambert Academic Publishing, Saarbrücken, Germany.

Shah, A., Cross, R.F., Palombo, E.A., 2004. Identification of the antibacterial component of an ethanolic extract of the Australian medicinal plant, *Eremophila duttonii*. *Phytotherapy Research* 18, 615–618.

Taeymans J., Clarys, P., Barel, A.O., 2014. Use of food supplements as nutricosmetics in health and fitness – A review. In: Barel, A.O., Paye, M., Maibach, H.I. (eds.), *Handbook of Cosmetic Science and Technology*. CRC Press, Boca Raton, FL.

Tan, A.C., Konczak, I., Ramzan, I., Zabaras, D., Sze, D.M., 2011. Potential antioxidant, anti-inflammatory, and proapoptotic anticancer activities of Kakadu plum and Illawarra plum polyphenolic fractions. *Nutrition and Cancer* 63, 1074–1084.

Vuong, Q.V., Hirun, S., Chuen, T.L.K., Goldsmith, C.D., Bowyer, M.C., Chalmers, A.C., Phillips, P.A., Scarlett, C.J., 2014. Physicochemical composition, antioxidant and anti-proliferative capacity of a lilly pilly (*Syzygium paniculatum*) extract. *Journal of Herbal Medicine* 4, 134–140.

Watson, R.E., Long, S.P., Bowden, J.J., Bastrilles, J.Y., Barton, S.P., Griffiths, C.E., 2007. Repair of photoaged dermal matrix by topical application of a cosmetic 'antiageing' product. *British Journal of Dermatology* 158, 472–477.

Wilkinson, J.M., Cavanagh, H.M.A., 2005. Antibacterial activity of essential oils from Australian native plants. *Phytotherapy Research* 19, 643–646.

Yamashita-Higuchi, Y., Sugimoto, S., Matsunami, K., Otsuka, H., Nakai T., 2014. Grevillosides J-Q arbutin derivatives from the leaves of *Grevillea robusta* and their melanogenesis inhibitory activity. *Chemical and Pharmaceutical Bulletin* 62, 364–372.

Zhao, J., Agboola, S., 2007. *Functional Properties of Australian Bushfoods*. Rural Industries Research and Development Corporation, RIRDC Publication No. 07/030, Canberra, Australian Capital Territory, Australia.

Zhou, K., Wang, H., Mei, W., Li, X., Luo, Y., Dai, H., 2011. Antioxidant activity of papaya seed extracts. *Molecules* 16, 6179–6192.

CHAPTER 22

Processing of Native Plant Foods and Ingredients

Yasmina Sultanbawa

CONTENTS

INTRODUCTION

The value of natural additives in the food and beverage industry is estimated to increase to 45 USD billions in the global market; this figure includes vitamins, minerals and functional food ingredients according to Leatherhead's Global Food Additives Market report, 2014. A certain growth percentage of this industry is attributed to consumer's concerns of possible toxic effects from synthetic additives such as antioxidants, chemical preservatives and colouring agents. As an example: alternative methods to retard lipid oxidation in foods have increased the need for natural antioxidants. Sources of these natural antioxidants include the use of herbs, spices, fruits, vegetables and by-products from the food industry. It is well known that the additive and synergistic effects of the complex phytochemical mix present in these plant sources are responsible for their enhanced antioxidant activity (Neacsu et al., 2015). Recent publications (Konczak et al., 2009, 2010b; Sakulnarmrat and Konczak, 2012) have reported about the high antioxidant capacities of commercially grown Australian native plant foods which are far greater than those previously reported for blue berries, well known for its high antioxidant levels. The enhanced antioxidant capacity of native plant foods such as Kakadu plum is mainly attributed to the high levels of vitamin C and phenolic compounds (Konczak et al., 2009).

The commercial potential for the use of native plant foods as natural additives in the food industry would be the potent bioactivity due to its complex blend of phytochemicals. The challenge to the industry is the retention of these bioactive properties during harvesting, processing, packaging and storage to promote them as functional foods and ingredients. Health Canada defines a functional food 'as similar in appearance to, or may be, a conventional food that is consumed as part of a usual diet, and is demonstrated to have physiological benefits and/or reduce the risk of chronic diseases beyond basic nutritional functions', that is they contain bioactive compounds. Functional ingredients are defined by Health Canada 'as standardised and characterised preparations, fractions or extracts containing bioactive compounds of varying purity that are used as ingredients by manufacturers of human and pet food'. This chapter will look at the traditional practices used by indigenous people to process and preserve food, address the issues of retaining the quality and bioactivity of commercial native foods during processing and storage and review functional properties and potential innovative applications as natural additives in the food industry.

INDIGENOUS AND CURRENT PRACTICES IN PROCESSING OF NATIVE FOODS

Till the 1830s, about 900 edible plants were known to have been used by indigenous people in Australia. It is also known that not all the edible species were used as food: for example products made palatable by boiling were not used, as indigenous culture did not have the technology to boil food (Anonymous, 2000). There are interesting examples reported in the literature referring to unique processing techniques used to preserve native plant food. For instance *Kunzea pomifera* F. Muell known as muntries has small pome-like fruits and grows in the sandy coasts of Western Victoria and South Australia. It is recorded that the Narrindjeri people of the Coorong in the south-east of South Australia use to pound this fruit into a paste and make it into large cakes and trade this product with other tribes (Gott, 2008). This fruit was not only eaten fresh, but dried and stored for the winter months. An indigenous community in the great sandy desert of North-western Australia, belonging to the Pintubi and Gugadja linguistic units lived a full hunter-gatherer existence in this part of Australia until the mid-1950s and early 1960s. This community processed acacia seeds (major species: *Acacia coriacea*, *A. holosericea*, *A. stipuligera*) to a milky liquid by first soaking in water and then squashing. The remaining residue was spread onto the surface of flat termites' nests and dried in the sun. The dried mixer from this process is roasted in hot sand, and then winnowed to get rid of any sand. The clean seeds from this process are then re-ground with water. The ground water paste is eaten raw, never cooked. This community also prepared dampers, a common term for wet-milled seed cakes and bread cooked on coals from seeds of different grasses mainly *Fimbristylis oxystachya* and *Panicum australiense* (Cane, 1987; Zeanah et al., 2015). This indigenous community used to store two species of bush tomato (*Solanum chippendalei* and *S. diversiflorum*) for several years, indicating a long storage life. The bush tomato was cooked in hot ashes without burning and removed from the fire and winnowed free of sand and foreign matter. The cooked flesh was mashed with water and rolled into a ball, covered with grass and wound up with twine. The bush tomato was sometimes mixed with *Ipomoea* tubers. These balls of food were traded with visiting tribes for meat (Cane, 1987).

CURRENT PROCESSING PRACTICES OF THE NATIVE FOOD INDUSTRY

The commercial species of native foods currently available in the market is made up of a great diversity of species, geographical areas, different products and uses. These include native herbs/spices, anise myrtle (*Syzygium anisatum*), bush tomato (*Solanum centrale*), lemon myrtle (*Backhousia citriodora*), Tasmanian pepper (*Tasmannia lanceolata*) and wattleseeds (*Acacia* spp.). Native fruits, Davidsons' plum (*Davidsonia* spp.), desert limes (*Citrus glauca*), finger limes (*Citrus australasica*), Kakadu plum (*Terminalia ferdinandiana*), lemon aspen (*Acronychia* spp.), quandong (*Santalum acuminatum*), muntries (*Kunzea pomifera*) and riberry (*Syzygium luehmannii*). Stakeholders in the supply chain of native foods are wild harvesters, specialist growers, a number of vertically integrated firms as well as firms that process and market Australian native food products (Clarke, 2012). The products range from conventional processed food products such as jams, chutneys, sauces, dried herb mixes to more functional food/beverage type products such as specialty teas, functional ingredients identified for a particular nutrient or bioactive property. There is a small amount of fresh produce that are bought by chefs for native food cuisine. The bulk of the domestic produce is dried, frozen or further processed to other value-added products or used as ingredients in other industries; refer to Figures 22.1 through 22.3. These are examples for a general production flow chart for native herbs and fruits, stored as dried or frozen products.

EFFECTS OF PROCESSING ON QUALITY AND BIOACTIVITY

The current need for natural additives in the food, feed and cosmetic industry has created a renewed interest in Australian native foods for cross-industry applications. However, the native food industry and non-native food industries are concerned about the effect on bioactivity when native foods are processed in to value-added products such as dried herbs and frozen fruits.

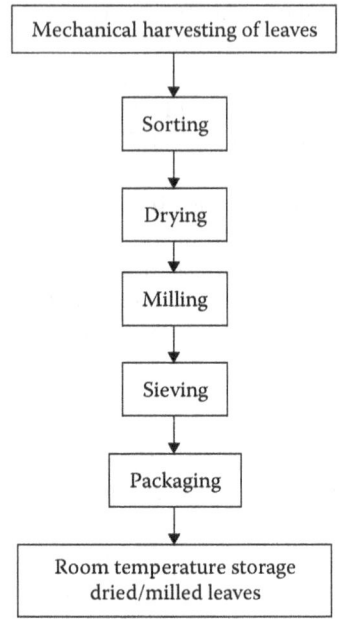

Figure 22.1   A general flow chart of dried Australian native herb leaves (lemon myrtle, anise myrtle and Tasmanian pepper leaves).

CONVENTIONAL PROCESSING TECHNOLOGIES

Thermal Processing

The application of heat is the most common method for processing food as it inactivates microorganisms and enzymes. Conventional thermal processing technologies include but are not limited to blanching, pasteurisation and sterilisation and thermal drying (dehydration) (Rawson et al., 2011). However, the heat treatment techniques used in the food industry to attain acceptable microbial safety to extend shelf life can lead to the loss of bioactivity in several ingredients including vitamins, antioxidants and proteins (George et al., 2013). A study on the impact of thermal processing on bioactive compounds in bush tomato and kakadu plum were investigated. Comparisons were made on the loss of bioactive compounds during commercial processing of bush tomato and kakadu plum. The commercial samples analysed were bush tomato chutney, bush tomato ketchup and kakadu plum chilli and ginger sauce. There was an increase in the amount of both lycopene and beta-carotene from the raw dried bush tomato to the processed product. This could be due to the release of these compounds from the food matrix during the heating process. Ascorbic acid in the kakadu plum was very sensitive to the heat, and there was a significant loss in the finished product in comparison to raw fruit, indicating that vitamin C is unstable when heated (Sommano et al., 2013). Wattle (*Acacia victoriae* Bentham) seed was

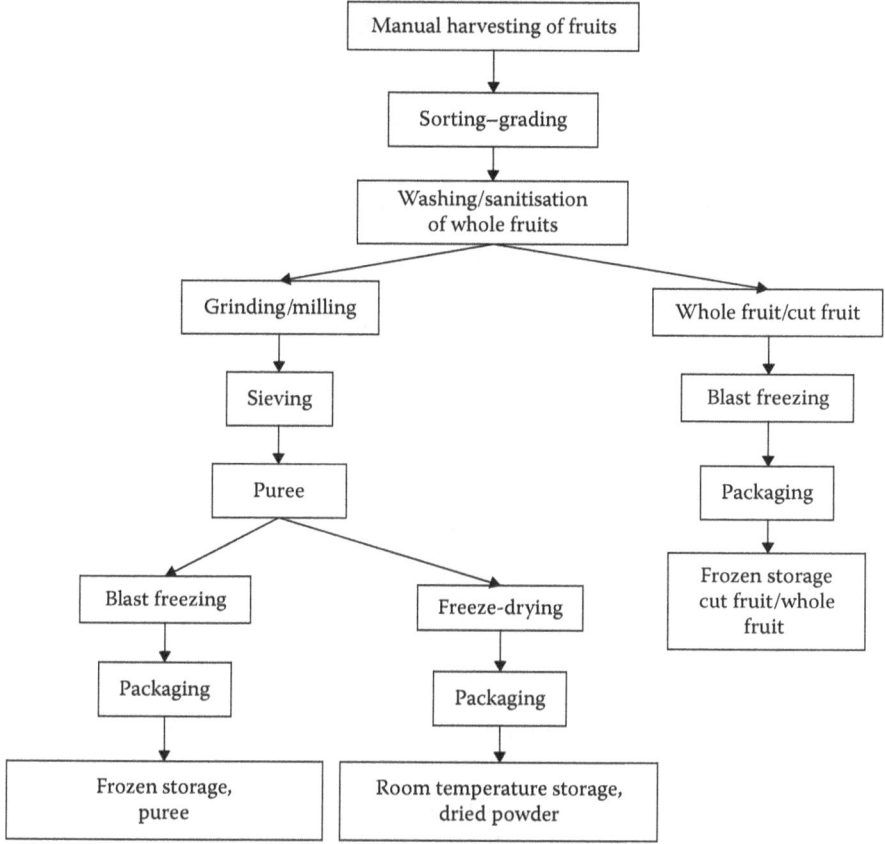

Figure 22.2 A general flow chart of frozen and dried Australian native fruits (Kakadu plum, Davidson plum and quandong).

extracted with water before or after soaking/heat treatment, to inactivate its protease inhibitors. This heating step reduced emulsion stability in comparison to the emulsion formed using extract from non-processed wattle seed which were very stable at both 20% and 50% oil levels (Ee et al., 2009).

Dehydration

Dehydration of fruits and vegetables by convective (hot air) drying is one of the most popular methods used to extend storage life of perishable produce. The removal of moisture prevents the growth of microorganisms and inhibits activity of enzymes that can result in quality losses during storage. In addition, drying transforms the fruit or vegetable into stable processed products that can be easily packaged and transported and stored at ambient temperature (Gamboa-Santos et al., 2014). Therefore, this type of dehydration is valued as a cost-effective method of

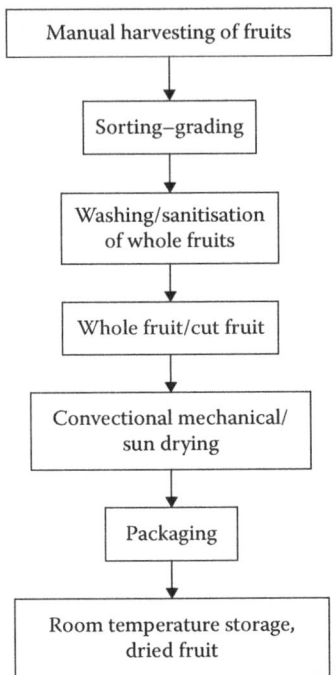

Figure 22.3   A general flow chart for drying of quandong and bush tomato.

food preservation. The effect of three drying methods including sun drying, hot air drying and freeze–drying on phytochemical compounds and antioxidant activity of premature fruit fall of citrus is reported. This example was selected to understand the effect of dehydration on different phenolic compounds and to demonstrate that the retention or degradation depended on the method of dehydration. The results showed that freeze–drying is good at retaining phenolic compounds, synephrine and antioxidants; hot air drying is good at retaining flavonoids and all three methods can be used for retaining limonoids (Sun et al., 2015). This clearly indicates that different drying methods affect different bioactive compounds in the tested plant material. Quandong fruit samples, cut in half, were dried using a hot air dryer at temperatures of 40°C, 50°C and 60°C. The dried quandong samples were packed and tested for total phenol content and antioxidant activity: ferric reducing anti-oxidant power (FRAP) over a 6-month period. This study concluded that the total phenolics and the FRAP values were similar in the quandong samples dried at these three temperatures and did not change significantly during storage at ambient temperature (Sultanbawa et al., 2015). Drying and milling of native herbs lemon myrtle, anise myrtle and Tasmanian pepper leaf indicated a loss of major volatiles during the milling process. The two major volatiles identified for the native herbs are lemon myrtle (Neral, Geranial), anise myrtle (Estragole, anethole) and Tasmanian pepper leaf (Eucalyptol, Eugenol). However, after drying, the major volatiles increased due to the reduction in moisture (Chaliha, 2013).

Freezing

Freezing as a preservation technique retains the quality of agricultural products over long storage periods. As a method of preservation of fruits and vegetables, freezing is generally regarded as a technique that retains the sensory attributes and nutritive properties in comparison to canning and dehydration (Fennema et al., 1973). Fruits are processed into purees, juices and nectars using various processing operations prior to freezing. The effect of freezing and frozen storage on Davidson's plum and kakadu plum whole fruit and puree was investigated. The freezing performance of kakadu plum indicated a good retention (in excess of 75%) of total phenolics and antioxidant activity (FRAP values) during the 6-month storage period in the whole fruit. The high vitamin C in kakadu plum and the low pH value of the fruit may have a positive effect on the stability of the phytochemicals. The retention of total phenolics and FRAP activities during storage in both the commercial and processed plum halves was over 80%, while in the puree, it was only 68%, indicating a greater loss of antioxidant activity in the frozen puree samples. The antimicrobial activity of both kakadu plum and Davidson's plum was stable over the frozen storage period, making it an attractive intermediate product for value addition to functional ingredients with these bioactive properties (Sultanbawa et al., 2015).

Processing of *Terminalia ferdinandiana* to frozen whole fruits, puree and freeze–dried powder with indigenous community engagement would be a good example of a native fruit with potential to becoming a mainstream agricultural produce. The envisaged plan is to develop the entire supply chain for Kakadu plum like it was done for the macadamia industry in Australia.

At present, the reported production volume of kakadu plum has been estimated to average 15–17 tonnes per annum from the Northern Territory and Western Australia combined (Clarke, 2012). There is far more fruit available in the wild in indigenous owned land that can be harvested in the future with more communities being involved in the harvesting. At present, kakadu plum is wild harvested in the Kimberly region of Western Australia and in parts of the Northern Territory; however, due to the increasing demand for kakadu plum, enrichment planting is being undertaken in Western Australia by indigenous producers. All fruits harvested in Northern Australia are sorted and frozen as part of the strict quality control measures in place to maximise quality and retention of vitamin C content (Cunningham et al., 2009). Coradji Pty Ltd., a private enterprise processing kakadu plum to value-added products (puree and freeze–dried powders), was the major buyer of kakadu plum. This company has now transferred ownership of equipment, patent (Pusateri et al. 2007) and intellectual property to the indigenous women of the Palngun Wurnangat Association and has relocated their processing plant to Wadeye in the Northern Territory. This transfer has made the harvesting and processing of kakadu plum a truly indigenous enterprise, and this would definitely add to the economic benefits of these communities.

NOVEL PROCESSING TECHNOLOGIES

These novel processing technologies include ohmic heating, dielectric heating/microwave heating/radio frequency heating which are grouped as novel thermal processing techniques which have been used in the processing of other food commodities, but not been looked for extensive use in the native food industry in Australia. Thermal processing of food in general leads to a loss of bioactive compounds (Rawson et al., 2011); if native food processors desire to produce products with maximum retention of bioactive compounds and long storage life, it would be worth exploring other novel non-thermal processing technologies. These would include but are not limited to high hydrostatic pressure processing (HHP), radiation processing, pulsed electric field (PEF) ozone and dense carbon dioxide and ultra-sound processing (Tiwari et al., 2009). Native fruits Davidson plum, riberry, quandong contain anthocyanins including pelargonidin, cyanidin, malvidin peonidin, petunidin and delphinidin (Konczak et al., 2009). HHP is a non-thermal food preservation technique for microbial and enzyme inactivation where pressures in excess of 100 MPa are applied to the food product with minimal effects on nutritional and quality parameters when compared with thermal processing, in particular on the anthocyanin content of fruits and vegetables (Tiwari et al., 2009). It is reported the stability of strawberry and red raspberry anthocyanins (pelargonidin-3-glucoside and pelargonidin-3-rutinoside) was most stable at 800 MPa for 15 minutes at moderate temperature (18°C–22°C). This may be due to complete inactivation of polyphenoloxidase. Enzymes such as polyphenoloxidase, peroxidise and β-glucosidase are known to cause degradation of anthocyanins (Garcia-Palazon et al., 2004). HHP would be suitable for some of the native fruits, in particular those which have bioactive compounds that could be retained under these processing conditions.

NATIVE FOODS AS POTENTIAL FUNCTIONAL INGREDIENTS

Australian native foods are been used in conventional food products such as jams, sauces, marinades and as spices/herb mixes to flavour food. However, one of the future market potentials for native foods in Australia lies in developing innovative functional ingredients targeting cross-industry applications as given in Figure 22.4. The challenge of using these ingredients in other applications such as food and beverages, nutraceuticals and cosmetics is the retention of the bioactivity during processing, packaging and storage of the value-added products. Towards achieving this goal, innovative delivery systems like microencapsulation and nanotechnology will have to be considered. The functional properties of commercial native foods are given in Table 22.1.

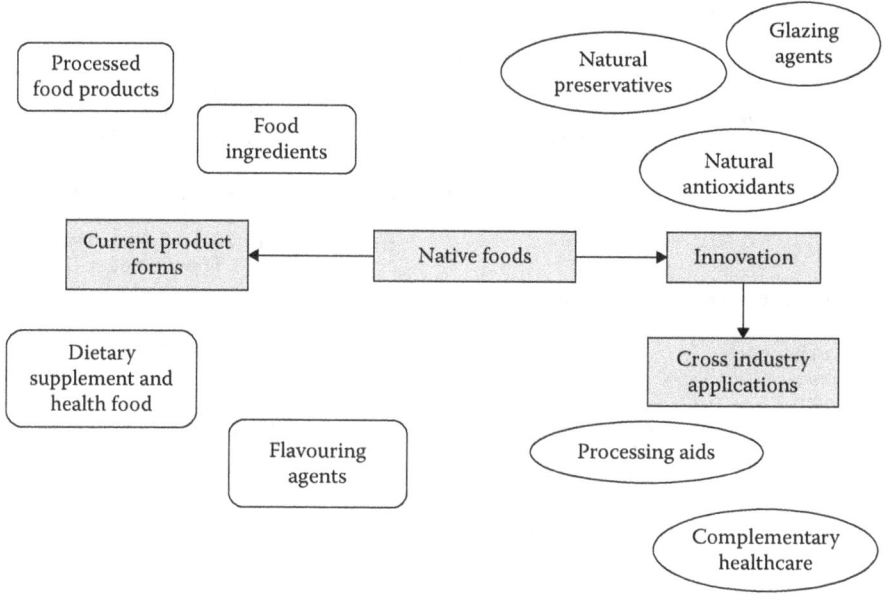

Figure 22.4 Creating a future for the native food industry in Australia through innovation.

INNOVATIVE DELIVERY SYSTEMS FOR PLANT BIOACTIVE COMPOUNDS

Challenges facing introduction of bioactive compounds into foods are not limited only to their inclusion in free-flowing powder or solutions. These bioactive compounds are often very sensitive to thermal and oxidative stress and are degraded during processing or storage. In addition, most of these bioactive compounds are not soluble in water; therefore, they need a vehicle to increase their solubility or dispersability in a food matrix (solid/liquid) which is mostly hydrophilic. In a food matrix, ideally the bioactive compounds should be stable and retain the bioactive properties during the expected storage period. If the benefits of the bioactive compounds are realised only after consumption, they could be affected by enzymatic degradation, acid conditions in the stomach, as well as changes in the osmolality of the intestinal fluids, and this could affect the properties of the bioactive compounds and their functionality (Shimoni, 2009). The ideal delivery system for bioactive compounds in foods should possess a wide array of controlled release properties. The manipulation of matter at a nanoscale represents an important discipline in the food industry. It is generally assumed that any particle between 1 and 100 nm can be considered a nanoparticle. A nanodelivery system under these circumstances would be most suitable as it would have properties of matter being organised at a molecular level. This would enable the precise control of material behaviour under temperature, pH, water activity and enzymatic environment (Gutiérrez et al., 2013; Shimoni, 2009).

Table 22.1 Functional Properties of Native Foods

Native Food	Functional Properties	References
Anise myrtle (*Syzygium anisatum*)	Antimicrobial and antioxidant activities, synergistic combinations with Tasmanian pepper leaf	Konczak et al. (2010b), Kubo and Himejima (1991), Sultanbawa et al. (2015)
Lemon myrtle (*Backhousia citriodora*)	Antimicrobial and antioxidant activities	Konczak et al. (2010b), Kurekci et al. (2013), Lazar-Baker et al. (2011), Sultanbawa et al. (2015), The Vien et al. (2008)
Bush tomato (*Solanum centrale*)	Antioxidant activity	Konczak et al. (2009)
Tasmanian pepper (*Tasmannia lanceolata*)	Antimicrobial and antioxidant activities, attachment inhibitor of blue mussel which causes marine fouling, improves the sensory acceptability of artificial sweeteners, enhances pungent or freshness notes	Ban et al. (2000), Himejima and Kubo (1993), Konczak et al. (2010b), Netzel et al. (2007), Starkenmann et al. (2011), Sultanbawa et al. (2015)
Wattleseeds (*Acacia* spp.)	Emulsifier, gelling and foaming properties	Ee et al. (2009)
Davidsons' plum (*Davidsonia* spp.)	Colour pigments, antimicrobial and antioxidant activity	Konczak et al. (2009, 2010a), Netzel et al. (2007), Sultanbawa et al. (2015)
Desert limes (*Citrus glauca*)	High folates	Konczak et al. (2009, 2010a), Netzel et al. (2007)
Finger limes (*Citrus australasica*)	Antioxidant activity	Konczak et al. (2010a), Netzel et al. (2007)
Kakadu plum (*Terminalia ferdinandiana*)	Antimicrobial and antioxidant activities	Konczak et al. (2010a), Netzel et al. (2007), Sultanbawa et al. (2015), Williams et al. (2014)
Lemon aspen (*Acronychia* spp.)	Antioxidant activity	Konczak et al. (2010a), Netzel et al. (2007)
Quandong (*Santalum acuminatum*)	Colour pigments and antioxidant activity	Konczak et al. (2009, 2010a), Netzel et al. (2007), Sultanbawa et al. (2015)
Muntries (*Kunzea pomifera*)	Antioxidant activity	Konczak et al. (2010a), Netzel et al. (2007)
Riberry (*Syzygium luehmannii*)	Colour pigments	Konczak et al. (2009, 2010a), Netzel et al. (2007)

Nanoencapsulated compounds were initially developed for the design of drug delivery systems. There is an increasing interest in this technology in the food industry. Several techniques can be used to nanoencapsulate bioactive compounds, some of them are self-assembly, high-pressure homogenisation, nanoemulsification, nanocomplexation, nanoprecipitation, coacervation, and lipid-based nanoencapsulation. An example of a nanoencapsulated bioactive compound is carotenoid lycopene, a European product which was claimed for use as a food supplement and food-fortifying agent. This product has been classified as generally regarded as safe (Gutiérrez et al., 2013).

An example of a native plant extract being incorporated into a microencapsulated delivery system is Lemon myrtle essential oil. It is not soluble in water due to this; its application has been limited to dry premixes and oil-soluble liquid applications. To enable the use of this essential oil as water-soluble additive microencapsulation has been used. Microencapsulation is the conversion of oils into solid and water-soluble forms, which can be used to extend the shelf life of the essential oil in other value-added product applications. The difference between the micro- and nanoencapsulation is the particle size, one is micro and the other is nano, due to the size difference, physico-chemical properties will be also be different. The lemon myrtle essential oil was incorporated in an oil-in-water emulsion which comprised of an aqueous phase, a solution of maltodextrin and Hi-Cap 100 (modified starch) or whey protein concentrate. The aqueous phase and the lemon myrtle essential oil were blended together to prepare coarse emulsions. The course emulsion was further homogenised using a microfluidiser. This system produces sub-micro emulsions. This emulsion was dried in a spray dryer to produce a powder. This type of microencapsulation is said to protect the functional properties of the lemon myrtle essential oil (The Vien et al., 2008).

CONCLUSIONS AND FUTURE DIRECTIONS

There is a growing market demand for botanicals and spices as natural additives. There is increase interest in the retention of bioactive compounds during harvesting, processing, packaging and storage.

Current knowledge indicates that in general, high-temperature treatments can affect the stability of bioactive compounds in native foods. There is an increase need to assess the potential of using novel non-thermal processing techniques to meet the need of nutritious food and functional food ingredients developed from native foods with bioactive compound retention. High hydrostatic pressure processing is of particular interest due to the retention of bioactive compounds and nutrients during processing. In particular for functional native food ingredients, it would be critical to evaluate different innovative delivery systems of the bioactive compounds to ensure efficacy, target delivery and stability during the storage of the value-added product.

REFERENCES

Anonymous, 2000. Inquiry into the utlization of Victorian native flora and fauna.

Ban, T., Singh, I.P., Etoh, H., 2000. Polygodial, a potent attachment-inhibiting substance for the blue mussel, *Mytilus edulis galloprovincialis* from *Tasmanianlanceolata*. *Bioscience Biotechnology and Biochemistry* 64, 2699–2701.

Cane, S., 1987. Australian aboriginal subsistence in the Western desert. *Human Ecology* 15, 391–434.

Chaliha, M., 2013. Effect of packaging materials and storage on major volatile compounds in three Australian native herbs. *Journal of Agricultural and Food Chemistry* 61, 5738–5745.

Clarke, M., 2012. *Australian Native Food Industry Stocktake*. Rural Industries Research and Development Corporation, RIRDC Publication No. 12/066, Canberra, Australian Capital Territory, Australia.

Cunningham, A.B., Garnett, S., Gorman, J., Courtney, K., Boehme, D., 2009. Eco-enterprises and *Terminalia ferdinandiana*: "Best laid plans" and Australian policy lessons. *Economic Botany* 63, 16–28.

Ee, K.Y., Rehman, A., Agboola, S., Zhao, J., 2009. Influence of heat processing on functional properties of Australian wattle seed (*Acacia victoriae* Bentham) extracts. *Food Hydrocolloids* 23, 116–124.

Fennema, O.R., Powrie, W.D., Marth, E.H., 1973. *Low-Temperature Preservation of Foods and Living Matter*. Marcel Dekker, New York.

Gamboa-Santos, J., Megías-Pérez, R., Soria, A.C., Olano, A., Montilla, A., Villamiel, M., 2014. Impact of processing conditions on the kinetic of vitamin C degradation and 2-furoylmethyl amino acid formation in dried strawberries. *Food Chemistry* 153, 164–170.

Garcia-Palazon, A., Suthanthangjai, W., Kajda, P., Zabetakis, I. 2004. The effects of high hydrostatic pressure on β-glucosidase, peroxidase and polyphenoloxidase in red raspberry (*Rubus idaeus*) and strawberry (Fragaria × ananassa). *Food Chemistry* 88, 7–10.

George, P., Kasapis, S., Bannikova, A., Mantri, N., Palmer, M., Meurer, B., Lundin, L., 2013. Effect of high hydrostatic pressure on the structural properties and bioactivity of immunoglobulins extracted from whey protein. *Food Hydrocolloids* 32, 286–293.

Gott, B., 2008. Indigenous use of plant in south-eastern Australia. *Telopea* 12, 215–256.

Gutiérrez, F.J., Albillos, S.M., Casas-Sanz, E., Cruz, Z., García-Estrada, C., García-Guerra, A., García-Reverter, J. et al., 2013. Methods for the nanoencapsulation of β-carotene in the food sector. *Trends in Food Science and Technology* 32, 73–83.

Himejima, M., Kubo, I., 1993. Fungicidal activity of polygodial in combination with anethole and indole against *Candida albicans*. *Journal of Agricultural and Food Chemistry* 41, 1776–1779.

Konczak, I., Zabaras, D., Dunstan, M., Aguas, P., 2010a. Antioxidant capacity and hydrophilic phytochemicals in commercially grown native Australian fruits. *Food Chemistry* 123, 1048–1054.

Konczak, I., Zabaras, D., Dunstan, M., Aguas, P., 2010b. Antioxidant capacity and phenolic compounds in commercially grown native Australian herbs and spices. *Food Chemistry* 122, 260–266.

Konczak, I., Zabaras, D., Dunstan, M., Aguas, P., Roulfe, P., Pavan, A., 2009. *Health Benefits of Australian Native Foods: An Evaluation of Health-Enhancing Compounds*. Rural Industries Research and Development Corporation, RIRDC Publication No. 09/133, Canberra, Australian Capital Territory, Australia.

Kubo, I., Himejima, M., 1991. Anethole, a synergist of polygodial against filamentous microorganisms. *Journal of Agricultural and Food Chemistry* 39, 2290–2292.

Kurekci, C., Padmanabha, J., Bishop-Hurley, S.L., Hassan, E., Al Jassim, R.A.M., McSweeney, C.S., 2013. Antimicrobial activity of essential oils and five terpenoid compounds against *Campylobacter jejuni* in pure and mixed culture experiments. *International Journal of Food Microbiology* 166, 450–457.

Lazar-Baker, E.E., Hetherington, S.D., Ku, V.V., Newman, S.M., 2011. Evaluation of commercial essential oil samples on the growth of postharvest pathogen *Monilinia fructicola* (G. Winter) Honey. *Letters in Applied Microbiology* 52, 227–232.

Neacsu, M., Vaughan, N., Raikos, V., Multari, S., Duncan, G.J., Duthie, G.G., Russell, W.R., 2015. Phytochemical profile of commercially available food plant powders: Their potential role in healthier food reformulations. *Food Chemistry* 179, 159–169.

Netzel, M., Netzel, G., Tian, Q., Schwartz, S., Konczak, I., 2007. Native Australian fruits – A novel source of antioxidants for food. *Innovative Food Science and Emerging Technologies* 8, 339–346.

Pusateri, D.J., Menon, G.R., Vergel de Dios, L.I., Schlipalius, L.E., 2007. Patent No. US 7,175,862 B2, 2007. Method of preparing Kakadu plum powder.

Rawson, A., Patras, A., Tiwari, B.K., Noci, F., Koutchma, T., Brunton, N., 2011. Effect of thermal and non thermal processing technologies on the bioactive content of exotic fruits and their products: Review of recent advances. *Food Research International* 44, 1875–1887.

Sakulnarmrat, K., Konczak, I., 2012. Composition of native Australian herbs polyphenolic-rich fractions and in vitro inhibitory activities against key enzymes relevant to metabolic syndrome. *Food Chemistry* 134, 1011–1019.

Shimoni, E., 2009. Nanotechnology for foods: Delivery systems. In: Barbosa-Cánovas, G., Motimer, A., Lineback, D., Spiess, W., Buckle, K., Colonna, P. (eds.), *Global Issues in Food Science and Technology.* Academic Press, San Diego, CA, pp. 411–424.

Sommano, S., Caffin, N., McDonald, J., Cocksedge, R., 2013. The impact of thermal processing on bioactive compounds in Australian native food products (bush tomato and Kakadu plum). *Food Research International* 50, 557–561.

Starkenmann, C., Cayeux, I., Birkbeck, A.A., 2011. Exploring natural products for new taste sensations. *CHIMIA International Journal for Chemistry* 65, 407–410.

Sultanbawa, Y., Williams, D., Chaliha, M., Konczak, I., Smyth, H., 2015. *Changes in Quality and Bioactivity of Native Foods during Storage.* Rural Industries Research and Development Corporation, RIRDC Publication No. 15/010, Canberra, Australian Capital Territory, Australia.

Sun, Y., Shen, Y., Liu, D., Ye, X., 2015. Effects of drying methods on phytochemical compounds and antioxidant activity of physiologically dropped un-matured citrus fruits. *LWT – Food Science and Technology* 60, 1269–1275.

The Vien, H., Caffin, N., Dykes, G.A., Bhandari, B., 2008. Optimization of the microencapsulation of lemon myrtle oil using response surface methodology. *Drying Technology* 26, 357–368.

Tiwari, B.K., O'Donnell, C.P., Cullen, P.J., 2009. Effect of non thermal processing technologies on the anthocyanin content of fruit juices. *Trends in Food Science and Technology* 20, 137–145.

Williams, D.J., Edwards, D., Pun, S., Chaliha, M., Sultanbawa, Y., 2014. Profiling ellagic acid content: The importance of form and ascorbic acid levels. *Food Research International* 66, 100–106.

Zeanah, D.W., Codding, B.F., Bird, D.W., Bliege Bird, R., Veth, P.M., 2015. Diesel and damper: Changes in seed use and mobility patterns following contact amongst the Martu of Western Australia. *Journal of Anthropological Archaeology* 39, 51–62.

CHAPTER 23

Quality Changes during Packaging and Storage of Australian Native Herbs

Mridusmita Chaliha

CONTENTS

INTRODUCTION

Australian native plants have been an integral part of the Aboriginal communities for centuries, and these plants have been used both as food and as traditional medicine (Cooper, 2004). Recently, these Australian native plants have gained significant interest at both national and international levels due to recent scientific discoveries linking many health-promoting properties with these plants (Wilkinson et al., 2002). Exploring the potential of these plants as functional food ingredients and/or novel pharmaceutical compounds is an emerging research area.

Out of the 13 currently commercialised Australian native plants, three can be classified as herbs, lemon myrtle, anise myrtle, and Tasmanian pepper leaf. This chapter will focus on the aforementioned native plants. These herbs are usually incorporated as the essential oil or as a milled form of the dried herb (Forbes-Smith and Paton, 2002). Like many other native plants, the three native herbs have unique sensory properties, which can be attributed to their unique volatile aroma profile.

Lemon myrtle (*Backhousia citriodora*), belongs to the Myrtaceae family, and native to eastern Australia (Hayes and Markovic, 2002). The characteristic lemon flavour of lemon myrtle is attributed to its predominant (95%) volatile compound – citral, which is the combination of the two isomeric aldehydes neral (Z-isomer) and geranial (E-isomer) (Brophy et al., 1995; Nhu-Trang et al., 2006; Penfold et al., 1951; Southwell et al., 1996). On their own, both neral and geranial impart a lemony aroma (Schieberle and Grosch, 1988). Leaves and flowers of lemon myrtle have lemon flavour (Smyth et al., 2012), and hence are incorporated in tea blends and beverages, dairy, cookies, breads, confectionery, pasta, syrups, liqueurs, flavoured oils, packaged fish and dipping and simmer sauces (Konczak et al., 2010).

Anise myrtle (*Syzygium anisatum*, Myrtaceae) is endemic to north-eastern New South Wales and Queensland. The characteristic aniseed flavour comes from the major (79%–90%) volatile anethole ((E)-1-methoxy-4-(10-propenyl) benzene) (Southwell et al., 1996). Anise myrtle leaves are incorporated in sweet and savoury dishes and cosmetic products (Smyth et al., 2012).

Leaves and berries of Tasmanian pepper (*Tasmanianlanceolata*, Winteraceae) have a strong heat and pungent flavour on the palate (Smyth et al., 2012), and could be attributed to the presence of the sesquiterpene polygodial (Drager et al., 1998).

The Australian native food industry is a newly emerging industry. Retention of the aroma and flavour of dried milled leaves of native herbs is a challenge to the industry. Deterioration of the sensory qualities, including aroma, occurs slowly during storage and the reduction in the perceived 'freshness' level can be detected by the consumers even at lower levels (Rizzo and Muratore, 2009).

Improvement of packaging materials is an integral part of processing in the agriculture and food industry. Packaging prevents contamination, protects product integrity and retains the desired sensory profile of the products (Mohney et al., 1988; Quezada-Gallo et al., 2000). Better understating of the ideal storage and packaging conditions are crucial for developing new market sectors and export opportunities for the Australian native food industry.

This chapter provides a brief overview of the current industry practices in terms of storage of the native herbs and recent research findings with regard to packaging conditions and its effect on product quality, most importantly aroma.

CURRENT PROCESSING CONDITIONS OF AUSTRALIAN NATIVE HERBS

Herbs and spices have traditionally been traded as dried products. For most of the commonly used herbs and spices, drying, grinding, packaging and storage constitute the most integral steps of the processing protocol (Figure 23.1). Depending on the type of the herb or spice, different parameters are employed.

The exact parameters involved in the processing of the three Australian native herbs are regarded as confidential by the native food industry. The general idea of the steps involved with the processing of these herbs is represented in the following flowcharts. Figures 23.2 and 23.3 depict the steps involved in the processing of lemon myrtle, anise myrtle and Tasmanian pepper respectively.

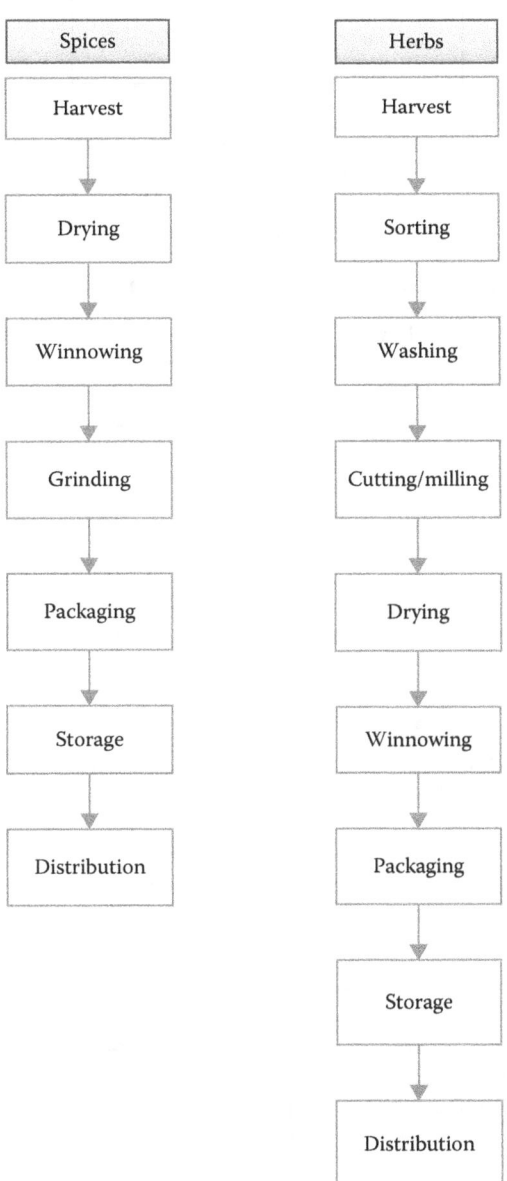

Figure 23.1 Steps involved in processing of common herbs.

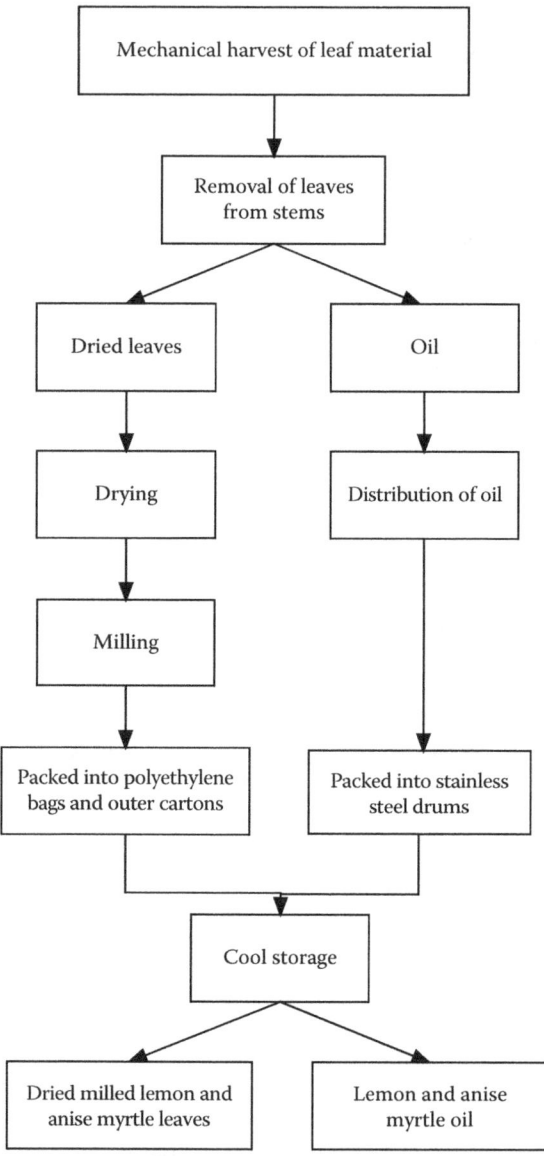

Figure 23.2 Processing of lemon and anise myrtle leaves. (Adapted from Sultanbawa, Y. et al., *Changes in Quality and Bioactivity of Native Foods during Storage*, Rural Industries Research and Development Corporation, RIRDC Publication No. 15/010, Canberra, Australian Capital Territory, Australia, 2015.)

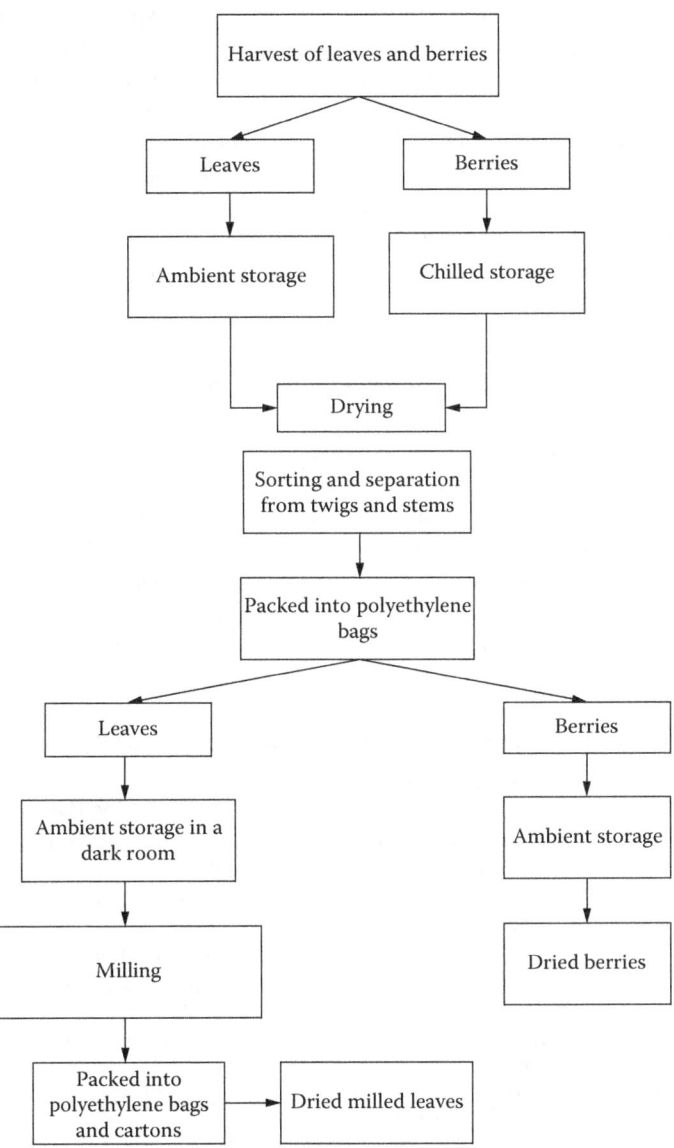

Figure 23.3 Processing of Tasmanian pepper leaves and berries. (Adapted from Sultanbawa, Y. et al., *Changes in Quality and Bioactivity of Native Foods during Storage*, Rural Industries Research and Development Corporation, RIRDC Publication No. 15/010, Canberra, Australian Capital Territory, Australia, 2015.)

EFFECT OF PROCESSING CONDITIONS ON VOLATILE COMPOUNDS

Drying is an integral part of processing of herbs and spices, as some herbs and spices can have a high moisture content of about 75%–80% (Díaz-Maroto et al., 2002) and can rapidly go through compositional change (Díaz-Maroto et al., 2004). The moisture levels are preferred to be less than 15% in dried herbs and spice (Díaz-Maroto et al., 2002). The moisture contents of the commercially available dried, milled lemon myrtle, anise myrtle and Tasmanian pepper leaves were 9%, 6% and 11%, respectively (Sultanbawa et al., 2015). This indicated that the current drying procedure employed by the native food industry was quite effective. Currently, there is no information on the loss of active volatiles during drying and milling of Australian native herbs.

IMPORTANCE OF PACKAGING IN RETAINING AROMA QUALITY

Contemporary packaging materials are developed by combining a number of different materials through lamination, coextrusion or coating, in order to provide better barrier function than just one material. Modern packaging materials combines different layers of foil, plastics like polyethylene (PE), polypropylene (PP), polyethylene terephthlate (PET), nylon, ethylene vinyl alcohol (EVOH), paper and adhesives. These multilayered packaging materials provide a much better barrier compared to the conventional PE materials. Table 23.1 summarises the barrier properties of some packaging materials that are used in the food industry.

Table 23.1 Water and Oxygen Transmission Rates of Some Food Packaging Polymers

Material (Layer Thickness in µm)	Transmission Rate of Water	Transmission Rate of Oxygen
LDPE (80)[a]	10–20 g/m²/24 h at 38°C (RH 90%)	6,500–8,500 cm³/(m²/24 hatm) at 38°C (RH 90%)
HDPE (80)[a]	7–10 g/m²/24 h at 38°C (RH 90%)	1,600–2,000 cm³/(m²/24 hatm) at 38°C (RH 90%)
OPP (35)	–	1,300 mL/(m²/24 hatm) at 23°C (RH 50%)
PVC (10)	>1000 g/m²/24 h	10,960 mL/(m²/24 hatm) at 23°C (RH 50%)
PVDC coated PET (12)/CPP (20)[a]	0.5–1 g/m²/24 h at 38°C at 38°C (RH 90%)	2–4 cm³/m²/24 h at 23°C (RH 50%)
PET (12)/PET (12)/ Foil (9)/LLDPE (65)[a]	0.25 g/m²/24 h at 38°C at 37°C (RH 98%)	0.02 at 25°C (RH 95%)

Abbreviations: LDPE, low-density polyethylene; HDPE, high-density polyethylene; OPP, oriented polypropylene; PVC, polyvinyl chloride; PVDC-coated PET/CPP, polyvinylidene chloride-coated polyethylene terephthalate/casted polypropylene; PET/PET/foil/LLDPE, polyethylene terephthalate/polyethylene terephthalate/aluminium foil/linear low-density polyethylene; RH, relative humidity.
[a] Effect of these packaging materials has been reported in major volatile retention of Australian native herbs.

Traditionally, lemon and anise myrtle industry used low-density polyethylene (LDPE) materials for storage, while Tasmanian pepper leaf industry used high-density polyethylene (HDPE) packaging. Storage of dried milled herbs in traditional polyethylene packaging was a problem for the Australian native food industry, as there was significant loss of aroma and flavours in the dried milled leaves after 1 month of storage in the polyethylene packaging material. Moreover, certain volatile compounds such as citral in lemon myrtle could migrate into the packaging material, causing disintegration of the packaging.

EFFECT OF PACKAGING MATERIALS AND STORAGE ON MAJOR VOLATILES OF AUSTRALIAN NATIVE HERBS

Until recently, the effect of packaging materials on the retention of active volatiles of the Australian native herbs when stored at room temperature was not known. Therefore, the Australian native food industry was unable to provide shelf life information for the commercially available dried herb products.

Our group published the first study that investigated the effectiveness of three packaging materials with various high-barrier properties in the prevention of loss of volatiles in three native Australian herbs stored at room temperature for 6 months (Chaliha et al., 2013). We compared the effectiveness of PE (LDPE and HDPE) with two high-barrier laminated packaging films, namely, polyvinylidene chloride–coated polyethylene terephthalate/ polyvinylidene chloride–casted polypropylene (PVDC-coated PET/CPP) and polyethylene terephthalate/polyethylene terephthalate/ aluminium foil/linear low-density polyethylene (PET/PET/Foil/LLDPE), in order to provide effective solutions to industry in terms of retaining aroma of herbs during long-term ambient storage.

The major target volatiles monitored in our study were neral and geranial for lemon myrtle, estragole and anethole for anise myrtle and eucalyptol and eugenol for Tasmanian pepper leaves. The concentrations of the major volatiles were determined using gas chromatography mass spectrometry (GCMS) in selected ion monitoring (SIM) mode. The detailed methods have been described previously (Chaliha et al., 2013).

Commercial samples of fresh leaves (with or without stems) and dried leaves (before and after milling) were analysed to assess the effect of processing on the volatile composition of the products.

It was observed that in lemon myrtle and Tasmanian pepper leaves, the concentration of major volatiles was highest in the dried leaves before milling (Tables 23.2 and 23.4). However, after milling, the concentrations of major volatiles for both lemon myrtle and Tasmanian pepper leaves decreased, indicating herb quality loss caused by the process of milling (Tables 23.2 and 23.4). There were no samples available of dried pre-milled anise myrtle (Table 23.3). Our study also indicated that inclusion or exclusion of stems with the leaves does not change the

Table 23.2   Concentrations of Major Volatiles (mg/kg) Present in Native Herb Samples at the Beginning of the Storage Trial (Month = 0) for Neral and Geranial in Lemon Myrtle (n = 4)

Storage Time (Months)	Sample Description	Neral (mg/kg)	Geranial (mg/kg)
0	Fresh whole leaves on stem	$1746.60^b \pm 337.31$	$1774.06^b \pm 363.10$
0	Fresh whole leaves without stem	$1426.07^b \pm 203.91$	$1438.38^b \pm 202.68$
0	Dried whole leaves	$3461.18^a \pm 193.34$	$3489.54^a \pm 127.08$
0	Dried, milled leaves (commercial product)	$794.70^c \pm 41.67$	$844.23^c \pm 44.35$

Source: Chaliha, M. et al., *J. Agric. Food Chem.*, 61, 5738, 2013.
Note: Average concentrations analysed using Student t test. Different letters (i.e. a, b, c) across sample types for each volatile denote significant differences between mean concentrations according to a Tukey–Kramer HSD.

Table 23.3   Concentrations of Major Volatiles (mg/kg) Present in Native Herb Samples at the Beginning of the Storage Trial (Month = 0) for Estragole and Anethol in Anise Myrtle (n = 4)

Storage Time (Months)	Sample Description	Estragole (mg/kg)	Anethol (mg/kg)
0	Fresh whole leaves on stem	$96.12^a \pm 5.00$	$126.87^b \pm 5.35$
0	Fresh whole leaves without stem	$22.15^c \pm 2.78$	$276.80^b \pm 35.72$
0	Dried, milled leaves (commercial product)	$60.30^b \pm 13.77$	$470.45^a \pm 101.80$

Source: Chaliha, M. et al., *J. Agric. Food Chem.*, 61, 5738, 2013.
Note: Average concentrations analysed using Student t test. Different letters (i.e. a, b, c) across sample types for each volatile denote significant differences between mean concentrations according to a Tukey–Kramer HSD.

Table 23.4   Concentrations of Major Volatiles (mg/kg) Present in Native Herb Samples at the Beginning of the Storage Trial (Month = 0) for Eucalyptol and Eugenol in Tasmanian Pepper Leaves (n = 4)

Storage Time (Months)	Sample Description	Euclyptol (mg/kg)	Eugenol (mg/kg)
0	Fresh whole leaves	$1.28^b \pm 0.03$	$7.01^b \pm 0.31$
0	Dried whole leaves	$13.06^a \pm 5.53$	$83.10^a \pm 33.21$
0	Dried, milled leaves (commercial product)	$5.16^b \pm 0.25$	$57.09^a \pm 2.29$

Source: Chaliha, M. et al., *J. Agric. Food Chem.*, 61, 5738, 2013.
Note: Average concentrations analysed using Student t test. Different letters (i.e. a, b, c) across sample types for each volatile denote significant differences between mean concentrations according to a Tukey–Kramer HSD.

overall aroma quality of the lemon and anise myrtle product (Tables 23.2 and 23.3). A sample of Tasmanian pepper leaf without stems was not available for comparison (Table 23.4).

Significant retention of the major volatiles for all the three herbs was observed in samples packed in the PVDC-coated PET/CPP and PET/PET/Foil/PE materials compared to those packed in the commercial LDPE and HDPE packages. Samples stored in the LDPE (or HDPE) packing material showed the most rapid decline in the concentration of key volatiles over the storage period. A reason behind this rapid decline may be the high gas permeability rate of LDPE and HDPE materials which allows the volatiles to migrate out of the sample matrix, whereas the higher barrier properties of the PVDC-coated PET/CPP and the PET/PET/Foil/PE materials have very low water and oxygen permeabilities and therefore better prevention of the loss of volatiles.

For lemon myrtle, PET/PET/Foil/PE packaging material performed significantly better than the PVDC-coated PET/CPP bags over storage time in retaining key volatiles (Table 23.5), whereas for anise myrtle and Tasmanian pepper leaf, there was no

Table 23.5 Change in Concentration of Volatiles (mg/kg) during 6 Months of Storage for Lemon Myrtle (Dried, Milled Leaves) in Packaging Materials – LDPE, PVDC-Coated PET/CPP and PET/PET/Foil/PE of Neral and Geranial ($n = 4$)

Packaging Material	Storage Months	Concentration of Neral (mg/kg)	Concentration of Geranial (mg/kg)
LDPE	1	627.15 ± 19.71	682.52 ± 23.29
	2	547.56 ± 20.62	618.24 ± 25.69
	3	319.75 ± 18.46	367.11 ± 28.43
	4	274.81 ± 41.19	317.09 ± 49.61
	5	221.35 ± 30.66	257.99 ± 27.58
	6	182.48 ± 20.57	214.82 ± 24.29
PVDC-coated PET/CPP	1	929.76[a] ± 45.32	941.60 ± 37.78
	2	859.94[b] ± 30.06	868.31 ± 34.04
	3	765.52[d] ± 6.89	766.18 ± 6.52
	4	816.38[bc] ± 43.10	813.58 ± 45.65
	5	644.27[e] ± 15.95	649.50 ± 17.12
	6	577.53[f] ± 33.88	588.31 ± 34.63
PET/PET/Foil/PE	1	1141.94[a] ± 96.49	1173.20 ± 112.40
	2	1089.30[a] ± 73.90	1097.50 ± 74.22
	3	962.81[b] ± 52.60	961.10 ± 53.81
	4	1174.59[a] ± 72.38	1164.14 ± 76.54
	5	1106.22[a] ± 70.28	1109.02 ± 69.60
	6	1082.31[ab] ± 132.54	1087.77 ± 131.32

Source: Chaliha, M. et al., J. Agric. Food Chem., 61, 5738, 2013.
Note: Different letters within a column (i.e., a, b, c, d, e, f) denote significant differences between means according to a Tukey–Kramer HSD.

Table 23.6 Change in Concentration of Volatiles (mg/kg) during 6 Months of Storage for Anise Myrtle (Dried, Milled Leaves) in Packaging Materials – LDPE, PVDC-Coated PET/CPP and PET/PET/Foil/PE of Estragole and Anethol ($n = 4$)

Packaging Material	Storage Months	Concentration of Estragole (mg/kg)	Concentration of Anethol (mg/kg)
LDPE	1	48.12 ± 7.31	525.49 ± 30.45
	2	25.95 ± 4.93	375.63 ± 63.55
	3	15.47 ± 1.05	294.63 ± 11.38
	4	8.98 ± 1.68	187.10 ± 42.96
	5	6.70 ± 0.44	153.32 ± 15.69
	6	4.14 ± 0.32	92.88 ± 5.68
PVDC-coated PET/CPP	1	110.01 ± 9.38	761.99 ± 68.19
	2	105.01 ± 4.65	727.47 ± 23.03
	3	95.66 ± 6.03	660.22 ± 41.38
	4	99.46 ± 14.86	676.20 ± 90.89
	5	101.94 ± 9.82	715.74 ± 74.33
	6	80.03 ± 5.30	561.18 ± 32.10
PET/PET/Foil/PE	1	114.52 ± 7.28	800.03 ± 44.77
	2	104.61 ± 9.56	726.02 ± 52.58
	3	92.46 ± 10.01	639.02 ± 60.61
	4	113.71 ± 13.35	750.40 ± 77.68
	5	106.60 ± 9.53	728.22 ± 67.97
	6	96.82 ± 11.48	668.25 ± 80.83

Source: Chaliha, M. et al., J. Agric. Food Chem., 61, 5738, 2013.

significant difference between samples stored in PET/PET/Foil/PE bags and those stored in PVDC-coated PET/CPP bags (Tables 23.6 and 23.7).

CONCLUSIONS AND RECOMMENDATIONS

High-barrier packaging materials such as PVDC-coated PET/CPP and PET/PET/Foil/PE compared to conventional HDPE and LDPE packaging materials showed significant improvement in retention of key volatiles in three Australian native herbs stored for 6 months. The native food industry can benefit from the improved packaging suggested in this study to target export markets, where a longer shelf quality is required when bulk herb products are shipped and transported. Enhanced opportunities will help in the growth of the Australian native food industry which in turn will increase cultivation, harvesting and processing of native foods in rural communities, generating much needed employment in remote areas.

Table 23.7 Change in Concentration of Volatiles (mg/kg) during 6 Months of Storage for Tasmanian Pepper Leaves (Dried, Milled Leaves) in Packaging Materials – LDPE, PVDC-Coated PET/CPP and PET/PET/Foil/PE of Eucalyptol and Eugenol (n = 4)

Packaging Material	Storage Months	Concentration of Eucalyptol (mg/kg)	Concentration of Eugenol (mg/kg)
LDPE	1	1.63 ± 0.06	68.87 ± 2.64
	2	8.90 ± 0.85	85.43 ± 3.98
	3	7.44 ± 1.41	78.27 ± 3.27
	4	0.69 ± 0.03	56.67 ± 3.96
	5	8.01 ± 0.71	68.43 ± 2.21
	6	7.47 ± 1.22	69.06 ± 8.61
PVDC-coated PET/CPP	1	0.47 ± 0.08	51.96 ± 1.27
	2	7.00 ± 1.41	59.87 ± 12.90
	3	7.56 ± 3.03	66.94 ± 9.27
	4	0.41 ± 0.08	46.46 ± 2.37
	5	4.47 ± 0.19	39.79 ± 1.23
	6	8.30 ± 0.45	69.71 ± 3.30
PET/PET/Foil/PE	1	0.21 ± 0.03	41.69 ± 7.17
	2	3.90 ± 0.15	40.82 ± 1.71
	3	6.40 ± 0.49	62.96 ± 1.83
	4	0.21 ± 0.01	32.07 ± 7.24
	5	4.88 ± 0.94	44.56 ± 14.45
	6	7.48 ± 1.02	64.52 ± 2.97

Source: Chaliha, M. et al., *J. Agric. Food Chem.*, 61, 5738, 2013.

REFERENCES

Brophy, J.J., Goldsack, R.J., Fookes, C.J., RandForster, P.I., 1995. Leaf oils of the genus *Backhousia* (Myrtaceae). *Journal of Essential Oil Reserach* 7, 237–254.

Chaliha, M., Cusack, A., Currie, M., Sultanbawa, Y., Smyth, H., 2013. Effect of packaging materials and storage on major volatile compounds in three Australian native herbs. *Journal of Agricultural and Food Chemistry* 61, 5738–5745.

Cooper, W., 2004. *Fruits of the Australian Tropical Forest.* Nokomis Editions Pty Ltd, Melbourne, Victoria, Australia.

Díaz-Maroto, M.C., Pérez-Coello, M.S., Cabezudo, M.D., 2002. Effect of drying method on the volatiles in bay leaf (*Laurus nobilis* L.). *Journal of Agricultural and Food Chemistry* 50, 4520–4524.

Díaz-Maroto, M.C., Sánchez Palomo, E., Castro, L., González Vinas, M.A., Perez-Coello, M.S., 2004. Changes produced in the aroma compounds and structural integrity of basil (*Ocimum basilicum* L) during drying. *Journal of Agricultural and Food Chemistry* 84, 2070–2076.

Drager, V.A., Garland, S.M., Menary, R.C., 1998. Investigation of the variation in chemicals composition of *Tasmanianlanceolata* solvent extract. *Journal of Agricultural and Food Chemistry* 46, 3210–3213.

Forbes-Smith, M., Paton, J.E., 2002. Innovative products from Australian native foods. A report for the Rural Industries Research and Development Corporation. RIRDC Publication No. 02/109, Canberra, Australian Capital Territory, Australia.

Hayes, A.J., Markovic, B., 2002. Toxicity of Australian essential oil *Backhousia citriodora* (Lemon myrtle). Part 1. Antimicrobial activity and in vitro cytotoxicity. *Food and Chemical Toxicology* 40, 535–543.

Konczak, I., Zabaras, D., Dunstan, M., Aguas, P., 2010. Antioxidant capacity and phenolic compounds in commercially grown native Australian herbs and spices. *Food Chemistry* 122, 260–266.

Mohney, S.M., Hernandez, R.J., Giacin, J.R., Miltz, J., 1988. Permeability and solubility of d-limonene vapor in cereal package liners. *Journal of Food Science* 53, 253–257.

Nhu-Trang, T.T., Casabianca, H., Grenier-Loustalot, M.F., 2006. Authenticity control of essential oils containing citronellal and citral by chiral and stable-isotope gas-chromatographic analysis. *Analytical and Bioanalytical Chemistry* 386, 2141–2151.

Penfold, A.R., Morrison, F.R., Willis, J.L., Mckern, H.G., Spies, M.C., 1951. The occurrence of a physiological form of *Backhousia citriodora* F Muell. and its essential oil. *Journal and Proceedings of the Royal Society of New South Wales* 85, 123–126.

Quezada-Gallo, J.A., Debeaufor, F., Voille, A., 2000. Mechanism of aroma transfer through edible and plastic packagings. In: Risch, S.J. (ed.), *Food Packaging Testing Methods and Applications*. American Chemical Society, Washington, DC.

Rizzo, V., Muratore, G., 2009. Effects of packaging on shelf life of fresh celery. *Journal of Food Engineering* 90, 124–128.

Schieberle, P., Grosch, W., 1988. Identification of potent flavor compounds formed in an aqueous lemon oil/citric acid emulsion. *Journal of Agricultural and Food Chemistry* 36, 797–800.

Smyth, H.E., Sanderson, J.E., Sultanbawa, Y., 2012. Lexicon for the sensory description of Australian native plant foods and ingredients. *Journal of Sensory Studies* 27, 471–481.

Southwell, I., Russel, M., Birmingham, E., Brophy, J., 1996. Aniseed myrtle – Leaf quality. *Australian Rainforest Bushfood Industry Association Newsletter* 4, 13–15.

Sultanbawa, Y., Williams, D., Chaliha, M., Konczak, I., Smyth, H., 2015. *Changes in Quality and Bioactivity of Native Foods during Storage*. Rural Industries Research and Development Corporation, RIRDC Publication No. 15/010, Canberra, Australian Capital Territory, Australia.

Wilkinson, J.M., Hipwell, M., Ryan, T., Cavanagh, H.M.A., 2002. Bioactivity of *Backhousia citriodora*: Antibacterial and Antifungal Activity. *Journal of Agricultural and Food Chemistry* 51, 76–81.

CHAPTER 24

Value Chains: Making the Connections between Producers and Consumers of Native Plant Foods

Anoma Ariyawardana, Ray Collins and Lilly Lim-Camacho

CONTENTS

NATIVE FOODS INDUSTRY IN AUSTRALIA

The Australian landscape with its rainforests, deserts, mountains, wetlands and tropical, sub-tropical and temperate regions provides rich and diverse climatic conditions for a wide spectrum of native flora species. For thousands of years, the Aboriginal communities of Australia have been using these species for food, medicine, essential oils, timber and wood products, seed for horticulture, crafts and craft material (Ahmed and Johnson, 2000; Morse, 2005; Ryder et al., 2009). Collection of *bush tucker* or *bush food* from the wild and its preparation into edible food is a skill that is inherited and embedded within Australian Aboriginal culture. In every region of the continent, native plant species were widely consumed by the Aboriginal population and to a limited extent by others who had the opportunity to share in their traditional knowledge. In more recent times, *bush foods* have attracted commercial interest as knowledge of their potential phyto-pharmaceutical, nutraceutical and other benefits have become available (Ahmed and Johnson, 2000; Konczak and Roulle, 2011). In the late 1990s, the commercialisation of unique Australian bush foods led to the recognition of an '*Australian Native Foods Industry*' (RIRDC, 2008). Recently, it has been estimated that the wholesale value of the Australian native foods industry excluding macadamia nuts is approximately $23 million a year (RIRDC, 2013; Spencer and Hardie, 2011). A driving force behind this industry has been the growing demand for food that can provide additional health benefits (Netzel et al., 2007; Sommano et al., 2011).

The unique nature of the industry, much of which is centered on knowledge originating from Aboriginal Australians, has the ability to generate livelihood opportunities for them while providing economic benefits to the country as a whole (Figure 24.1). However, as a number of studies have pointed out, developing employment opportunities and improving livelihoods through commercialisation of native species requires the development of markets as well as best practice guidelines (Ahmed and Johnson, 2000; Merne Altyerre-ipenhe (Food from the Creation time) Reference Group et al., 2011; Spencer and Hardie, 2011). Some studies have demonstrated the benefits of tackling this challenge from the perspective of the whole chain from producer to consumer, an approach referred to as adopting a 'value chain perspective' (Bryceson, 2008; Lee, 2012). Thus, the aim of this chapter is to explore the key concepts associated with adopting 'value chain thinking' and to suggest how they might be applied to developing the Australian native food industry.

CHAINS THAT LINK PRODUCERS AND CONSUMERS OF NATIVE FOOD

All native food products, whether harvested from the wild or produced on farms, must be transferred along a chain of businesses that link the harvester or producer with the final consumer, and many products are processed into different forms as they move along this chain. Ideally, the businesses that make up such chains will distribute products to consumers as efficiently and effectively as possible, and in doing so, they usually rely on the support of other businesses such as third party logistics providers, financial service providers and consultants, and other stakeholders such as government organisations (Figure 24.2). Achieving chain efficiency and effectiveness requires not only that individual firms be managed with this goal in mind, but also that the chain itself is managed. Originally, 'managing the chain itself' meant managing the physical product flows from initial suppliers through to final customers, a practice defined as 'supply chain management'

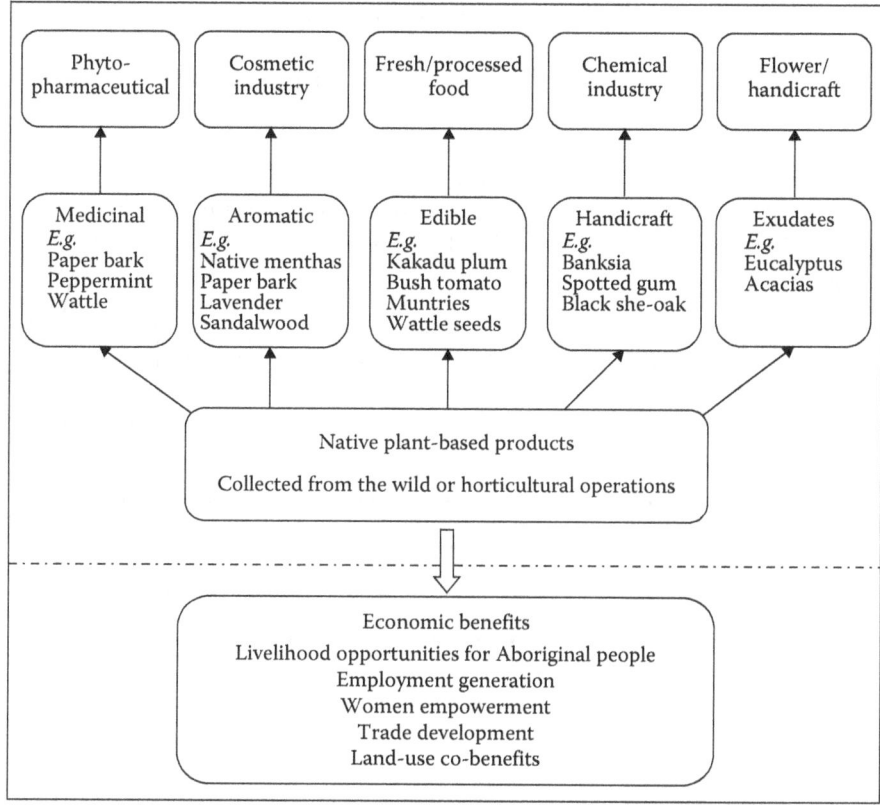

Figure 24.1 Uses of Australian native plant species (Acacias (*Acacia* spp.), Banksia (*Banksia* spp.), Bush Tomato (*Solanum central*), Black She-oak (*Allocasuarina littoralis*), Eucalyptus (*Eucalyptus* spp.), *Kakadu Plum (Terminalia ferdinandiana)*, Lavender (*Lavandula angustifolia*), *Muntries (Kunzea pomifera)*, Mountain pepper (*Tasmanianlanceolata*), Native menthas (*Mentha australis*), Native pepper *(Tasmanian* spp.), Paper Bark (*Melaleuca alternifolia*), Peppermint plant (*Mentha x piperita*), Spotted Gum (*Corymbia Maculata*), Quandong (*Santalum acuminatum*), Wattle seed (*Acacia* spp.)).

(Lambert et al., 1998). This concept was later expanded to incorporate other critical dimensions of a chain's operations, such as the flow of information, the management of relationships and governance issues within and across the businesses in a chain (Boehlje, 1999; Braziotis et al., 2013; Lusch et al., 2010).

In highly competitive business environments, supply chain management was recognised as a strategy that could give businesses a competitive edge, but it meant that businesses needed to develop specialised skills in managing relationships with other chain members (Barney, 1991). Over time, the adoption of supply chain management as a strategic tool changed the nature of competition: from competition between firms to competition between chains (Boehlje, 1999; Dyer, 2000; Fearne et al., 2012). Thus, for any business, engaging in supply chain management means recognising

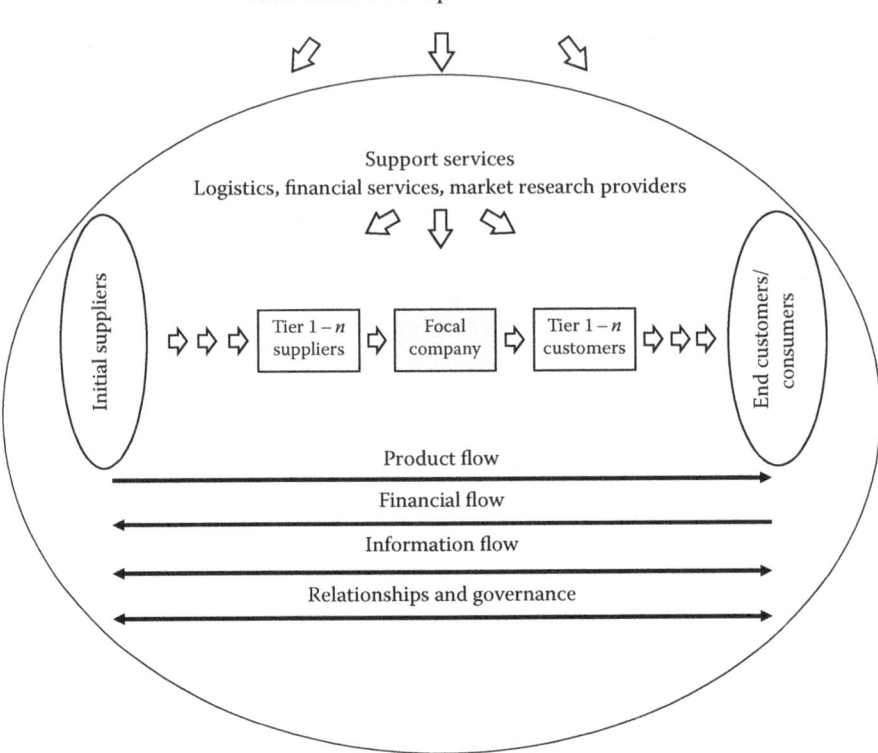

Figure 24.2   The supply chain as an integrated system. (Developed based on Lambert, D.M. et al., *Int. J. Logist. Manag.*, 9, 1, 1998; Boehlje, M., *Am. J. Agric. Econ.*, 81, 1028, 1999.)

what needs to be managed, and how (Lambert et al., 1998). According to Anderson, Britt and Favre (2007), this process begins by identifying customer service needs and determining how to meet these needs.

FROM SUPPLY CHAINS TO VALUE CHAINS

The concept of value was first introduced by Porter (1985), who argued that a company's competitiveness revolves around its ability to provide superior value to its customers, and that within a company, there is a chain of activities, called a value chain, that create and deliver this value. By incorporating a supply chain management perspective, Porter's concept became extended from how a single business might operate to how a whole chain might operate.

Defining a producer-to-consumer chain as a value-driven system means understanding what constitutes value to customers and consumers, how and where value is generated along the chain and what processes within each firm are responsible for generating value (Bonney et al., 2007; Hines and Rich, 1997; Taylor, 2005). In contrast to supply chain approaches which primarily view the chain as a supply driven system with waste reduction and efficiency as its key objectives, the value chain perspective is driven by the need to create and deliver *value* to customers and consumers. Thus, value chain approaches are more demand driven; supply chain approaches are more supply driven.

The concept of value is multifaceted. Ultimately, it revolves around the consumer's perspective of the benefits of a product, because this determines willingness to pay for that product, and paying for the product creates the revenue stream that flows back to the businesses in the chain that produce and deliver the product (Priem, 2007). How consumers perceive a product is complex, and Kotler et al. (2010) have argued that consumer perceptions involve the whole human mind, heart and spirit, incorporating functional, emotional and spiritual aspects in assessing what constitutes value in a product. They therefore emphasised the importance of consumer research in understanding the complex nature of value satisfaction in the mind of the consumer.

Value chain management thus combines the need for efficiency drawn from supply chain management principles with the need to be focused on what consumers' value and are willing to pay for. The underlying assumption is that if the businesses that make up a chain can target consumers by delivering what they value, and doing so as efficiently as possible, then the superior performance of the whole chain will confer competitive benefits on each business within the chain.

NATIVE FOOD PRODUCTS AND CONSUMER VALUE

What does this mean for native food products? Clearly, understanding what consumers value is the starting point. Identifying attributes that distinguish native food products in the eyes of the consumer provides essential information for building competitive value chains. Every product can be thought of as having four different types of attributes, as shown in Figure 24.3. Beyond the core product, consumers may have expectations such as taste, smell and texture; they may respond to how the product is packaged and branded and they may value its provenance, its originality or its contribution to a local economy. Knowing the dimensions and relative importance of these groups of attributes enables consideration of how and where the value in each attribute can be created in a chain – because as stated earlier, when a chain can deliver what consumers value and are willing to pay for, businesses in that chain can improve their competitiveness.

UNDERSTANDING THE MARKET VALUE OF
AUSTRALIAN NATIVE FOODS

Understanding what customers and consumers value, and being able to deliver that value, is crucial for new products as there are higher risks of failure in under- or

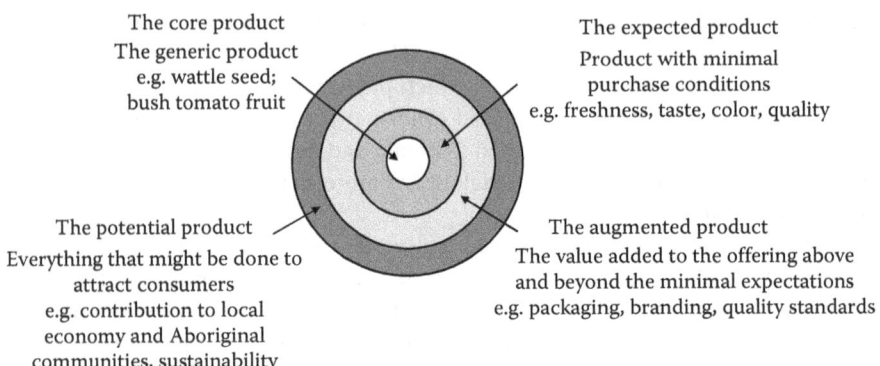

The core product
The generic product
e.g. wattle seed;
bush tomato fruit

The expected product
Product with minimal
purchase conditions
e.g. freshness, taste, color, quality

The potential product
Everything that might be done to
attract consumers
e.g. contribution to local
economy and Aboriginal
communities, sustainability

The augmented product
The value added to the offering above
and beyond the minimal expectations
e.g. packaging, branding, quality standards

Figure 24.3 Sources of consumer value in native food products. (Developed based on Levitt, T., *Harvard Bus. Rev.*, January–February, 1, 1980.)

undeveloped markets. Research that is aimed at identifying marketing opportunities or potential problems for existing or new native food products will enable companies to assess market potential so as to inform business decisions along the value chain (Lim-Camacho and Dunne, 2006; Malhotra, 2010). Marketing research can be performed at all four stages of the new product development process, from opportunity identification to development, testing and launch (Suh, 1990; Urban and Hauser, 1993; van Kleef et al., 2005). These approaches are common to new product development across a wide range of food categories, and apply equally to the introduction of new native food products in order to increase the chances of success in a market.

Marketing Research for Australian Native Foods

Marketing research for Australian native foods can have multiple objectives. As a starting point, there are three core areas to explore, as identified by Malhotra (2010): (1) who are the customer groups, (2) what is the ideal marketing mix or product offering and (3) what are the key environmental factors influencing success to market?

Customer Groups

Successful marketing of new products is not solely based on a healthy consumer demand; the supply chain actors that define the path to consumers, such as processors, wholesalers and retailers, can act as facilitators or barriers to accessing consumer demand. Identifying and understanding the wants and needs of these actors is vital to successful delivery of native food products to market. For example, the bush tomato and wattle seed supply chains are comprised of wholesalers, processors and manufacturers, distributors, restaurants and, finally, consumers (Bryceson, 2008). Each step of the chain, all the way to consumers, should be understood and explored, as each would have a varying set of needs that need to be met before they can work together to take a product through the chain to consumers.

A 'walking-the-chain' exercise of current customers or potential customers will allow suppliers to understand who the actors are and what role they will play in delivering full product value to consumers. Walking-the-chain involves desktop research to identify key players, and a primary data collection phase where suppliers and other stakeholders are recommended to be physically present to observe how the different stages of the chain operate. Such exercises also offer the potential to initially gauge consumer perceptions of the product and identify different consumer segments that may exist. Walking-the-chain is also a cost-efficient method of conducting sound marketing research using qualitative approaches. It provides detailed information that can be easily actionable (tactical approaches), as well as provide long-term direction (strategic approaches) for new product introduction.

CONTRIBUTORS TO THE PRODUCT OFFERING

There are many facets to understanding the potential of a new product offering, and being able to answer the who, how, when, why and what of the product are starting points.

1. Who are the consumers of the product? Walking the chain exercises as described earlier should enable the initial identification of potential consumer segments for the product. Further research should enable the profiling of consumers – gender, age groups, location and socio-economic status are some basic descriptors. Qualitative research in the form of focus groups and/or taste tests will allow initial exploration, while larger surveys will enable identification of the potential size of the market.
2. How and when will the product be used? Exploring this question relates to understanding what the potential consumer experience will be like – from purchasing to post-consumption. Identifying where they would purchase the product, what are the potential substitutes and what they would use it for should be explored. In the case of wattle seed, for example, consumers might expect to purchase this only at health food stores and use it primarily for baking or as a topping to breakfast cereals, as they might with chia seeds. It is worthwhile to explore *why* these are the top-of-mind uses, and what aspects of the product lead consumers to think of this. Doing so would uncover underlying product attributes that could serve as either limitations or marketing levers.
3. What core consumer needs are met by the product? Associated with the *why* questions described earlier is understanding core consumer needs. Often, these are difficult to articulate, and there are various research techniques that can be used to uncover them (projection exercises, idea generation, etc.). Understanding these core needs can be powerful levers in an age where new food products are constantly introduced. Some possible needs that can be met by native foods might be the desire to be different (to rebel against big companies, to try something exotic), desire to be healthy (as in the case of nutraceuticals) or the desire to be more 'connected' (ethical purchasing, going back to grass roots). These needs are often addressed in new products through promotion and packaging – the common pathways for communicating with consumers.

4. What are the basic requirements of market entry? It is important to identify what key customer requirements exist for chain actors to be able to take a product to the market. For native foods destined for international markets, quarantine requirements, minimum volumes and shelf life specifications are common concerns. In addition, quality standards, packaging requirements and labelling specifications are also areas that become stumbling blocks for new products if not met. A challenge for new native food products is that often, standards and requirements are not established, and companies at the forefront may have to work with various institutions to arrive at these standards. This can be both a challenge and a benefit, but it is a process that cannot be avoided.

KEY ENVIRONMENTAL FACTORS

Linked to the final point is the need to understand what external factors could influence success to market. Environmental factors are those that are beyond the control of the chain and therefore can be difficult to manage. Food safety standards, truth in labelling laws, environmental performance reporting and ethical sourcing requirements are some examples of factors that can influence success to market. Some of these regulations might be easy to adhere to, while others, such as environmental performance, might be costly to report against.

An example of such factors at play is action taken by the group People for the Ethical Treatment of Animals (PETA) to discourage the use of kangaroo skins as a leather product. This placed pressure on large sporting goods manufacturers to demonstrate the sustainability of their suppliers, with some, such as Adidas, eventually taking the decision to discontinue the use of kangaroo skins for its prestige boots.

Native foods are not immune to such pressures, and it is important to address ethical considerations while understanding their markets. In considering the implications of consumer research findings to native food value chain strategies, it is important to note the goals that may underlie commercialisation of products. It is not uncommon for native foods around the world to be developed for commercialisation with community economic participation, livelihood development and gender equity as priorities above profitability (Merne Altyerre-ipenhe (Food from the Creation time) Reference Group et al., 2011). If these are indeed priorities in particular contexts, it would be pertinent to explore the value of these objectives for markets and consumers. Likewise, exploring consumer value trade-offs for core product attributes *versus* social benefits of commercialisation would enable identification of the extent to which consumers are willing to sacrifice certain features such as consistency of quality for broader, but less visible, benefits, such as remote community livelihood development or women's empowerment.

NATIVE FOOD PRODUCT VALUE CHAINS

Mapping any supply chain network is product and company specific. However, given the diversity of native food products that are currently being marketed and have the potential to be marketed, a generic native food chain can be mapped based on previous studies. This mapping exercise provides valuable information such as the key stages, activities and players along the chain (Figures 24.4 and 24.5). Accordingly, four main types of chains could be identified: (1) harvester–trader–retailer–consumer; (2) harvester–trader–processor–retailer–consumer; (3) harvester–trader–processor–consumer and (4) harvester–processor–retailer–consumer. Traditionally, the first three chains dominated the native foods industry with traders playing a crucial role in linking harvesters with other chain members. A fourth chain that has emerged is the processor-led value chain, examples of which are the chains led by Robins Foods. This company has a direct supply chain partnership with Indigenous Australian Foods Ltd (IAF) which is owned and operated by Aboriginal communities. This is a not-for-profit company supplying native food products to Robins Foods with an endorsement that ensures authenticity of the product, integrity of the partnership and fair prices paid to Aboriginal communities. Similarly, the company has contractual arrangements with Aboriginal communities who are supported by the Outback Spirit Foundation – a public benevolent institution established by Robins Foods whose aim is to develop the livelihoods of rural Aboriginal communities. Robins Foods processes products under the 'Outback Sprit' brand, marketed exclusively through Coles, one of the biggest food retail chains in Australia (Robins, 2007). From each sale, Coles and Robins Foods make a donation of five cents each to Coles Indigenous Food Fund (CIFF), an initiative of Coles to support Aboriginal communities and to promote native foods (Brain, 2014).

More recently, a number of other initiatives have been established as harvester-led value chain models that directly link Aboriginal communities with consumers. Murri Munchies Ltd in Queensland and Aboriginal Bush Traders in the Northern Territory are not-for-profit initiatives that aim to support Aboriginal people in engaging in economic activities. They promote bush harvesting and marketing and provide members with training on entrepreneurship, business planning, marketing, quality assurance and research. Their products are sold under the brands '*Murri Munchies*' and '*Aboriginal Bush Traders*'.

The '*Mayi Harvests*' model can be seen as a variation of the aforementioned models which has emerged from a community development programme in Western Australia. This business model was developed by one of the seven Aboriginal families who are involved in a cooperative that markets Kullari plums through the Indigenous Harvest Australia Co-operative Limited (IHA) in Western Australia. Currently, *Mayi Harvests* produces high-value, gourmet native food products for restaurants, tourist outlets and health food shops and is operating with the support of the local Community Development Employment Program (CDEP).

Wild or Commercial Harvesting

Although products of native plant species are commonly collected from natural stands in the wild, there is a growing interest in cultivating commercially important crops such as bush tomato, quandong, *Acacia*, *Citrus* and muntries in horticultural enterprises. This interest is driven by the irregular supply and inconsistent quality of wild harvested products (Ryder et al., 2009). As mapped in Figure 24.5, it is

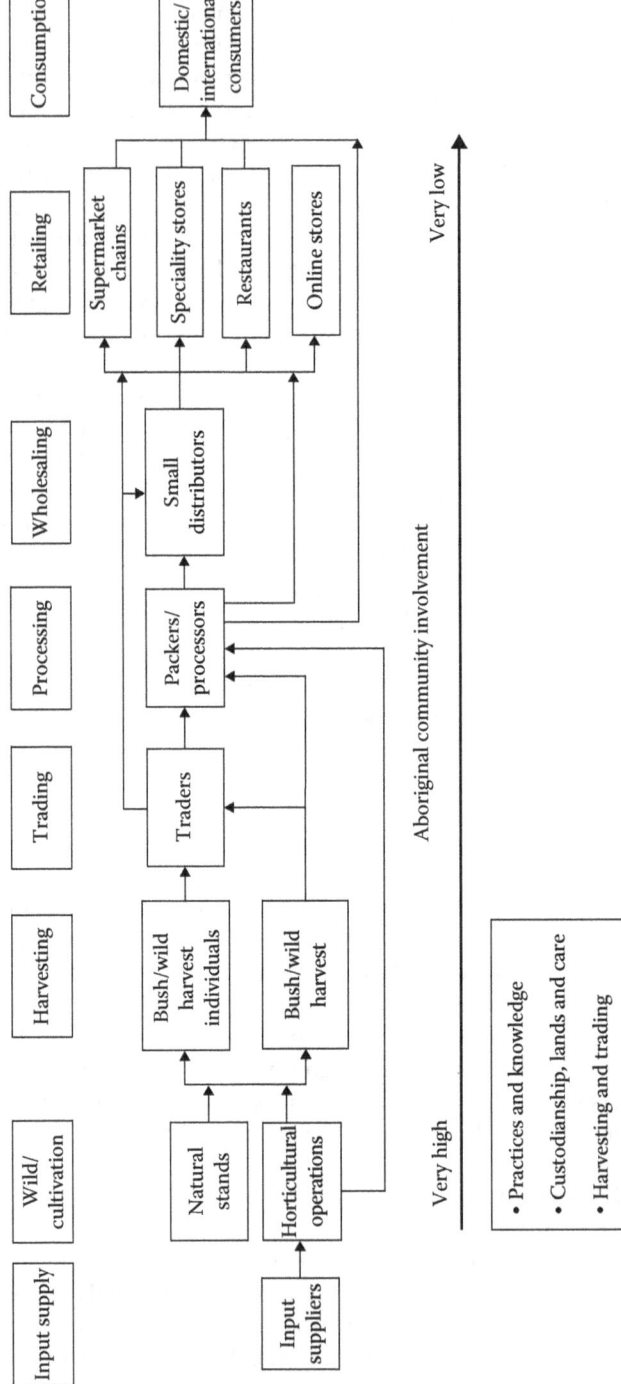

Figure 24.4 Generic native food supply chains and their links with the Aboriginal communities of Australia. (Based on Bryceson, K.P.. Value chain analysis of bush tomato and wattle seed products, DKCRC research report 40, Desert Knowledge Cooperative Research Centre, Alice Springs, Northern Territory, Australia, 2008; Merne Altyerre-ipenhe (Food from the Creation time) Reference Group et al., Aboriginal people, bush foods knowledge and products from central Australia: Ethical guidelines for commercial bush food research, industry and enterprises, DKCRC report 71, Ninti One Limited, Alice Springs, Northern Territory, Australia, 2011, and other studies on native food.)

Natural stands/cultivation	Harvesting	Trading	Processing	Retailing	Consumption
• Management of natural stands (N) • Propagation, cultivation and management of horticultural operations (V) • Burning practices and species burn responses (N) • Selective breeding and translocation (V)	• Long-distance travel to harvest (N) • Species recognition and maturity (V) • Expertise in harvesting (V) • Sorting, cleaning and drying (N) • Basic processing and preparation such as detoxification (V) • Post-harvest handling (V) • Integrating Aboriginal knowledge of bush foods management, harvest, preparation and trade (V) • Integrating Aboriginal knowledge, traditions and culture (V)	• Order and receive produce from harvesters (N) • Long-distance travel to pick up produce (N) • Species recognition by Aboriginal and English species names (V) • Clean and sterilise produce (N) • Roast and grind depending on the product (V) • Sort and store (N) • Package and label produce (V) • Supply to processors or consumers (N) • Product experimentation (V) • Monitor seeding or fruiting of produce (N)	• Sorting and cleaning (N) • Roast, grind and mill depending on the product (V) • Package and label produce (V) • Further processing depending on the product types such as chocolate, jams and sources (V)	• Storage and display (N) • Store demonstrations (N)	*Consumption related* • Palatability including flavor, aroma and texture • Appearance • Consistency • Convenience *Product* • Price • Availability • Percentage composition • Packaging • Labelling *Standards* • Nutritional information • Food safety information (e.g. HACCP) • Ethical standards (e.g. Fairtrade) • Origin • Authenticity *Others* • Cooking instructions • Uses • Traceability

Figure 24.5 Key functions and potential consumer value-adding opportunities for native food. (Developed based on Ryder, M. et al., Sustainable bush produce systems: Progress report 2004–2006, DKCRC Working Paper No. 31, Desert Knowledge CRC, Alice Springs, Northern Territory, Australia, 2009, and other studies on native food). *Note*: V, potential value-adding; N, necessary but not value-adding.

apparent that any improvement in management practices has the potential to add consumer value by improving the core product attributes of native products such as appearance, texture and palatability. It would also improve the consistency of supply, ultimately enhancing value to customers as well as to the final consumer.

Products from the wild are generally harvested by women, both Aboriginal and non-Aboriginal. Although not scientifically tested, Merne Altyerre-ipenhe (Food from the Creation time) Reference Group et al. (2011) pointed out that the quality of native food products hand-picked by Aboriginal women is extremely high compared to products harvested by non-Aboriginal women or by using mechanised techniques. As documented previously, this could be due to the integration of traditional Aboriginal knowledge and practices during harvesting, preparation and trade by Aboriginal women (Spencer and Hardie, 2011). Depending on the product, these items are subjected to sorting, drying and basic processing and then are aggregated ready for sale to traders.

By examining the possible dimensions of consumer value, it is evident that harvesting is one of the crucial stages in the chain that can add a number of potential consumer value attributes. For example, harvesters' knowledge in species recognition and maturity and basic processing techniques, such as detoxification, sorting, cleaning and drying, may contribute significantly to the product's flavour, aroma, palatability and texture. Similarly, assurance of authenticity and origin, and providing a fair share to the Aboriginal people for their traditional knowledge, traditions and culture during harvesting, would provide higher-order consumer value attributes.

Trading

Traders generally travel long distances to meet with Aboriginal communities to collect native food products. These usually small-scale operators play a significant role in linking harvesters with the rest of the chain (Cleary, 2012; Ryder et al., 2009). Based on a study conducted in Central Australia, Ryder et al. (2009) highlighted that these traders not only act as consolidators of native food products but also extend their services as merchants and network facilitators to support people and hence, they suggest that these merchants be called traders, not wholesalers. Traders usually pay cash for the raw materials collected. They also perform key activities such as species recognition and naming, cleaning, sorting, packaging and basic labelling upon consolidation. As highlighted in Figure 24.5, these activities contribute strongly to expected consumer value attributes. Value-added products in bulk are then sold to different channels depending on demand. Cleary (2012), by applying net chain analysis to the native foods industry, highlights that value associated with traders in developing social relationships with harvesters goes beyond simply moving products to other chain members.

Processing

Except for a small amount of fresh produce channelled to restaurants, the bulk of native food products are dried, frozen or further processed, often in combination with non-native food ingredients (RIRDC, 2008). The number of processors involved in the native foods industry is limited. Currently, processors such as Robins Foods as well as Oz Tukka and Kurrajong Native Foods produce a wide range of value-added products including jams, sauces, chutneys and syrups for both domestic and international markets. They have their own input supply networks and distributions channels and market processed native food products under their own brands. Bryceson (2008) provides a detailed description of two dominant chains with respect to bush tomato and wattle seeds with links to Australia's two major supermarket chains, Coles' Indigenous Food Fund and Woolworths' Outback Café series. As highlighted in Figure 24.5, the processing stage is crucial in augmenting core product attributes and providing consumer value attributes related to convenience, nutrition, food safety and compliance with ethical production standards.

WHOLESALE/RETAIL AND CONSUMPTION

Native food products are available through some supermarkets, speciality stores, online and some food service businesses that specialise in uniquely Australian food. Limited availability coupled with poor consumer awareness on uses and nutritional value has hindered the development of a domestic market for native food products. Ahmed and Johnson (2000) highlight the value of in-store demonstrations as a mechanism for developing the industry. Similarly, by studying the food service and manufacturing sectors of the native food industry, Cherikoff (2000) concludes that chefs and food media can play a significant role in promoting native food consumption. His argument is to position the native foods industry in the minds of consumers as related to a uniquely Australian dining experience coupled with prestige, modernity, food interest and flavour appeal. It is well accepted that the development of stronger domestic food markets is a stepping-stone to developing more competitive industries (Porter, 1998).

MANAGING NATIVE FOOD PRODUCT VALUE CHAINS

Growing consumer appeal for natural and nutritionally rich food supplements has created a potentially vibrant market for native food and potential to provide economic benefits to Aboriginal communities in Australia. To achieve these results, it is vital to link the two ends of the chain with a view to providing consumer value attributes as efficiently as possible. Responding to market signals or creating consumer value is a chain-wide task that cannot be achieved by businesses in isolation. Both upstream and downstream players must work together to efficiently deliver value to consumers. Thus, having chain management competencies that leverage chain members' capabilities is the key to achieve sustainable competitive advantage (Spekman et al., 2002).

While this kind of value creation improves the performance of all chain players and enhances each company's competitive position, Porter and Kramer (2011) argue that the value thus generated also needs to be shared across the broader communities within which a chain operates. This argument applies well to the native foods industry where there is a significantly greater involvement of Aboriginal communities, their traditions, knowledge and custodianship of land in the upstream part of the chain, while the downstream elements are dominated by non-Aboriginal businesses (Bryceson, 2008; Merne Altyerre-ipenhe (Food from the Creation time) Reference Group et al., 2011). Consistent with this view, Lee (2012) and Spencer and Hardie (2011) point out that the development of the native foods industry should take place so as to safeguard the hereditary rights of traditional custodians while ensuring equitable benefits to traditional owners for their intellectual property. According to Porter and Kramer (2011), government could take a leading role in this process by facilitating and encouraging companies to share the value they create across the broader society. Although processor-led chains such as Robins Foods have already voluntarily shared this value with Aboriginal communities through their philanthropic initiatives, Spencer and Hardie (2011) argue for a formalised indigenous certification and labelling programme that would help benefit the Aboriginal population who are the custodians of land, traditional knowledge and traditional practices. At the same time, there is a growing interest in cultivating native plant species in horticultural enterprises so as to increase production and improve reliability of supply. Although these strategies would add value to downstream players, limited engagement with traditions and lower levels of involvement could potentially reduce the benefits flowing back to Aboriginal communities. Cunningham (2009) pointed out that horticultural production of muntries, mountain pepper and quandong has generated more benefits to non-Aboriginal producers than Aboriginal communities. This clearly highlights the challenge of integrating Aboriginal communities in the commercialisation of native food species so that they receive an equitable share of returns for their expertise, cultural knowledge, values and traditions.

As highlighted in Figure 24.5, there are multiple dimensions of consumer value that could be created along a native food value chain. However, to capture this value, the native foods industry needs greater knowledge of its existing and potential consumers, and 'chain champions' who can drive entrepreneurial initiatives using a value chain framework – or 'value chain thinking'. With few exceptions, native food chains are currently fragmented and based on arm's length transaction-by-transaction relationships. In this environment, small operators such as harvesters compete against each other, price is king and information is withheld as a source of power. The key to building stronger value chains lies in strengthening the capacity of harvesters to link with each other and with other chain members. Clusters of harvesters, for example, could share knowledge and technology, accumulate supplies of products, work to agreed quality standards and negotiate with buyers who are interested in higher volumes and values of product. Such clusters have a poor track record, because many fail to achieve the levels of trust and commitment required for effective collective action. Trust plays a crucial role in developing value chain relationships, so it is important to have governance mechanisms in place that enable trust-building (Morgan and Hunt, 1994) and help to overcome difficulties associated with power and conflict (Mentzer et al., 2000). Ultimately, the level of trust and commitment in a value chain directly influences its ability to create value and distribute it equitably. Native food value chains based on effective relationships can deliver what consumers want and will pay for, while at the same time distributing returns equitably. For this way of doing business to become the norm, peak industry bodies such as the Australian Native Food Industry Limited (ANFIL) have an important role to play. A first step would be to develop and implement native food certification schemes that authenticate the value associated with Aboriginal knowledge and cultural heritage.

REFERENCES

Ahmed, A.K., Johnson, K.A., 2000. Horticultural development of Australian native edible plants. *Australian Journal of Botany* 48, 417–426.

Anderson, D.L., Britt, F.F., Favre, D.J., 2007. The best of supply chain management review: The 7 principles of supply chain management. *Supply Chain Management Review* 11, 57.

Barney, J.B., 1991. Firm resources and sustained competitive advantage. *Journal of Management* 17, 99–120.

Boehlje, M., 1999. Structural changes in the agricultural industries: How do we measure, analyze and understand them? *American Journal of Agricultural Economics* 81, 1028–1041.

Bonney, L., Clark, R., Collins, R., Fearne, A., 2007. From serendipity to sustainable competitive advantage: Insights from Houston's Farm and their journey of co-innovation. *Supply Chain Management* 12, 395–399.

Brain, C., 2014. Supermarket grows support for indigenous bush food producers. http://www.abc.net.au/news/2014–05–23/coles-indigenous-food-fund/5462888 (accessed February 27, 2015).

Braziotis, C., Bourlakis, M., Rogers, H., Tannock, J., 2013. Supply chains and supply networks: Distinctions and overlaps. *Supply Chain Management: An International Journal* 18, 644–652.

Bryceson, K.P., 2008. Value chain analysis of bush tomato and wattle seed products. DKCRC research report 40. Desert Knowledge Cooperative Research Centre, Alice Springs, Northern Territory, Australia.

Cherikoff, V., 2000. *Marketing the Australian Native Food Industry*. Rural Industries Research and Development Corporation, RIRDC Publication No. 00/61, Canberra, Australian Capital Territory, Australia.

Cleary, J., 2012. Business exchanges in the Australian desert: It's about more than the money. *Journal of Rural and Community Development* 7, 1–15.

Cunningham, A.B., Garnett, S.T., Gorman, J., 2009. Policy lessons from practice: Australian bush products for commercial markets. *GeoJournal* 74, 429–440.

Dyer, H.J., 2000. *Collaborative Advantage: Winning through Extended Enterprise Supplier Networks*. Oxford University Press, New York.

Fearne, A., Martinez, M.C., Dent, B., 2012. Dimensions of sustainable value chains: Implications for value chain analysis. *Supply Chain Management: An International Journal* 17, 575–581.

Hines, P., Rich, N., 1997. The seven value stream mapping tools. *International Journal of Operations and Production Management* 7, 46–64.

Konczak, I., Roulle, P., 2011. Nutritional properties of commercially grown native Australian fruits: Lipophilic antioxidants and minerals. *Food Research International* 44, 2339–2344.

Kotler, P., Kartajaya, H., Setiawan, I., 2010. *Marketing 3.0: From Products to Customers to the Human Spirit*. Wiley & Sons, Inc., Hoboken, NJ.

Lambert, D.M., Cooper, M.C., Pagh, J.D., 1998. Supply chain management: Implementation issues and research opportunities. *The International Journal of Logistics Management* 9, 1–20.

Lee, L.S., 2012. Horticultural development of bush food plants and rights of Indigenous people as traditional custodians – The Australian Bush Tomato (*Solanum centrale*) example: A review. *The Rangeland Journal* 34, 359–373.

Levitt, T., 1980. Marketing success through differentiation – Of anything. *Harvard Business Review* January–February, 1–10.

Lim-Camacho, L., Dunne, A., 2006. Market research methodologies for Australian native cutflowers: The case of Grevillea sp. varieties to Japan. In: *Seventh Australian Native Flower Conference*, Bardon, Queensland, Australia.

Lusch, R.F., Vargo, S.L., Tanniru, M., 2010. Service, value networks and learning. *Journal of the Academy of Marketing Science* 38, 19–31.

Malhotra, N.K., 2010. *Marketing Research: An Applied Orientation*. Pearson Education, Upper Saddle River, NJ.

Mentzer, J.T., Min, M., Zacharia, Z.G., 2000. The nature of interfirm partnering in supply chain management. *Journal of Retailing* 76, 549–568.

Merne Altyerre-ipenhe (Food from the Creation time) Reference Group, Douglas, J., Walsh, F., 2011. Aboriginal people, bush foods knowledge and products from central Australia: Ethical guidelines for commercial bush food research, industry and enterprises. DKCRC report 71. Ninti One Limited, Alice Springs, Northern Territory, Australia.

Morgan, R.M., Hunt, S.D., 1994. The commitment-trust theory of relationship marketing. *Journal of Marketing* 58, 20–38.

Morse, J., 2005. Bush resources: Opportunities for aboriginal enterprise in central Australia. Desert Knowledge Cooperative Research Centre report, Alice Springs, Northern Territory, Australia.

Netzel, M., Netzel, G., Tian, Q., Schwartz, S., Konczak, I., 2007. Native Australian fruits – A novel source of antioxidants for food. *Innovative Food Science and Emerging Technologies* 8, 339–346.

Porter, M.E., 1985. *Competitive Advantage: Creating and Sustaining Superior Performance*. The Free Press, New York.

Porter, M.E., 1998. *The Competitive Advantage of Nations: With a New Introduction*. Macmillan Press Ltd, London, U.K.

Porter, M.E, Kramer, M.R., 2011. Creating shared value. *Harvard Business Review* 89, 62–77.

Priem, R.L., 2007. Consumer perspective on value creation. *The Academy of Management Review* 32, 219–235.

RIRDC, 2008. *Native Foods R&D Priorities and Strategies 2007–2012*. Rural Industries Research and Development Corporation, Canberra, Australian Capital Territory, Australia.

RIRDC, 2013. *Annual Report 2012–2013*. Rural Industries Research and Development Cooperation, Canberra, Australian Capital Territory, Australia.

Robins, L., 2007. *Outback Spirit Bush Foods: A Learning Model in Marketing and Supply Chain Management*. Rural Industries Research and Development Corporation, RIRDC Publication No. 07/037, Canberra, Australian Capital Territory, Australia.

Ryder, M., Walsh, F., Douglas J. et al., 2009. Sustainable bush produce systems: Progress report 2004–2006. DKCRC Working Paper No. 31. Desert Knowledge CRC, Alice Springs, Northern Territory, Australia.

Sommano, S., Caffin, N., McDonald, J., Cocksedge, R., 2011. Food safety and standard of Australian native plants. *Quality Assurance and Safety of Crops and Foods* 3,76–18.

Spekman, R.E., Spear, J., Kamauff, J., 2002. Supply chain competency: Learning as a key component. *Supply Chain Management: An International Journal* 7, 41–55.

Spencer, M., Hardie, J., 2011. *Indigenous Fair Trade in Australia – Scoping Study.* Rural Industries Research and Development Corporation, RIRDC Publication No. 10/172, Canberra, Australian Capital Territory, Australia.

Suh, N.P., 1990. *The Principles of Design.* Oxford University Press, New York.

Taylor, D.H., 2005. Value chain analysis: An approach to supply chain improvement in agri-food chains. *International Journal of Physical Distribution and Logistics Management* 35, 744–761.

Urban, G.L., Hauser, J.R., 1993. *Design and Marketing of New Products.* Prentice–Hall, Englewood Cliffs, NJ.

van Kleef, E., van Trijp, H.C.M., Luning, P., 2005. Consumer research in the early stages of new product development: A critical review of methods and techniques. *Food Quality and Preference* 16, 181–201.

CHAPTER 25

New Market Opportunities for Australian Indigenous Food Plants

Vic Cherikoff

CONTENTS

A SHORT HISTORY OF THE MODERN USE OF WILD FOODS

Australian wild foods are an opportunity waiting to happen, though ironic that even in the middle of 2015, the indigenous foods and medicines that sustained the world's longest living culture for over 60,000 years are almost the final frontier of plant-based resources. My work on the nutritional value of Australian Aboriginal Bushfoods at the Human Nutrition Unit, University of Sydney, since the early 1980s and my concurrent commercial activities as a pioneer in the development of indigenous plants as foods, flavours, nutritionals, antimicrobials, medicines, cosmetic actives and other functional applications are proofs of its potential.

At the time, my view was that in 3–5 years, we should see a population of creative chefs designing innovative food concepts that would rock food culture around the world. Australian cuisine would impact culinary schools and feature on menus from takeaways to fine dining establishments, and chefs would blossom into professionals showcasing our unique flavours to their peers, patrons and food critics.

This array of a few dozen deliciously aromatic herbs, pungent spices, uniquely interesting fruits, low fat nuts, delicious grains and even some food adjuncts (e.g. paperbark) and infused smoke and flavoured oils would be assimilated by thousands of chefs. Surely more innovative cooks and creative food technologists would experiment with these new ingredients and adopt and adapt them for new and highly marketable flavour variants. An Australian cuisine would evolve and become well established in just a few years and spread internationally in no more than a decade...

But no. It didn't happen.

Today, more than 30 years have passed after the commercialisation of the initial launch of 24 wild foods in 1980, but there is still only a murmur of interest in an authentically Australian cuisine. Our country is still deeply entrenched in 'MediterrAsian' food styles with a strong representation of U.S. fast food and drive-throughs, a snapshot of Indian and Mexican, a rare taste of North African and perhaps a handful of some other ethnic food styles, mostly confined to our major cities. Out in rural centres more than 3 hours drive from any metropolis and the food is still typically trapped in the 1960s instant 'coffee' reigns supreme and not a cappuccino, latté or macchiato in sight.

How similar is this to the classic story of the two shoe salesmen coming home from a market assessment in Africa as they report their findings? One says: forget the African market. No one wears any shoes. The other is excitedly optimistic and says, 'Quick, the market is huge. No one has any shoes'.

We do need to be grateful that there is still a wild food resource to offer contemporary Australians. It is only due to the careful, devotional and expert management of the land by each of the 600 clans who made up the World's longest living culture that we still have a remnant of the wild foods this land once supported. Interestingly, this 'residual' resource is a valuable window into the nutrition of this most ancient culture and provides clues to our future survival.

MOVING FORWARD

There are numerous opportunities for wild foods to make their mark as the essential elements of an authentic Australian food style that spreads globally. Two approaches are evolving: first, foraging chefs present a menu that changes daily and features highly seasonal and local ingredients. Many wild foods would need to be enhanced for supply to meet the demand with appropriate management for sustainability and expanding harvest.

The second path is already more developed where indigenous ingredients are marketed through mainstream and specialty food distributors for chefs across the food industry including takeaway food outlets, food trucks and truck stops, clubs and pubs, restaurants and caterers from private chefs to venue and event caterers, institutional caterers (schools, military establishments, airlines, cruise ships, railways, hospitals, hospices, aged care facilities) and so on.

Both of these approaches require better resource management than has occurred to date in Australia where we have one of the worst records for species extinction of any country. Perhaps a growing demand for wild foods could reinstate the indigenous Australian environmental micro-management and in doing so, create a new opportunity for Australian biota.

It is becoming more widely accepted that indigenous Australians micro-managed every piece of this country. They turned the soil-harvesting tuberous plants; protected some ecosystems from disturbance; or appropriately in frequency and intensity, burned their traditional lands in mosaic patterns. This technique of firestick farming not just facilitated the mineralisation/mobilisation of nutrients ensuring healthy plant communities and soil microflora but also maximised the productivity for both humans as well as the animals and birds they hunted for food or held as totems back through time.

In some low-lying areas, channels were worked to inundate expansive flat country resulting in sufficiently large productions of 'crop' species such that regular, large gatherings of people from many clans were possible and the social and technical exchanges accelerated the spread of innovations and discoveries.

The harvests of key species such as *Portulaca oleracea* (pigweed), *Eragrostis eriopoda* (love grass) or *Microlena stipoides* (Alpine meadow rice) or even *Marsilea drummondii* (nardoo) were clearly, managed productions.

Other species were natural bonanzas and were harvested over wide districts including the yam daisy (*Microseris scapigera*), vanilla and chocolate lilies (*Anthropodium milleflorum* and *A. stricta*, respectively) and various ground orchids (e.g. *Pterostylus nutans*). Numerous grass and shrub seeds can be included in this group of bountiful harvest foods. These were only possible because of the land management practices used. For example, the physical process of harvesting with some seed or tubers left for regrowth, turning and mulching the soil, fire regimes and fallow periods.

There is little doubt that had the invading British arrived with a more noble cause than to dump their convict criminals as far from British soil as possible, Australia may have progressed along a better path. Had the first explorers chosen to observe, learn, study and adopt the methods of plant and land resource management of the world's longest living culture, Australia may have been in a far different place now, agriculturally. Instead, we rely on a handful of species to provide for our failing nutrition. We consume one-tenth of the number of foods than did Australian hunter-gatherers and at less than one-sixth the nutritional quality and falling annually.

We suffer from soil erosion and fertility declines in impoverished ecosystems that are no longer environmentally sustainable over the long term. Biodiversity has always been the key to survival and not only have modern cultures depleted the original biological reserves but we have disrupted entire eco-systems and micro-climates compromising the resilience of natural processes to recover.

A LIGHT AT THE END OF THE TUNNEL

As a way to redress our position, I would like to present some other areas of potential for our natural resources as functional ingredients in the food and other industries and will touch on some traditional Indigenous remedies which fall in the divide between food and medicine. Some threats will also be addressed.

But before this, another side of the production/economics equation also exists and this embraces a wide range of opportunities for our wild food resources.

SOCIAL GOALS

Foraging food in a sustainable way or managing existing stands for a more co-ordinated production or setting up enriched production areas along traditional forage-farming practices can provide a host of tangible and intangible benefits.

Indigenous and non-indigenous community's involvement in production and supply or that of individuals wild harvesting within or across traditional Aboriginal lands are areas of opportunity.

There is a significant social contribution from the management, harvesting and renewed consumption of foraged foods in the community. It rekindles the ties to country and wild foods contribute valuable nutrients that are increasingly absent from 'agricultured' foods. Economically, the wider picture might expand nursery offerings beyond the ubiquitous azaleas, agapanthus and other inappropriate weeds and lead to more interesting, productive and biologically diverse indigenous Australian landscapes in domestic urban and community gardens. These might even create habitats for indigenous reptiles and birds rather than introduced snails and pests.

Nursery plants of native edibles for urban gardens have been attempted before and proved a challenge in the past when a South Australian firm unsuccessfully attempted to release a range of wild food plants to the nursery market. Problems of distribution thwarted their plans and a lack of public interest or poor marketing failed to motivate householders to embrace their botanic offerings. Admittedly, they also included several hybridised limes (branded as blood limes and sunrise limes) which could not be called wild foods for those passionate about them and were more horticultural curiosities from a now defunct, CSIRO cross-breeding programme.

Times change and perhaps there are more foraging gardeners (there are certainly more foraging chefs) today that might make this opportunity more successful particularly if backed up with one of those popular backyard makeover or gardening shows on television. The opportunity for wild food plants in another trend, that of cityscapes of vertical gardens is exciting with many species coming to mind such as warrigal greens (*Tetragonia tetragonoides*), midyim (*Austromyrtus dulcis*), scrambling lily (*Geitonoplesium cymosum*), zigzag vine (*Melodorum leichhardtii*), munthari (*Kunzea pomifera*) and licorice leaf (*Smilax glyciphylla*), to name just a few with food or foraging potential.

The development of this range of species might urge research into some plants that need to be understood better in terms of potential food value. For example, Daintree nut (*Omphalea queenslandii*) has a golf ball–sized nut with an eggshell-thin shell and a nut with a flavour blend of macadamia, cashew and chestnut. The downside is that some people find it a laxative. There are interesting parallels with cashew nuts which contain irritant compounds that necessitate machine processing rather than hand preparation to render the nuts edible and not highly allergenic. A simple investigation of the laxative agent(s) in Daintree nuts and strategies to remove or inactivate them (appropriate roasting) might yield a new nut that could match the macadamia in economic value. The vine is extremely hardy with a single specimen in its natural habitat of wetlands south of Cairns, spreading over 3 ha. Even vines introduced to a Sydney garden adapted to a wild food garden setting, producing an annual bounty of cannonball-sized fruits containing from one to four nuts each. The vigorous habit necessitates regular pruning as the plant's growth rate is somewhat akin to Jack's beanstalk, and there is a possibility that the vine could also be cut and used as canes for furniture and other items.

ECONOMICS OF PRODUCTION

The challenge of developing new food species in Australia is the cost of land and that for labour and post-harvest freight for plantation production. Even for foods that are wild harvested from remote regions, there is the cost of labour and transport of the resources pre-harvest, then freight post harvest. Wild plants in this country are also adapted to our ancient soils and many produce heavier fruit crops every second or third year with local climatic conditions and global warming being other disruptions to regular production.

Even in good seasons, fruits often ripen over far wider windows than conventional produce requiring the extended picking times and storage of regular smaller harvests, adding more cost in labour, refrigeration and fuel. However, rather than promoting or inducing fruit set with chemical (e.g. potassium nitrate sprays on mangoes) or physical methods (e.g. tree shakers used to increase nut production of pecans, walnuts and other species), wild characteristics should be maintained where nutritional value is the prime market consideration. This may not be as important if new foods are to simply be produced as indigenous Australian alternatives to the nutritionally compromised produce we eat today. It is possible to breed bigger and sweeter variants of the wild fruits with concomitantly less nutritional density and flavour.

However, it must be appreciated that a global opportunity exists for the nutritional contribution that wild foods offer for our on-going health and well-being.

The simple, regular addition of a number of wild foods to the diet can be enormously beneficial.

NEW USES FOR EXISTING PRODUCTS

Developing secondary products and the extraction of functional elements from by-products of the main harvest of food plants is an important strategy to shift the economics of wild food production.

Quandong

For example, the use of quandong seeds supplements the returns on fruit harvests and offers multiple opportunities. The kernels can be extracted from the crushed seeds and a product made which reflects the traditional use of the kernel oils as anti-itch treatment for scalp conditions. Santalbic acid is a cyclic fatty acid in the kernel oil which appears to have anti-prostaglandin activity, suggesting a role in hair and skin care preparations. Last, the spent shells can be milled as an abrasive inclusion (exfoliant) in soaps and scrubs. As with the Daintree nut, further research might reveal the way the quandong kernel could be treated (a controlled roasting again) to break the cyclic fatty acid and render the roasted kernel edible. Years ago, this was being prepared and used as a thin layer in a specialty ice cream to good effect. The toasted kernel is deliciously rich in Maillard products and aromatics.

Kalari or Kakadu Plum

The Kalari or Kakadu plum is world-famous following the discovery and proof of its vitamin C content during my early work at the Human Nutrition Unit, University of Sydney (Cherikoff, 2000). The seeds of the fruits are treated in a similar way as the seeds of quandong. Additionally, extracts from the plum are in the process of being integrated into packaging to extend the shelf life in products such as farmed prawns, other seafood and meat. This species has been the target of U.S. patent attempts with corporations such as Mary Kay attempting to control the use of the plum in their highly competitive market. The United States allows natural extracts and processes to be patented even when there is nothing new or novel in the patent. International objection and lobbying thwarted the validity of the anti-competitive action and for the time being, at least, the Kalari or Kakadu plum is still a free for all. It does remain an area of vigilance for anyone investing in the opportunities that wild plants offer.

Illawarra Plum

Research at the CSIRO Nutrition Unit in Adelaide some years ago looked at the Illawarra plum as a potential weight loss food (pers. Comm.). Data discussed at a conference on nutrition and innovation suggested that the Illawarra plum contained an as-yet unidentified component(s) which inhibited the development of beta-adipose cells (immature fat storage cells) to mature functional storage tissue. The *in vitro* studies pose an opportunity for more research into the burgeoning field (pun intended) of obesity control and healthy diets.

Unpublished graduate work at the University of Sydney around 2010 looked at the lethal effect of indigenous plant extracts on tissue cultures of various cancer cell lines. Many herbs, spice and other food species yielded extracts which killed the cancer cells in less than an hour from administration, suggesting more work is needed to understand the effect better. It also underpins the absence of many cancers in indigenous Australians in traditional times despite the longevity of the elders who lived well into their 70s and 80s.

Acacia (Wattle Seed)

Several promising extracts come from the seed pod, stem and leaf coating of dryland *Acacia* species. *A. tenuissima* yields a powerful anti-inflammatory resin which is still used in the bush and has a 6000-year history at least. *A. victoriae* contains a well-researched (Gutterman et al., 2005) collection of avrins and avrini-cins which have been shown to be powerfully anti-cancer. A suggested approach for the commercialisation of these functional compounds is the bulk tissue culture of non-differentiated cells in tanks rather than sourcing extractable material from field crops. However, where the same species are grown for food, the resin may be a valu-able on-farm production as a by-product.

Additionally, *A. victoriae* also produces a gum exudate which is very similar in properties to gum Arabic (*Senegalia senegal* – formerly *Acacia senegal*). Wild harvesting of the gum in Australia may not be economically feasible but plantation trees might be utilised, even if only when seed production falls as the relatively short-lived trees reach maturity and prior to replacement planting.

An unexplored by-product from a range of *Acacia* species are the arils. These are fleshy structures which attach the seeds to the pods and many are highly coloured. In fact, it is known that red arils are generally harvested by ants which are the main seed spread vectors for these *Acacia*, whereas yellow to orange arilate seeds attract birds which then spread the seeds. Many arils are highly fragrant and some were used as food adjuncts. For example, the seeds of *A. coriacea*, *A. oswaldii*, *A. tenuissima* and *A. Adsurgens*, each with their arils attached, were worked in water. This imparted a rich characteristic flavour to the water and still yielded the seeds which were heat-parched and dry-milled to a seed cake flour. Analyses (Brown et al., 1987) showed that the arils are high in polyunsaturated oils and the colour comes from beta-carotenes. The long stability of the arils over time suggests that vitamin E activity is marked and this means a range of antioxidants are in play.

While *Acacia* arils may be an expensive commodity due to their low weight, their functional value may be equal to saffron and their role in particular disease states is yet to be determined. Saffron is now famous for its preventative action in macular degeneration and *Acacia* arils may be similarly effective in this condition or others. More work is needed.

MICRO-SUGARS

An emerging area of research in the health and wellness industries is that of glycation and its regulation. This is clearly relevant as every one of our trillions of cells and those of the bacteria in our gut is coated with the peachy fuzz that is the product of sugars with other sugars, proteins or lipids. This layer is involved in how cells communicate with each other and with their environment and is crucial in the development of disease and ageing and in the maintenance of health.

One particular sugar that has only recently entered the food industry as a (good) sugar is trehalose. This non-reducing, di-glucose sugar has a host of seemingly 'magic' properties (Cherikoff, 2015), the most functional being the ability to preserve proteins and lipids from damage in a number of situations. The most obvious and the phenomenon that led to the common name of Resurrection Sugar is the water-holding capacity of this uniquely structured disaccharide. As a plant rich in trehalose desiccates, the trehalose forms a glass (a gel) of a sugar and water matrix which protects surrounding structural proteins and lipids. On rehydration, the plant is able to recommence photosynthesis often within 48 hours imparting a significant evolutionary advantage for a dryland plant species.

Trehalose is being shown to be far more ubiquitous than previously assumed and although the work has not yet been done, Australian flora are expected to be rich in trehalose with even many rainforest species able to withstand prolonged drought conditions. Trehalose-rich foods are a valuable addition to a healthy diet and the essential nature of the wide range of micro-sugars is only now being appreciated as we saturate our diet with sucrose and begin to learn why fructose deserves the title, bad sugar (cf. bad fats). Micro-sugars are associated with many antioxidants as structural components and as we eliminate these vital functional ingredients through breeding fruits and vegetables low in antioxidants, we reduce our intake of micro-sugars. Micro-sugars may possibly be as important nutritionally, as any of the alphabet vitamins.

FOOD AND COSMETIC SCIENCE AND TECHNOLOGY

Secondary products from a host of food species can yield a wide variety of functional ingredients for food, skin and hair care and a range of other industries. These products include cosmetic extracts and inclusions, ingredient extenders and flavour enhancers, flavours, fragrances and extracts, emulsifiers, colours, thickeners, waxes, dietary fibres, antimicrobials, antioxidants, anti-inflammatories, micro-sugars, adaptogens and enzyme regulators.

Essential oils distilled from eucalypts, tea trees, citrus and other indigenous Australian species were recognised as valuable since the days of the white invasion and are already widely used in cosmetics, perfume and food applications. There is a mix of plantation and foraged source materials.

THERAPEUTICS IN CONVENTIONAL AND
ALTERNATIVE MEDICINES

This future opportunity for Australian wild foods is for those with a long-term vision with new compounds taking the path of tests, refinements, trials (from bio-modelling to human trials but hopefully with minimal or no animal testing). The range here is huge, including anti-cancer and anti-tumour agents, adaptogens, anti-septics, anti-hypertensives and functionals in medical devices (e.g. actives in personal lubricants, toothpaste additives, eye moisturisers and anti-fungals).

The downside of prospecting for therapeutics is the enormous and usually prohibitive cost of development, the competition from the pharmaceutical industry and the challenges of patenting natural product derivatives. This is the reason behind the strategy of using the actives as cosmetic ingredients where they generally get listed as fragrances (parfum). This sneaks them under the radar and creates a cosmetic range that can be more than just a skin cream or liquid full of synthetic chemicals that are biologically inert, poorly absorbed or, in some cases, even deleterious.

Some warnings need to be noted. For example, lemon myrtle cannot be recommended in skin care products where exposure to sunlight can occur as lemon myrtle can induce photosensitisation and a painful, recurring, persistent allergy-like reaction can be debilitating once contracted.

MARKETING AND MORE

Marketing of new indigenous-inspired products is an opportunity for remote communities to also license art in label designs or be pro-actively associated with the branding and supply of raw materials. Aspects of traditional resource management are marketable promotional values and more products need to be presented with this as a major benefit to the end user.

Government organisations are often directed to provide indigenous Australian businesses the opportunity to tender for a significant proportion of the supplies and services they need annually and some corporations have a similar ethic. More can be made of this in marketing with the providers promoting the mutual association.

An opportunity also exists to re-introduce the traditional forage-farming methods of some wild foods (Gammage, 2012). This was a sustainable production system in the true sense of the words. Numerous species were enriched through managed annual harvests, water farming, maintaining healthy soils and appropriate fire regimes while encouraging multi-purpose, resilient mixed eco-systems. Different clans maintained productive seral stages in selected habitats around their traditional Country, maximising the food availability, range and abundance. Some Australian National Parks are attempting to utilise the fire management practices guided by local indigenous elders but much of the country continues to be mismanaged in an attempt to mitigate the intensity of wildfires rather than improve the health of the land.

Perhaps an increasing demand for wild foods, medicines and other resources might lead to forage-farming projects giving benefits that may spill over to conventional agriculture.

Last, too few non-indigenous Australians include any original Australians in their immediate circle of friends. An opportunity exists to find a value-niche and trade on one's Aboriginality as we still have a lot to learn from the World's longest living culture.

REFERENCES

Brown, A.J., Cherikoff, V., Roberts, D.C.K., 1987. Fatty acid composition of seeds from Australian Acacia species. *Lipids* 22, 490–494.

Cherikoff, V., 2000. *Marketing the Australian Native Food Industry.* Rural Industries Research and Development Corporation, RIRDC Publication No. 00/61, Canberra, Australian Capital Territory, Australia.

Cherikoff, V., 2015. Resurrection sugar. http://cherikoff.net/wp-content/uploads/Resurrection-Sugar1.pdf (accessed January 2015).

Gammage, B., 2012. *The Biggest Estate on Earth: How Aborigines Made Australia.* Allen & Unwin, Sydney, New South Wales, Australia.

Gutterman, J.G., Lai, H.T., Yang, P., Haridas, V., Gaikwad, A., Marcus, S., 2005. Effects of the tumor inhibitory triterpenoid avicin G on cell integrity, cytokinesis, and protein ubiquitination in fission yeast. *Proceedings of the National Academy of Sciences of the United States of America* 102(36), 12771–12776 (accessed October 2008).

Appendix: Australian Native Food Recipes

Jude Mayall

Today there is a growing excitement about our native food. These easy-to-follow recipes are a great way to start your 'bushfood' journey, I hope you enjoy them.

BUSH TOMATO PASTA SAUCE

This is a basic sauce recipe, serve with your favourite pasta or to accompany any meat dish.

Ingredients

- 5 large fresh ripe tomatoes, chopped into small chunks
- 1 brown onion and finely chopped garlic
- 2 dessertspoons of ground bush tomato
- 1 dessertspoon of tomato paste
- Mountain pepper to taste
- Murray River salt to taste
- Fresh basil
- Oil

Method

1. In a frying pan, heat the olive oil and then add the brown onion.
2. Add the chopped fresh tomatoes.
3. Add the crushed or chopped garlic.
4. Add the tomato paste.
5. Cook, stirring frequently, until the mixture is soft and pureed.
6. Add the bush tomato; this will thicken the sauce.
7. Add some stock or water if needed.
8. Add the mountain pepper and the Murray River salt to taste.
9. Add the finely chopped fresh basil.
10. Keep cooking until the desired consistency is achieved.
11. If you want a thinner sauce, add more stock.
12. For a thicker sauce, keep on heat until the sauce reduces and thickens.

Optional Additions

- Olives
- Finely chopped ham
- Grated zucchini
- Grated carrot

BUSH TOMATO PESTO

This quick and easy pesto is full of flavour that gives a multitude of options: tossed with pasta, added as toppings in fish or chicken recipes or just enjoyed on bread sticks.

Ingredients

- ½ cup of roasted macadamia nuts
- ½ cup of parmesan cheese
- ½ cup of Australian extra virgin olive oil
- 1 cup of fresh chopped basil leaves
- 1 cup of fresh chopped parsley
- 1 teaspoon of bush tomatoes
- A good grind of pepper berries

Method

1. Place the nuts, parmesan cheese and ¼ cup of oil in a blender and puree until smooth.
2. Add the remaining ingredients and blend to a chunky paste; add more oil if needed.
3. Serve on bread sticks.

DAVIDSON PLUM ICE CREAM

Ingredients

- 1 cup of milk
- 1 cup of cream
- 1 cup of Davidson Plum puree
- ¾ cup of sugar
- 5 egg yolks

Method

1. Heat together the cream, milk and Davidson Plum puree until almost boiling.
2. Whisk the egg yolks and sugar until pale and creamy.
3. Continue whisking and add the hot cream mixture a little at a time until evenly combined.
4. Put the mixture into the saucepan and place it back on medium heat, but don't allow the sauce to boil or it will separate. Keep stirring until the mixture thickens slightly and clings on the back of the spoon.
5. Put in fridge to chill.
6. Once chilled, add to an ice cream maker and churn.
7. Place in a container and freeze.

SNAPPER WITH DESERT LIME BUTTER SAUCE

Ingredients

- 2 good sized pieces of snapper (of fish of your choice)
- ½ cup of finely chopped desert limes
- 1 tablespoon of butter
- ¼ cup of olive oil
- A good grind of pepper berry
- 1 teaspoon of roasted garlic (see the following section, 'Method')
- Salt to taste
- Chilli (optional)

Method

1. Heat the butter and olive oil in a pan, and then add the roasted garlic and the pepper berry.
2. Add the desert limes, salt and chilli, if desired.
3. Mix them together and put the mixture aside.
4. Pan-fry the snapper and place it on a plate.
5. Top with the desert lime butter sauce.

Roasted Garlic

- Halve a head of garlic; leave skin on.
- Place it on foil and then sprinkle with olive oil and season with salt and pepper leaf.
- Close the foil and cook it on a moderate oven for approx. 1 hour.
- Take it out of the oven and allow it to cool (cloves are now soft).
- Squeeze out the cloves into a bowl, and then cover them with olive oil.
- Mix them together and store in a glass jar in a refrigerator.

GREEN SALAD WITH FINGER LIMES

Crisp and crunchy, the iceberg lettuce combined with finger limes and riberry dressing makes a fresh and healthy salad to accompany any meal.

Ingredients

- 1 iceberg lettuce
- 1 cucumber, finely sliced
- 5 finger limes, cut in half and caviar squeezed out

Riberry Dressing

Mix together

- 3 tablespoons of olive oil
- 1 dessertspoon of balsamic vinegar

- 1 teaspoon of honey
- 1 dessertspoon of riberries, roughly chopped
- Salt and pepper to taste

Method

1. Arrange the lettuce leaves on a plate.
2. Add the cucumber slices.
3. Mix 3 of the caviar finger limes through the salad.
4. Drizzle dressing over the salad.
5. Add the remaining caviar finger limes to the top.
6. Finger limes come in all colours and flavours from bright green to red.
7. The bright green limes are quite tart: the pinker the colour, the softer the flavour. Use a combination of these or choose whichever is in season; they all taste amazing and add a real tang to this salad.

KAKADU PLUM AND APPLE CRUMBLE

Ingredients for Fruit Mixture

- 1 kg of green apples (peeled and chopped into small pieces) or (quick way) 800 g can of apples
- ½ cup of sugar
- 1 cup of Kakadu Plums
- 1 tablespoon of lemon juice
- ½ teaspoon of cinnamon

Method

1. Mix the apples with the lemon juice and Kakadu Plum pulp (see the following section, 'Method for Kakadu Plums'), and then add sugar and cinnamon. Put them into a baking dish.
2. Top with crumble mix.

Method for Kakadu Plums

Place Kakadu Plums into a saucepan with ½ cup of water and cook them until soft. Then push through a sieve to get the pulp and discard skins and stones.

Macadamia Crumble Topping

Ingredients for the Crumble

- ¾ cup of plain flour
- 80 g of butter, chilled and chopped into pieces

- 100 g of chopped macadamia nuts, (chop these by hand is used to get finely chopped and large nuts, making the final product more interesting)
- ½ cup of rolled oats
- ½ cup (firmly packed) of brown sugar
- ½ teaspoon of cinnamon

Method

1. Place the flour in a bowl or on a bench, and then rub the chilled butter into the flour with your fingertips until it resembles breadcrumbs.
2. Add the nuts, oats and sugar, and mix them together.
3. Place the mixture on top of the kadadu plum and apples and bake in an oven at 180°C, approx. 20 minutes.

ASPARAGUS RISOTTO WITH LEMON ASPEN

Ingredients

- 2 tablespoons of olive oil
- 1 large onion, finely chopped
- 2 cups of Arborio rice
- 500 mL of chicken stock (hot)
- Approx. 2½ cups of water (extra, boiling)
- 1 cup of white wine
- 400 g of chopped fresh asparagus, cooked but with a slight crunch
- A good handful of chopped parsley
- A zest of 1 lemon
- ½ cup of grated tasty cheese
- ½ cup of sour cream
- Salt
- A good grind of Australian pepper berries
- ¼ cup of finely sliced lemon aspen
- ¼ cup of salted roasted pepitas
- ¼ cup of roughly chopped coriander (optional)

Method

1. Cook the onion in the oil until soft, but don't brown it. Add the rice and stir until covered in oil and slightly toasted.
2. Heat in a separate saucepan the wine and chicken stock.
3. Add the heated stock about ½ cup at a time into the rice allowing it to be absorbed. Continue stirring until all stock/wine is absorbed.
4. Add the asparagus, salt and ground pepper berries.
5. Add the cheese and cream and stir in.
6. Add the boiling water a little at a time until the rice is cooked. Here you can add a lot of water to make your risotto 'soupy' or a little only until the rice is cooked and the risotto is creamy.

7. When you have your risotto the way you like it, stir through the lemon aspen put in a serving dish and then sprinkle on top the pepitas and coriander.

Note: Leave the lemon aspen until last to make sure you make the most of this fantastic flavour. Also, add it a little at a time to ensure you get the taste that's right for you.

BIRCHER MUESLI WITH LEMON MYRTLE

This is a magic recipe that is definitely open to suggestions.

You can also add some grated green ginger, cinnamon or nutmeg of your choice. If you have a cup of quandongs handy, add those. It would also be amazing to add muntries. Let your imagination run free with this recipe. This is a perfect pick-me-up in the morning or at any time of the day.

Ingredients

- 2 cups of rolled oats
- ⅓ cup of pepitas
- ⅓ cup of sunflower kernels
- ½ cup of chopped macadamia nuts
- ½ teaspoon of lemon myrtle
- 1½ cups of fresh orange juice
- 1 grated apple (Granny Smith)

Method

1. Mix the ingredients together, and then leave it for an hour or overnight for the flavours to merge.
2. Add some honey or sweetener to taste.
3. Serve with natural yoghurt and enjoy.

MUNTRIE MUFFINS

Ingredients

- 3½ cups of plain flour
- 1 cup of sugar
- 190 g of butter
- ½ teaspoon of baking soda
- 1 heaped dessertspoon of baking powder
- 1½ cups of low fat milk
- 2 eggs
- A pinch of salt

- A pinch of cinnamon
- 1½ cups of muntries

Method

1. Set the oven temperature to 200°C.
2. Prepare a 12-cup muffin tray.
3. Mix the flour, sugar, salt and cinnamon into a bowl.
4. Rub the butter into the flour until it resembles fine breadcrumbs.
5. Set aside 1 cup of this mixture for the crumble toppings.
6. Add to the mixture the baking powder and baking soda.
7. Whisk them together in a separate bowl and mix them with the milk, eggs and vanilla.
8. Mix them into the flour mixture.
9. Fold in muntries.
10. Divide them evenly into muffin tins or patty pans.
11. Top with the crumble toppings.
12. Bake them for approx. 25–35 minutes until golden brown and skewer comes out clean.

PEPPERBERRY BEER DAMPER

The flavour of the beer makes this damper exceptional!

Traditionally made over an open fire with just flour and water, beer came later. This method made a really tough damper and the following recipe has been modified with a few extra touches to make a lighter damper, which lasts for a few days.

Ingredients

- 3 cups of self-raising flour
- 1 teaspoon of ground pepperberries
- 1 tablespoon of finely chopped chives
- 1 teaspoon of baking soda, or sodium bicarbonate
- A pinch of salt
- 50 g approx. or 3 tablespoons of butter
- 2 cups of beer (of your choice)

Method

1. Mix the flour, soda and a pinch of salt together.
2. Mix in the pepper berries and chives.
3. Rub the butter into the flour with your fingers until it resembles breadcrumbs; don't overdo this (few big bits are fine).
4. Finally, mix in the beer.

5. Put the mixture in a greased pan or a billy over the camp fire and cook it.
6. The damper can also be wrapped around the end of a stick and cooked holding over the fire, just like marshmallows. Extra flour might be needed to make the dough a little firmer.
7. If cooking at home, put the mixture into a pan of any shape and place it onto a baking paper.
8. Then in a fan forced oven set at 200°C or 180°C, cook it for approx. 25–30 minutes.
9. When cooked wrap in a tea towel to keep the moisture in.
10. Damper will last a couple of days.

PRAWNS WITH ANISE MYRTLE AND PEPPERLEAF

Ingredients

- 1 cup of dry white wine
- ¼ cup of finely chopped spring onions
- ½ teaspoon of Tasmanian Pepperleaf
- 500 g of green prawns, peeled with head on
- 1 level teaspoon of anise myrtle
- 1 tablespoon of butter
- 1 tablespoon of olive oil
- Salt

Method

1. Heat the butter and olive oil in a pan, and then add prawns and cook quickly. Once cooked, take the prawns out from the pan and place them on a dish.
2. Add the white wine and spring onions to the pan; bring it to boil and allow it to reduce.
3. Add ½ teaspoon of anise myrtle and taste the dish, and then add the other half.
4. Add ½ teaspoon of Tasmanian Pepperleaf.
5. Add salt to taste.
6. Add prawns to the pan and toss to cover with sauce, and then serve.

QUANDONG COMPOTE

These tart-tasting fruits are a real star; you can add these to your favourite breakfast cereal and use in conjunction with other fruits and berries for pies or baked goods or just enjoy them for the unique fruits that they are.

Ingredients

- 2 cups of dried quandongs
- 4 cups of water

- 1 star anise
- 1 cup of brown sugar
- 1 teaspoon of vanilla

Method

1. Put all these ingredients into a saucepan and bring them to boil; reduce to the lowest as possible heat. Then put lid on the saucepan and let it simmer until quandongs have softened; they will grow to approx. 3 times their size.
2. Don't overcook; take off heat when the quandongs are soft but still have their shape. Discard the star anise, mix in vanilla and allow it to cool.
3. This is deliciously served with cream, yoghurt or ice cream.

RIBERRY AND PEAR SLICE

Filling Ingredients

- ¼ cup of riberries
- Approx. 6 pear halves, each cut into half again
- 125 g of butter
- 125 g of cream cheese
- ½ cup of caster sugar
- 1 generous teaspoon of vanilla
- 1½ cups of ground almonds
- Approx. ½ cup of flaked almonds
- 2 eggs

Filling Method

1. Set the oven temperature to 150°C.
2. Beat the butter, cream cheese, vanilla and sugar together.
3. Add the eggs and continue beating the mixture until smooth.
4. Stir in the ground almonds and riberries.
5. Pour the mixture into a prepared/cooked pastry case.
6. Top with pears and then sprinkle flaked almonds on top.
7. Cook for approx. 50 minutes or until golden brown.

Pastry Ingredients

- 1 1/2 cups plain flour
- A pinch of salt
- 125 g of butter
- ⅓ cup of caster sugar
- 1 egg

Pastry Method

1. Rub the flour, butter, sugar and salt together using your hand until it resembles fine breadcrumbs.
2. Add the egg and mix in until smooth; don't overmix. Wrap them in a paper and refrigerate for approx. 30 minutes.
3. Roll them out onto a board and put them into a prepared tin (36 cm × 12 cm), and then bake blind for approx. 20 minutes.

CHOCOLATE MOUSSE WITH WATTLESEED

Here's one for you chocoholics and it's guilt-free; it contains all the good stuff and it's dairy-free, egg-free and gluten-free.

Ingredients

- 1 ripe avocado
- 1 ripe banana
- 1½ dessertspoons of quality cocoa powder (not sweetened)
- 1 dessertspoon of honey or maple syrup (sweetened to taste)
- 1 teaspoon of wattleseed syrup

Method

1. Blend all ingredients together in a processor (hey presto, it's done!)
2. You can eat it as is or freeze it for a 'supercool' treat.
3. The wattleseed adds a lovely coffee flavour to the mousse.

WATTLESEED SYRUP

Ingredients

- 2 tablespoons of roasted and ground Wattleseed
- ¾ cup of water

Method

1. Bring water and Wattleseed to boil.
2. Let it simmer to reduce the syrup to about ⅓ of the volume.
3. Strain out Wattleseed.
4. Chill in fridge before using.

WATTLESEED PAVLOVA

The coffee/chocolatey flavour of the Wattleseed adds another dimension to a spectacular dessert.

Ingredients

- 3 egg whites, stored at room temperature
- 1 cup of castor sugar
- 1 teaspoon of white vinegar
- 1 teaspoon of vanilla
- 1 dessertspoon of ground and roasted Wattleseed
- 1 teaspoon of cornflour

Method

1. Whip the egg whites until they stand in peaks.
2. Mix the Wattleseed in with the caster sugar.
3. Add a small quantity of the sugar/Wattleseed; mix a little at a time into the egg whites and keep whipping until the mixture becomes firm.
4. Fold in the cornflour, and then gently fold in the vanilla and vinegar.
5. Prepare a tray linked with the baking paper and then sprinkle some cornflour over.
6. Spread the pavlova mixture onto the paper into a round shape, approx. 20 cm.
7. Oven at 150°C.
8. Cook for approx. 45 minutes and then turn the oven off. Leave it in the oven to cool down; it might crack a bit (it often happens), but don't worry no one else will.
9. Once already cooled, top pavlova with whipped cream and fruit mix (of your choice).

Index